D1067642

PRINCIPLES OF GEOCHEMISTRY

PRINCIPLES OF GEOCHEMISTRY

THIRD EDITION

BRIAN MASON

Curator, Department of Mineral Sciences
U. S. National Museum
Washington, D. C.

John Wiley & Sons, Inc. NEW YORK LONDON SYDNEY

10 9 8 7 6 5 4 3

Library of Congress Catalog Card Number: 66-26752

Printed in the United States of America

PREFACE

This book has developed from a course of lectures originally given at the University of Canterbury in New Zealand and later at Indiana University and Columbia University. These lectures were designed not only for geology students but also for students in other sciences, especially chemistry. The aim has been to summarize the significant facts and ideas concerning the chemistry of the earth and to synthesize these data into a coherent account of the physical and chemical evolution of the earth. Such an aim is fraught with difficulties—difficulties of collecting and evaluating information from many fields, difficulties of interpretation and correlation of apparently unrelated phenomena, and finally difficulties of presentation of the material in a logical manner. The presentation followed in this book is adapted fairly closely from a one-semester course of the same title.

The book begins with a chapter describing the scope and subject matter of geochemistry and giving a brief account of its development. The next chapter deals with the earth as a planet and its relationship to the solar system and universe as a whole and is followed by one discussing the internal structure of the earth and its composition. From the data presented in these two chapters an account is given of the relative abundances of the elements and isotopes, both in the earth and in the universe as a whole, and on the basis of the evidence an attempt is made to present a logical account of the probable pregeological history of the earth. The aim of these first chapters is to provide a background for the remainder of the book, which is concerned with materials and processes at and near the surface of the earth. Because geochemistry is to a large extent the application of physicochemical principles to processes on and within the earth, some account of these principles is given in the next chapter, with special reference to the chemistry of the solid state. The geochemistry of igneous rocks is then dealt with, followed by a chapter on sedimentation and sedimentary rocks. Subsequent chapters deal in turn with the geochemistry of the hydrosphere, the atmosphere, and the biosphere. The geochemistry of the hydrosphere is essentially the geochemistry of sea water and poses such fundamental questions as the mode of origin and evolution of the ocean. The geochemistry of the atmosphere is concerned with the nature of the primeval atmosphere and the changes it has undergone through interaction with the hydrosphere, the biosphere, and the lithosphere. The geochemistry of the biosphere involves a discussion of the amount of organic matter, its composition, and the role of organisms in the concentration and deposition of individual elements. Then follows a chapter on metamorphism and metamorphic rocks, and the final chapter is a brief summary and synthesis in terms of the geochemical cycle.

Throughout this book temperatures are expressed in degrees Celsius, unless otherwise stated; tons are metric tons (1000 kg).

The emphasis throughout is on interpretation rather than description, on what is yet to be learned as well as what is already known. It is assumed that the reader is conversant with the fundamental concepts of geology and the principles of physics and chemistry and is familiar with the standard literature on these subjects. Although the book is written primarily from the point of view of a geologist, I hope that it will also be interesting and stimulating to chemists and physicists who are attracted by the fundamental problems of the earth. These problems can be solved only by the combined efforts of workers in many fields. Large gaps exist in our knowledge and understanding, but even a cursory survey reveals the possibilities of research which could contribute to reducing these gaps—and might also bring out others now unsuspected. There is an urgent and continuing need for more precise chemical and physical data on materials of the earth, for which the geologist is largely dependent on the chemist and physicist.

This book is necessarily a compilation from many sources, sources which include not only published books and papers but also correspondence and conversations with many people. I have been greatly helped by discussing many of the ideas presented with my colleagues. I am especially indebted to Professor J. Verhoogen, who read the manuscript for the first edition, and to Professor G. J. Wasserberg and Professor S. Epstein, who read that for the second edition; their comments and suggestions have been of the greatest value.

Because the literature of geochemistry is extremely scattered, a selected bibliography is appended to each chapter. The choice of these references may appear somewhat capricious; however, the policy has been to include comprehensive texts, significant recent papers (especially those in journals not usually referred to by geologists), and review articles with comprehensive bibliographies. The abbreviations used for journals and periodicals are those of *Chemical Abstracts*. It is hoped that the reader will be guided thereby to the most useful literature in the field from which he can proceed to further bibliographic research as the spirit moves him.

In the eight years since the second edition of this book was published vigorous research in geochemistry has resulted in a great increase in data and considerable progress in concepts and theories. Particularly noteworthy have been the advances in techniques for age determination of rocks, the keen interest in the nature and composition of the earth's interior (and the study of phase changes at temperatures and pressures previously unattainable in the laboratory), the large amount of work on minor and trace elements in geological materials, and the further development of isotope geochemistry. The significance and applications of geochemistry have achieved wide recognition.

Under these circumstances a revision of *Principles of Geochemistry* has become highly desirable. Many of the data in the second edition have been superseded by more reliable figures, and there has been a concomitant development in the theoretical framework of the subject. This revision retains the original organization of the book, but the changes in detail are numerous. The number of tables, and especially of figures, has been considerably increased in order to incorporate more information in a readily comprehended form. The bibliographies accompanying each chapter have been thoroughly revised to include new sources of information and to eliminate other references which have been largely incorporated in the newer works. At the request of many teachers, a section of questions and problems has been added as an appendix.

WASHINGTON, D. C. *Brian Mason*
August, 1966

CONTENTS

PRINCIPLES OF GEOCHEMISTRY

CHAPTER ONE

INTRODUCTION

THE SUBJECT OF GEOCHEMISTRY

In the simplest terms geochemistry may be defined as the science concerned with the chemistry of the earth as a whole and of its component parts. At one and the same time it is both more restricted and also more extensive in scope than geology. Geochemistry deals with the distribution and migration of the chemical elements within the earth in space and in time. The science of the occurrence and distribution of the elements in the universe as a whole is called cosmochemistry.

Clarke, in *The Data of Geochemistry*, defined the subject in a more restricted form:

> Each rock may be regarded, for present purposes, as a chemical system in which, by various agencies, chemical changes can be brought about. Every such change implies a disturbance of equilibrium, with the ultimate formation of a new system, which, under the new conditions, is itself stable in turn. The study of these changes is the province of geochemistry. To determine what changes are possible, how and when they occur, to observe the phenomena which attend them, and to note their final results are the functions of the geochemist. . . . From a geological point of view the solid crust of the earth is the main object of study; and the reactions which take place in it may be conveniently classified under three heads—first, reactions between the essential constituents of the earth itself; second, reactions due to its aqueous envelope; and third, reactions produced by the agency of the atmosphere.

V. M. Goldschmidt (1954) described geochemistry in the following terms:

> The primary purpose of geochemistry is on the one hand to determine quantitatively the composition of the earth and its parts, and on the other to discover the laws which control the distribution of the individual elements. To solve these problems the geochemist requires a comprehensive collection of analytical data on terres-

trial material, such as rocks, waters, and the atmosphere; he also uses analyses of meteorites, astrophysical data on the composition of other cosmic bodies, and geophysical data on the nature of the earth's interior. Much valuable information has also been derived from the laboratory synthesis of minerals and the investigation of their mode of formation and their stability conditions.

The main tasks of geochemistry may be summarized thus:

1. The determination of the relative and absolute abundances of the elements and of the atomic species (isotopes) in the earth.
2. The study of the distribution and migration of the individual elements in the various parts of the earth (the atmosphere, hydrosphere, crust, etc.), and in minerals and rocks, with the object of discovering principles governing this distribution and migration.

THE HISTORY OF GEOCHEMISTRY

The science of geochemistry has largely developed during the present century; nevertheless, the concept of an autonomous discipline dealing with the chemistry of the earth is an old one, and the term "geochemistry" was introduced by the Swiss chemist Schönbein (discoverer of ozone) in 1838. The history of geochemistry naturally includes much of the history of chemistry and geology. Because geochemistry is basically concerned with the chemical elements, their discovery and recognition are landmarks in the history of the subject. The modern concept of an element can be said to date from Lavoisier's definition in his "Traité élémentaire de Chimie" (1789), although some seventeenth- and eighteenth-century scientists certainly understood the distinction between elements and compounds. Lavoisier recognized the following thirty elements: O, N, H, S, P, C, Cl, F, B, Sb, Ag, As, Bi, Co, Cu, Sn, Fe, Mn, Hg, Mo, Ni, Au, Pt, Pb, W, Zn, Ca, Mg, Ba, Al, Si. Of these, Au, Ag, Cu, Fe, Pb, Sn, Hg, S, and C were already known to the ancient world. The last decade of the eighteenth century saw the discovery of U, Zr, Sr, Ti, Y, Be, Cr, and Te. The discovery or isolation of the elements during the nineteenth century can be summarized as follows:

1800–1809: Na, K, Nb, Rh, Pd, Ce, Ta, Os, Ir
1810–1819: Li, Se, Cd, I
1820–1829: Br, Th
1830–1839: V, La
1840–1849: Ru, Tb, Er
1850–1859:
1860–1869: Rb, In, Cs, Tl
1870–1879: Sc, Ga, Sm, Ho, Tm, Yb
1880–1889: Ge, Pr, Nd, Gd, Dy
1890–1899: He, Ne, Ar, Kr, Xe, Po, Ra, Ac

The data show an interesting pattern; during the first decade a considerable number of elements were discovered or isolated for the first time, reflecting the theoretical developments from Lavoisier and Dalton. The rate of discovery then fell off, until in the decade 1850–1859 no new elements were discovered. In 1860 Bunsen and Kirchoff demonstrated the utility of the spectroscope in the detection and identification of the elements, and in rapid succession Cs, Rb, Tl, and In were discovered with this new instrument. The 1870's and 1880's were marked by the discovery of several new lanthanides and the elements predicted by Mendeleev on the basis of the periodic table: eka-aluminum (Ga), eka-boron (Sc), and eka-silicon (Ge). In 1894 argon, the first of the inert gases, was discovered, followed rapidly by the remaining members of the group—Ne, He, Kr, Xe. The discovery of radioactivity by Becquerel in 1896 resulted in the recognition of polonium and radium by the Curies in 1898, and actinium by Debierne in 1899.

By 1900, therefore, the periodic table was essentially complete, except for some short-lived radioactive elements and for Eu (1901), Lu (1907), Hf (1923), and Re (1925). Proof of this, however, was first provided in 1914, when Moseley demonstrated the correlation between X-ray spectra and the atomic numbers of the elements.

Throughout the nineteenth century geochemical data were mainly the byproduct of general geological and mineralogical investigations and comprised more and better analyses of the various units—minerals, rocks, natural waters, and gases—making up the accessible parts of the earth. For many years this work was largely confined to European laboratories, but with the organization of the U. S. Geological Survey and the appointment of F. W. Clarke as Chief Chemist in 1884 a center devoted to the chemical investigation of the Earth was founded on the American continent.

Clarke was Chief Chemist for 41 years, until he retired in 1925. He was responsible for a vast and ever-growing output of analyses of minerals, rocks, and ores collected by the field staff or submitted for examination; moreover, he was always concerned with the fundamental significance of the mass of factual data thus acquired. In 1889 he published a classic paper, "The relative abundance of the chemical elements," which represented the first attempt to utilize the accumulated rock analyses to determine the average composition of the earth's crust and the relative abundances of the elements. It is interesting to note that even with the inadequate data then available, Clarke was able to draw some remarkably prescient conclusions. He wrote "If . . . we assume that the elements have been evolved from one primordial form of matter, their relative abundance becomes suggestive . . . the process of evolution seems to have gone on slowly till oxygen was reached. At this point the process exhibited its maximum energy, and beyond it the elements forming stable oxides were the most readily developed, and in the largest amounts. On this supposition the scarcity of the elements above iron becomes somewhat intelligible; but the theory does not account for everything and is to be regarded as merely tentative." However, in his great compendium, *The Data of Geochemistry*, which was first published as U. S. Geological Survey Bulletin 330 in 1908 and passed through five editions in less than twenty years, Clarke did not pursue this theory of the origin of the elements.

In many respects the appearance of the fifth edition of *The Data of Geochemistry* in 1924 marks the end of an era. During the preceding hundred years geochemical research was largely synonymous with the analysis of those parts of the earth accessible to visual inspection and chemical assay. From the nature of things it could be little more; interpretative geochemistry, the creation of a philosophy out of the mass of factual information, had to wait upon the development of the fundamental sciences, especially physics and chemistry. A single illustration serves to demonstrate this: the failure of all attempts to explain adequately the geochemistry of the silicate minerals before the discovery of X-ray diffraction provided a means for the determination of the atomic structure of solids.

The development of geochemistry in new directions was greatly advanced by the establishment of the Geophysical Laboratory by the Carnegie Institution of Washington in 1904. The policy followed in this laboratory of careful experimentation under controlled conditions, and the application of the principles of physical chemistry to geological processes, was an immense step forward. Previously geologists and chemists had been skeptical of the possibility of applying the techniques and principles of physics and chemistry to materials and processes as complex as those on and within the earth.

At the same time the original staff of the Geophysical Laboratory was beginning work in Washington, a new school of geochemistry was growing up in Norway. Fathered by J. H. L. Vogt and W. C. Brøgger, it attained world-wide distinction through the work of V. M. Goldschmidt and his associates. Goldschmidt graduated from the University of Oslo in 1911, and his doctor's thesis, *Die Kontaktmetamorphose im Kristianiagebiet*, was a basic contribution to geochemistry. It applied the phase rule, recently codified by the work of Bakhuis Roozeboom, to the mineralogical changes induced by contact metamorphism in shales, marls, and limestones, and it showed that these changes could be interpreted in terms of the principles of chemical equilibrium. During the next ten years his work was devoted largely to similar studies on rock metamorphism. These studies stimulated related research in other Scandinavian countries and led eventually to the enunciation of the principle of "mineral facies" by Eskola in a paper published from Goldschmidt's laboratory.

In many ways 1912 can be considered a critical date in the development of geochemistry. In that year von Laue showed that the regular arrangement of atoms in crystals acts as a diffraction grating towards X rays and thus made the discovery that enabled the atomic structure of solid substances to be determined. Since the geochemist is largely concerned with the chemistry of solids, the significance of this discovery can hardly be overestimated. However, some years elapsed before the impact of this new development was felt in geochemistry. In the fifth edition of *The Data of Geochemistry*, published in 1924, no mention is made of it. It is a tribute to Goldschmidt's insight that he not only realized the significance of crystal structure determinations for geochemistry, but also devised a plan of research which led to a maximum of results in a minimum of time. Between 1922 and 1926 he and his associates in the University of Oslo worked out the structures of many compounds and thereby established the extensive basis on which to found general laws governing the distribution of elements in crystalline substances. The results were published in a series of papers entitled *Geochemische Verteilungsgesetze der Elemente*, which in spite of the title dealt largely with the crystal structures of inorganic compounds. In these publications Goldschmidt's name is associated with T. Barth, W. H. Zachariasen, L. Thomassen, G. Lunde, I. Oftedahl, and others, all of whom have since had notable careers.

In 1929 Goldschmidt left Oslo for Göttingen, where he began investigations on the geochemistry of the individual elements, applying the principles discovered in the previous years and making use of the current development in quantitative spectrographic methods for rapidly determining small amounts of many elements with a high degree of precision. The results are well summarized in the seventh Hugo Müller lecture of the Chemical Society of London "The principles of distribution of chemical elements in minerals and rocks" (1937). Because of conditions in Germany Goldschmidt returned to Oslo in 1935. Unfortunately, after the German invasion of Norway in 1940 he was able to do little more work. He was forced to flee Norway in 1942 to avoid deportation to Poland, and escaped to Sweden and thence to England. His health was seriously affected as a result of imprisonment in concentration camps in Norway, and he never recovered, dying in 1947 at the comparatively early age of 59.

Thanks largely to the work and the stimulus of Goldschmidt, the last 30 years have seen geochemistry develop from a somewhat incoherent collection of factual data to a philosophical science based on the concept of the geochemical cycle in which the individual elements play their part according to established principles. Geochemical speculation has extended beyond the accessible parts of the earth to the nature and constitution of the interior, the development of the earth throughout geological time, and ultimately to its pregeological history and to the history of the solar system as a whole.

An important school of geochemistry has developed in Russia, especially since 1917. Its greatest names have been V. I. Vernadsky and his younger colleague A. E. Fersman. Its productivity has been immense, including a very large work on geochemistry by Fersman, published in four volumes between 1933 and 1939. Unfortunately, practically all this literature is in Russian, and much of it is not readily available in other countries. However, this situation has been greatly improved in recent years with better abstracting and translation services. Geochemistry in Russia has been particularly directed towards the search for and exploitation of mineral raw materials, evidently with considerable success.

THE LITERATURE OF GEOCHEMISTRY

Geochemical facts, theories, and fantasies are widely scattered throughout the scientific literature. The first source of such data is naturally geological publications, but their number is legion and geochemical papers may be found in any of them. Chemical publications contain, of course, much that has geochemical significance, and occasionally classical material appears therein, as, for example, the symposium entitled "The physical chemistry of igneous rock formation," published by the Faraday Society in 1925. Fortunately, the chemical literature is relatively accessible through abstract journals. The literature of astronomy, biology, and of physics also carries information of importance to the geochemist.

Two journals devoted to geochemistry are *Geochimica et Cosmochimica Acta*, which began publication in 1950 under direction of an international board of editors, and *Geokhimiya*, published since 1956 by the Academy of Sciences of the U.S.S.R. The second has been published in translation by the Geochemical Society under the title *Geochemistry* (until 1963; in 1964 *Geochemistry* was replaced by *Geochemistry International*, comprising selected articles from *Geokhimiya* along with translated articles from other languages). Although these journals attract many specifically geochemical papers, many data of significance for geochemistry continue to appear as incidental material in publications concerned with mineralogy, petrology, economic geology, inorganic and physical chemistry, and other sciences.

Three outstanding books are standard reference works in the field of geochemistry, one of which, the 1924 edition of Clarke's *The Data of Geochemistry*, is still the finest and most readily available collection of analytical data on geological material, and on this account alone is an essential work. In addition, it documents excellently an enormous mass of literature, much of which still has significance. The second is *Geochemistry* by Rankama and Sahama, published in 1950. This authoritative and comprehensive book not only treats the general aspects of the subject, but also gives a detailed account of the geochemistry of each element. The third is *Geochemistry* by V. M. Goldschmidt. Goldschmidt had prepared a great deal of manuscript for this work when he died in 1947, and thanks to the conscientious editorial work of Dr. A. Muir, and contributions to missing sections by other people, it was finally published in 1954. Although similar in scope to the book of Rankama and Sahama, it bears the distinctive stamp of Goldschmidt's genius for arriving at broad generalizations which correlate a large number of unconnected observations.

SELECTED REFERENCES

Clarke, F.W. (1924). The data of geochemistry (fifth edition). 841 pp. *U. S. Geol Surv. Bull.* 770.

Fleischer, M. (ed.) (1962–). The data of geochemistry (sixth edition). *U. S. Geol. Surv. Prof. Paper 440.* The new edition of this great work is a cooperative effort of many authors, and is being published in parts.

Goldschmidt, V. M. (1954). *Geochemistry.* 730 pp. Clarendon Press, Oxford.

Green, J. (1959). Geochemical table of the elements for 1959. *Bull. Geol. Soc. Amer.* **70,** 1127–1184. This is a very useful compilation of up-to-date geochemical data, compactly presented on a large periodic table.

Hawkes, H. E., and J. S. Webb (1962). *Geochemistry in mineral exploration.* 415 pp. Harper & Row, New York. Besides giving a comprehensive account of the application of geochemistry to the discovery of ore deposits, this book contains a great deal of useful information in general geochemistry.

Rankama, K., and Th. G. Sahama (1950). *Geochemistry.* 912 pp. University of Chicago Press, Chicago.

Smales, A. A., and L. R. Wager (ed.) (1960). *Methods in geochemistry.* 464 pp. Interscience Publishers, New York. An excellent collection of articles on the techniques of geochemical research, with special emphasis on modern developments; good bibliographies.

Weeks, M. E. (1956). *Discovery of the elements* (sixth edition). 910 pp. J. Chem. Educ., Easton, Pa. A fund of interesting and significant data related to the history of geochemistry.

CHAPTER TWO

THE EARTH IN RELATION TO THE UNIVERSE

THE NATURE OF THE UNIVERSE

The earth is a unit within the solar system, which consists of the sun, the planets and their satellites, the asteroids, the comets, and the meteorites. The sun itself is only one star within our galaxy, which comprises probably more than 10^{11} stars and has a lens-like form with a diameter of about 70,000 light-years (1 light-year $\simeq 10^{13}$ km). Beyond our own galaxy there is a very large number of other systems of stars of approximately the same size. These systems, the extragalactic nebulae, are scattered fairly uniformly through space, the nearest to us being the Andromeda nebula at a distance of about 1,000,000 light-years. The spectra of these extragalactic nebulae show a displacement of the lines towards the red end of the spectrum, this red shift being approximately proportional to their distance. The red shift is regarded as a Doppler effect due to recession of the nebulae with velocities approximately proportional to their distance, and leads to the picture of an expanding universe. A large part of the theory of the expanding universe is still highly speculative; for this reason conclusions based on the theory also partake of its speculative character.

THE AGE OF THE UNIVERSE

The theory of an expanding, dynamic universe implies that the universe has been and is in a state of evolution. If we extrapolate backward in time we arrive at a state when the universe was "contracted to a point," or when all the matter in the universe was concentrated into a very small region. It is customary to regard this as its primitive state and to reckon the astronomical age of the universe from this epoch. With certain assumptions as to the rate of expansion the astronomical age of the universe is computed to be about 11 $\times$ 10^9 years. An alternative view—the steady-state theory—is that the universe has always been much as it is now, with new matter being continuously created between the receding galaxies.

THE AGE OF THE SOLAR SYSTEM

The question about which of these theories—the expanding universe or the steady-state universe—better explains the facts is yet unresolved. However, the solar system can be treated as an isolated unit, and its age considered independently of the rest of the galaxy and the universe as a whole. The solar system is essentially a closed system, and its elemental composition the same as when it formed, except insofar as it has been modified by the conversion of hydrogen to helium in the sun and by the decay of radioactive elements. The underlying assumption is that the material of the solar system was segregated at some definite time (the zero point of this time scale) and that the primitive constitution was subsequently modified by radioactive decay. The age of the elements is then reckoned from the time when the naturally radioactive series consisted entirely of the parent elements (non-radiogenic "daughter" elements may have been present). The natural radioactive series of importance for this dating are given in Table 2.1.

Data on the abundances of the elements have shown that those of atomic number greater than 40 are of approximately equal abundance, indicating that, whatever the process of formation, nuclei of approximately the same complexity were formed in approximately equal amounts. Since the abundance of U^{238}, with a half-life of 4.5 $\times$ 10^9 years, is about the same as that of the stable elements bismuth and mercury, the elements cannot have formed more than a few thousand million years ago, as otherwise U^{238} would have largely disappeared through disintegration. On the other hand, the low abundance of the naturally radioactive potassium isotope, K^{40}, with a mean half-life of 1.3 $\times$ 10^9 years, indicates that the formation of this nucleus occurred at least a few thousand million years ago, giving it sufficient time to decay to a large extent. Similar reasoning explains the almost complete absence of Np^{237} and Pu^{239}, whose half-lives are 2.25 $\times$ 10^6 and 2.4 $\times$ 10^4

Table 2.1. Some Natural Radioactive Series of Significance for Geological Dating

Parent Elements	Total Half-lives	End Products
U^{238}	4.5 $\times$ 10^9 years	Pb^{206} + $8He^4$
U^{235}	7.1 $\times$ 10^8 years	Pb^{207} + $7He^4$
Th^{232}	1.4 $\times$ 10^{10} years	Pb^{208} + $6He^4$
Rb^{87}	5.1 $\times$ 10^{10} years	Sr^{87}
K^{40}	1.3 $\times$ 10^9 years	A^{40}, Ca^{40}

years. An independent solution is provided by consideration of the relative abundances of U^{235} and U^{238}. If these two isotopes were originally formed in approximately equal amounts, the present ratio of U^{235} to U^{238} (1:140) is due to the shorter life of U^{235}, which causes it to disappear much more rapidly than U^{238}. The half-life of U^{235} is 7.1×10^8 years; hence to reduce its amount to $1/140$ that of U^{238} would take about 5×10^9 years. Evidence of this sort, although qualified by uncertainties in the primordial abundances of the elements, indicates that a unique value can be assigned to the age of the solar system.

Other arguments also support the view that the solar system as we know it dates from a few thousand million years ago. Thus, of the isotopes of lead (Pb^{204}, Pb^{206}, Pb^{207}, Pb^{208}), Pb^{204} is not of radioactive origin, and its amount should be the same now as when it was originally formed; the other isotopes are presumably partly original and partly derived from the decay of U^{238}, U^{235}, and Th^{232}. In material containing both uranium and lead the isotopic composition of the lead has undergone a progressive change during geological time; the relative amounts of the radiogenic isotopes have increased with respect to the nonradiogenic Pb^{204}. Meteorites provide us with a particularly satisfactory illustration of this fact. Iron meteorites contain no uranium, and the small amount of lead present has the highest relative amount of Pb^{204} of any natural material. Stony meteorites contain measurable amounts of uranium, and the lead present shows the effect of the continual addition of radiogenic lead in much higher Pb^{206}/Pb^{204} and Pb^{207}/Pb^{204} ratios than those for lead extracted from iron meteorites. A mathematical analysis of the data gives an age for meteorites of 4.5×10^9 years, which has been independently confirmed by potassium-argon and rubidium-strontium age determinations on stony meteorites. This is the time since the iron and the stony meteorites were differentiated, and it can be plausibly equated with the time of planet formation in the solar system.

Another approach to this general problem of the age of the solar system is provided by a consideration of the age of terrestrial materials. The change in the isotopic composition of lead with time has been investigated in lead ores of known geological ages; as predicted, it has been shown that Pb^{204} is progressively less abundant in relation to the other lead isotopes in ores of progressively younger age. A statistical study of the data provides a number of independent solutions for the time at which the primeval lead began to be modified by the addition of isotopes generated from uranium and thorium. These solutions show a significant concentration of values at about 3.5×10^9 years. Since the data are determined on specimens from the earth's crust, this age is that of the separation of the crust from the mantle. This evidently postdates the appearance of the earth as an individual body in the solar system, which is probably best given by the age of meteorites, viz., 4.5×10^9 years. The significance of this billion-year interval between the age of the earth and the age of the crust is discussed in the next chapter.

The age of 3.5×10^9 years for the consolidation of the earth's crust is corroborated by the information available from the dating of individual rocks of the crust by one or another of the radioactive decay schemes. On all the continents except Antarctica rock ages of around 2700 million years are well established. It has been much more difficult to extend the record further back, but the oldest ages from Africa are 3000 to 3300 million years; from North America, 3100 to 3500 million years; from Europe, 3500 million years; and from Australia, 3000 million years. It seems unlikely that the age record for crustal rocks will be extended much beyond 3500 million years. The time of formation of the primordial crust and the age of the earth as an individual body must, of course, be greater than these dates. The hiatus of 1000 million years between the probable age of the earth and that of the oldest crustal rocks was evidently occupied by the evolution of a stable crust.

THE NATURE OF THE SOLAR SYSTEM

In the study of geochemistry the solar system is of primary importance, although it is inconspicuous within our own galaxy and insignificant in relation to the universe as a whole. Data on the solar system are given in Table 2.2. Any satisfactory theory of the origin of the solar system must explain its regularities, the most important of which are the following:

1. The sun contains over 99.8% of the mass of the system.
2. The planets all revolve in the same direction around the sun in elliptical orbits, and these orbits all lie in practically the same plane.
3. The planets themselves rotate about their axes in the same direction as their direction of revolution around the sun (except Uranus, which has retrograde rotation); most of their satellites also revolve in the same direction.
4. The planets show a rather regular spacing as expressed by Bode's law,[1] and they form two contrasted groups: an inner group of small planets

[1] Bode's law is an empirical series which closely approximates the relative distances of the planets from the sun and can be formulated as follows:

Mercury	Venus	Earth	Mars	(Vacant)	Jupiter	Saturn	Uranus	Neptune	Pluto
4	4	4	4	4	4	4	4		4
0	3	6	12	24	48	96	192		384
4	7	10	16	28	52	100	196		388
Actual distances of the planets from the sun in terms of the Earth's distance as 10									
3.9	7.2	10	15.2	..	52	95	192	301	395

The law thus gives a good agreement with the measured distances, except for Neptune; the gap between Mars and Jupiter is occupied by the asteroids, at a mean distance of 29.

Table 2.2. Data on the Solar System

	Sun	Mercury	Venus	Earth	Mars	Jupiter	Saturn	Uranus	Neptune	Pluto
Mass (Earth = 1)	332,000	0.05	0.81	1.00	0.11	318	95.3	14.5	17.2	0.03?
Radius (in kilometers)	695,000	2400	6100	6371	3400	69,900	57,500	23,700	21,500	2900
Volume (Earth = 1)	1,300,000	0.05	0.87	1.00	0.15	1320	736	51	39	0.1
Density	1.41	5.33	5.15	5.52	3.97	1.35	0.71	1.56	2.47	2?
Maximum surface temperature	5500	350	600?	60	30	−138	−153	−184	−200	−220
Gases in atmosphere*	Many	None	CO_2	Many	CO_2, H_2O	CH_4, NH_3	CH_4, NH_3	CH_4	CH_4	?

* Nitrogen, hydrogen, and helium are probably abundant in the atmospheres of the larger planets, although undetectable by present methods.

(Mercury, Venus, Earth, and Mars), which are called the terrestrial planets, and an outer group of large planets (Jupiter, Saturn, Uranus, and Neptune), which are called the major planets.

5. The major part of the angular momentum of the solar system is concentrated in the planets, not in the sun, in spite of the concentration of mass in the sun.

THE ORIGIN OF THE SOLAR SYSTEM

To be acceptable, any theory of the origin of the solar system must account for these regularities. Two main schools of thought exist, both with ancient and respectable antecedents. Both consider the solar system as derived from an ancestral sun or solar nebula. They differ essentially in that one prescribes the action of an external force to form the planets from the sun, whereas the other rejects the idea of an external force and finds the energy required to form the planets within an ancestral solar nebula. The first school of thought dates back to the French philosopher Buffon in 1749, who suggested that the planets were torn from the body of the sun by a collision with another star. The other originated with the speculations of Kant in 1755. Kant suggested that within an original solar nebula, regions with slightly higher density than the mean would act as sinks for matter and the planets would thereby grow at the centers of these regions. Laplace in 1796 visualized the original state of the sun as a rotating tenuous mass of gas occupying the entire volume of the present solar system, from which contraction, accompanied by increasing rotational speed, led to the disengaging of a series of gaseous rings by centrifugal force, these rings then condensing to form the planets.

The Laplace hypothesis held the field for some sixty years, until the physicist Clerk Maxwell showed that the physics of the solar system is inconsistent with the mode of origin postulated by Laplace. The major difficulties lay in the concentration of angular momentum in the planets, and not in the sun, and in the mechanism by which the annular rings of gas might have condensed into planets. These considerations brought the Laplace theory into disfavor, and Buffon's theory was revived and given more precise form, first by Chamberlin and Moulton in the United States and later by Jeans and Jeffreys in England. These theories had in common the formation of the earth and the other planets from material torn from the sun by the impact or close approach of another star. The Chamberlin-Moulton hypothesis visualized the formation of the planets by the aggregation of small solid particles (planetesimals); the Jeans-Jeffreys hypothesis considered that the planets were formed by the condensation of masses of incandescent gas. These impact theories were favored for a number of years but have been abandoned since it was shown that material torn from the sun by the impact of another body could under no circumstances condense to form planets but would be completely dissipated throughout space in a very short time.

In recent years the origin of the earth and of the solar system as a whole has been the subject of intense speculation. Variants of the nebular theory have been suggested by a number of astronomers. The version proposed by von Weizsäcker seems to fit the facts best. It pictures the primitive sun as a rapidly rotating mass surrounded by an extended lens-shaped envelope consisting of solid particles and gas in turbulent motion. In this lens-shaped envelope eddy-like vortices would form, causing local accumulations of matter which aggregated to form the planets. Von Weizsäcker gives reasons for the regular spacing of these vortices, which would explain Bode's law. Von Weizsäcker's theory also explains the remarkable difference in size and density between the inner and the outer planets. In the lens-shaped envelope from which the planets condensed the temperature would decrease with distance from the center according to the inverse square law. Because of this falling-off of temperature more material could condense in the outer parts than in the inner parts. In the region of the inner planets only compounds of low volatility condensed, whereas in the outer regions the condensation products contained much material of low critical temperature. Because of this difference in the amount of condensation the outer planets also grew faster and larger than the inner ones. Hence the inner planets are small and dense, whereas the outer ones are large and have a low specific gravity (Pluto is exceptional). Von Weizsäcker's theory incorporates part of the Chamberlin-Moulton hypothesis, in that it considers that the planets were built up by the aggregation of solid particles, that is, planetesimals, rather than by condensation of incandescent gas. This theory, especially in its chemical aspects, has been extensively developed by Urey (1952).

Other ideas on the mode of formation of the solar system are linked up with the nature of double stars. All the double stars that have been observed have a large amount of angular momentum. Most of the single stars, such as the sun, have very little angular momentum, as far as we know. It is therefore reasonable to suppose that a double star may reach a condition of greater stability by evolving into a single star with a system of planets revolving about it, the planetary system carrying the major portion of the angular momentum. Hoyle suggested that the evolution of a double star into a single star with a planetary system may result from the disintegration of one component of the double star with an accompanying supernova outburst. Most of the material of the supernova would be dispersed into space, but sufficient matter was left within the sun's sphere of influence to condense into the planets.

THE COMPOSITION OF THE UNIVERSE

Our knowledge of the chemical composition of the universe is obtained by spectroscopic examination of solar and stellar radiation, by the analysis of meteorites, and by what we know of the composition of the earth and other planets. Spectroscopic observation indicates the elements responsible for the radiation, and by careful analysis of the intensities of the spectral lines rough estimates can be made of the relative amounts of the different elements present in the outer layers of the radiating body. The data are consistent with the belief that the universe consists throughout of the same elements, and despite local variations which may generally be readily explained the relative abundances of the different elements are everywhere much the same. Only once has an element not previously known to occur on the earth been discovered elsewhere; this was helium, first detected in the sun's spectrum by Lockyer in 1868.[1] Nearly 30 years later, in 1895, helium was identified on the earth by Ramsay as the gas evolved when uraninite is heated with a mineral acid (Hillebrand a few years previously noticed this evolution of an inert gas from uraninite but thought it to be nitrogen).

THE COMPOSITION OF THE SUN

Spectroscopic studies of the sun have been made over many years and many data have been accumulated. The major limitations of this method of study are (*a*) some elements either do not give detectable spectra, or their strong lines have wavelengths less than 2900 A and are absorbed by the atmosphere of the earth and cannot be observed (this limitation is being overcome by spectrographic data obtained at high altitudes by rockets); (*b*) the spectra are produced in the outer part of the sun and give the composition of the solar atmosphere. Whether this composition is really representative for the sun as a whole depends on the effectiveness of convection to stir the material into a homogeneous mixture. These limitations must be borne in mind when considering the following information. Sixty-six of the 92 elements have been recognized in the sun's spectrum, and there is no reason to conclude that any element is really absent, the presence of the others being unobservable on account of their small abundance or the limitations previously stated. The relative abundances of the more common elements in the solar atmosphere are given in Table 2.3. The most striking feature is the extreme abundance of hydrogen and helium, a feature that is probably characteristic of the larger planets, Jupiter, Saturn, Neptune, and Uranus, also.

[1] Technetium, which does not occur naturally on the earth, has been identified in stellar spectra.

Table 2.3. Abundances of Elements in the Solar Atmosphere

Element	Atomic number	Abundance (atoms/10^4 atoms Si)
H	1	3.2×10^8
He	2	0.5×10^8
C	6	166,000
N	7	30,000
O	8	290,000
Na	11	630
Mg	12	7900
Al	13	500
Si	14	10,000
S	16	6300
K	19	16
Ca	20	450
Sc	21	0.2
Ti	22	15
V	23	1.6
Cr	24	33
Mn	25	25
Fe	26	1200
Co	27	14
Ni	28	260
Cu	29	35
Zn	30	8

After Aller, *The abundance of the elements*, 1961.

THE COMPOSITION OF THE PLANETS

Visual inspection and spectroscopic examination of the surfaces of the planets can tell us little about their bulk composition, since they are inhomogeneous, and their interior is undoubtedly different from their surface. However, data on their densities and analogies with the earth do provide some guide. Of the inner planets, Mercury has no atmosphere, and its density indicates that it is probably similar in composition to the earth. Venus is our nearest neighbor, and has a very dense atmosphere, consisting almost entirely of carbon dioxide and nitrogen which conceals its surface. The size and mass of Venus suggest that its composition is probably much like that of the earth. Mars, the next planet beyond the earth, has an atmosphere that does not obscure the surface of the planet and is therefore rarefied; however, clouds and dust storms have been observed on the face of Mars, and polar ice caps form in winter and melt in summer, showing that the atmosphere must contain some water vapor. The reflecting power, or albedo, of Mars indicates that its surface is largely composed of reddish rock, which Kuiper believes is probably rhyolitic material. The size and mass of Mars indicate a composition probably similar to that of the earth. However, the oblateness of Mars suggests that it is of uniform chemical composition throughout and is not differentiated into an iron core and silicate mantle, as is the earth.

The major planets, Jupiter, Saturn, Neptune, and Uranus, have many features in common, in particular low densities and thick atmospheres that completely obscure their surfaces. The low densities and thick atmospheres are explained by an abundance of hydrogen and helium probably comparable with that in the sun. Much of the hydrogen is evidently present as methane and ammonia. It has been shown that the rings of Saturn probably consist of ice particles, and the albedos and densities of some of the satellites of these planets suggest that they consist largely of ice also. The data we have on the major planets indicate that they have interiors similar to that of the earth but are covered with a great thickness of ice and condensed gases and have atmospheres containing hydrogen, helium, nitrogen, methane, and ammonia.

Pluto, the distant planet similar in size to Mercury, is an enigmatic body of which we know very little; it appears to be without an atmosphere, and its albedo is like that of the moon, suggesting a surface of dark-colored rock, possibly similar to basalt. It may have been a satellite of Neptune or a body captured by the solar system.

THE COMPOSITION OF METEORITES

Spectrographic evidence tells us nothing about the composition of the interiors of the planets. We must fall back on analogies with our own planet and with the evidence provided by meteorites, which are parts of the solar system (possibly fragments of disrupted asteroids) that eventually land on the earth. There are presumably millions of meteorites of all sizes in the solar system, from the finest dust particles up to those that are miles in diameter (if we include the asteroids, which appear to be similar to meteorites in many respects). Meteoritic matter is continually falling on earth, mostly in the form of dust undetectable except by special means; it is estimated that the rate of meteoritic infall is between 1000 and 10,000 tons daily. Our knowledge of the composition of meteorites comes from the larger and more spectacular ones that are seen to fall or from objects that are recognized as meteorites by the special characters distinguishing them from terrestrial rocks.

Meteorites consist essentially of a nickel-iron alloy, of crystalline silicate, mainly olivine or pyroxene, or of a mixture of these; glassy bodies, called tektites, are also found and may be meteorites. None resembling sedimentary or metamorphic rocks has been found. Many systems of classification have been devised for meteorites, but for our purpose they may be grouped as follows:

1. Siderites or irons (average 98% metal).
2. Siderolites or stony irons (average 50% metal, 50% silicate).
3. Aerolites or stones.
4. Tektites.

The siderites, or iron meteorites, consist essentially of a nickel-iron alloy (Ni is usually between 4 and 20%, rarely greater), generally with accessory troilite (FeS), schreibersite $(Fe, Ni, Co)_3P$, and graphite. Additional accessory minerals, such as daubreelite ($FeCr_2S_4$), cohenite (Fe_3C), and lawrencite ($FeCl_2$) occur more rarely. These accessory minerals are present as small rounded or lamellar grains scattered through the metal. The metal generally shows a definite structure known as Widmanstatten figures, which is brought out by etching a polished surface with an alcoholic solution of HNO_3. This structure consists of lamellae of kamacite (a nickel-iron alloy with about 6% Ni) bordered by taenite (a nickel-iron alloy with about 30% Ni). The lamellae are parallel to the octahedral faces of an originally homogeneous crystal of nickel-iron, and meteorites showing Widmanstatten structure are therefore known as octahedrites. This structure is typical of exsolution in an alloy that has cooled very slowly from a high temperature. Hexahedrites are irons consisting entirely of kamacite, nickel-rich ataxites irons with more than 14% Ni and consisting largely of taenite.

The siderolites, or stony-iron meteorites, are made up of nickel-iron and silicates in approximately equal amounts. Two distinct groups, the pallasites and the mesosiderites, of different chemical and mineralogical composition, are recognized. The pallasites are made up of a continuous base of nickel-iron enclosing grains of olivine which often show good crystal forms. In the mesosiderites the metal phase is discontinuous and the silicates are mainly plagioclase feldspar and pyroxene, sometimes with accessory olivine.

On the basis of texture the aerolites or stones are divided into two groups, the chondrites and the achondrites. The chondrites are so named because of the presence of chondrules or chondri, which are small rounded bodies (averaging 1 mm in diameter) consisting of olivine and/or pyroxene. Chondrules seem to be unique to these meteorites and have never been observed in terrestrial rocks, hence are probably significant in terms of the origin of such meteorites. The average composition of chondrites is about 40% olivine, 30% pyroxene, 10–20% nickel-iron, 10% plagioclase, and 6% troilite. One group of chondrites, the carbonaceous chondrites, is unique among meteorites in consisting largely of hydrated iron-magnesium silicate (serpentine or chlorite) and containing up to 10% of complex organic compounds. The origin of these compounds, whether the remains of extraterrestrial organisms or the products of nonbiological synthesis, has been argued for over a century and is still actively controversial.

The achondrites are a diverse group of stony meteorites which do not contain chondrules and which are usually much more coarsely crystalline than the chondrites. Many achondrites resemble terrestrial igneous rocks in composition and texture, hence have probably crystallized from a silicate melt.

Tektites consist of a silica-rich glass (average about 75% SiO_2) resembling obsidian, yet distinct from terrestrial obsidians in composition and texture. They have an unusual chemical composition, which consists of the conjunction of high silica and comparatively high alumina, potash, and lime with low magnesia and soda; this composition resembles a few granites and rhyolites, and some silica-rich sedimentary rocks. Tektites are found, generally as small (up to 200–300 g) rounded masses, in areas that preclude a volcanic origin. Unlike the other meteorite types, tektites have not been observed to fall, and their identification as meteorites is disputed; some authorities regard them as the product of the impact of comets or gigantic meteorites on the Earth. In view of their enigmatic origin and their aberrant composition they are not further considered here. However, a comprehensive account is provided by O'Keefe (1963).

Many chemical analyses have been made of meteorites. The irons form a rather homogeneous group, differing from one to another essentially in nickel content, and an average composition is easily derived. The stony-irons and the achondrites are very diverse groups, and are comparatively rare, so averages of their compositions have little significance. The chondrites are abundant and show a remarkable homogeneity in chemical composition. Table 2.4 gives a selection of the available compositional data. The close correspondence between the average composition of iron meteorites with the average composition of the metal from chondrites strongly suggests a common source. The iron meteorites probably represent metal segregated by the partial or complete melting of material of chondritic composition.

There is general agreement that meteorites provide us with the best sample from which to derive the absolute abundances of the nonvolatile elements; several tables of elemental abundances have been consequently compiled from the analytical data on meteorites. A principal difficulty in such compilations has been the selection of the analytical data, since meteorites differ greatly in composition and in relative abundance. This is shown in Table 2.5, which divides meteorites into two groups, the *finds* (those collected but not seen to fall) and the *falls* (those collected after having been seen to fall). The figures in Table 2.5 show a remarkable reversal in

Table 2.4. The Composition of Meteorite Matter

(weight per cent)

	Metal* (from irons)	Metal† (from chondrites)	Silicate† (from chondrites)	Average Chondrite†
O			43.7	33.24
Fe	90.78	90.72	9.88	27.24
Si			22.5	17.10
Mg			18.8	14.29
S				1.93
Ni	8.59	8.80		1.64
Ca			1.67	1.27
Al			1.60	1.22
Na			0.84	0.64
Cr			0.38	0.29
Mn			0.33	0.25
P			0.14	0.11
Co	0.63	0.48		0.09
K			0.11	0.08
Ti			0.08	0.06

* Brown and Patterson, *J. Geol.* **56,** 87, 1948.
† Mason, *Amer. Museum Novitates,* No. 2223, 1965.

*Table 2.5. Frequency of Meteorite Finds and Falls**

	Finds		Falls	
Type	Number	Per Cent	Number	Per Cent
Irons	545	58.1	33	4.6
Stony-irons	53	5.7	11	1.5
Achondrites	7	0.7	56	7.8
Chondrites	333	35.5	621	86.1
Total	938	100.0	721	100.0

* After Mason, 1962.

proportions between the finds and the falls. The reason is not far to seek. The relative abundance of irons as finds is due to their being easily recognized as meteorites, whereas a stony meteorite unless seen to fall could easily be overlooked as such. A truer indication of the relative abundance of the different meteorite types is therefore given by the relative proportions of those seen to fall. Such a compilation completely reverses the situation and indicates that the chondrites are far more abundant than all other types. The composition and structure of the chondrites favor the hypothesis that they may well represent the planetesimals which aggregated to form the planets;

the other meteorite types can plausibly be developed by the partial or complete melting and differentiation of material of chondritic composition. On this account the chemical composition of the chondrites has been the primary source of information regarding the absolute or cosmic abundances of the elements.

Determination of the ages of meteorites by the Rb^{87}—Sr^{87}, the K^{40}—Ar^{40}, and the Pb^{206}—Pb^{207} methods has given results indicating that both the irons and the stones crystallized about 4.5×10^9 years ago. Thus meteorites are much older than any terrestrial rocks. Earlier figures based on the helium content of iron meteorites gave widely varying ages, ranging up to 7×10^9 years. It is now clear, however, that these high ages are spurious. The impact of cosmic rays on meteorites causes nuclear disruptions accompanied by the production of helium; the increment in helium content results in an erroneously high figure for the apparent age deduced by this method.

THE COSMIC ABUNDANCE OF THE ELEMENTS

On the basis of data on the composition of meteorites and of solar and stellar matter, Goldschmidt in 1937 compiled the first adequate table of cosmic abundances of elements and isotopes. The data on hydrogen and helium were derived largely from examination of the sun and stars, and the figures for most of the other elements were based on their relative abundances in meteoritic material. Suess and Urey (1956) have published a revised table (Table 2.6), using the more extensive and accurate data accumulated since 1937; the major features of Goldschmidt's abundance figures are not altered, although there are numerous differences in detail.

In general, the agreement between the relative abundances determined in different regions of the universe seems reasonably good. Of the differences observed, most are explicable in terms of present or past physical conditions in the locale involved. For example, meteoritic and terrestrial matter differs from stellar material in the relative scarcity of the gaseous elements, but these disparities can readily be explained in terms of the physical conditions accompanying the evolution of the solar system. Variations in the abundances of hydrogen, helium, lithium, beryllium, boron, carbon, and nitrogen in different parts of the universe are almost certainly due to the participation of these elements in thermonuclear transformations responsible for energy production in the stars. In recent years it has been found that the abundance of heavy elements varies from star to star and is related to stellar age. This finding strongly suggests that these elements are synthesized within the stars and are distributed by stellar explosions.

Table 2.6 and Figure 2.1 show that the relative abundance of the different elements, especially the lighter ones, varies considerably. An element may be a hundred or a thousand times more or less abundant than its immediate

Table 2.6. Cosmic Abundances of the Elements

(atoms per 10,000 atoms Si)*

Z	Element	Abundance	Z	Element	Abundance
1	H	4.0×10^8	44	Ru	0.015
2	He	3.1×10^7	45	Rh	0.002
3	Li	1.0	46	Pd	0.007
4	Be	0.20	47	Ag	0.003
5	B	0.24	48	Cd	0.009
6	C	35,000	49	In	0.001
7	N	66,000	50	Sn	0.013
8	O	215,000	51	Sb	0.002
9	F	16	52	Te	0.047
10	Ne	86,000	53	I	0.008
11	Na	440	54	Xe	0.040
12	Mg	9100	55	Cs	0.005
13	Al	950	56	Ba	0.037
14	Si	10,000	57	La	0.020
15	P	100	58	Ce	0.023
16	S	3750	59	Pr	0.004
17	Cl	90	60	Nd	0.014
18	A	1500	61	Pm	—
19	K	32	62	Sm	0.007
20	Ca	490	63	Eu	0.002
21	Sc	0.28	64	Gd	0.007
22	Ti	24	65	Tb	0.001
23	V	2.2	66	Dy	0.006
24	Cr	78	67	Ho	0.001
25	Mn	69	68	Er	0.003
26	Fe	6000	69	Tm	0.0003
27	Co	18	70	Yb	0.002
28	Ni	270	71	Lu	0.0005
29	Cu	2.1	72	Hf	0.004
30	Zn	4.9	73	Ta	0.0007
31	Ga	0.11	74	W	0.005
32	Ge	0.51	75	Re	0.001
33	As	0.04	76	Os	0.010
34	Se	0.68	77	Ir	0.008
35	Br	0.13	78	Pt	0.016
36	Kr	0.51	79	Au	0.001
37	Rb	0.07	80	Hg	0.003
38	Sr	0.19	81	Tl	0.001
39	Y	0.09	82	Pb	0.005
40	Zr	0.55	83	Bi	0.001
41	Nb	0.01	90	Th	0.0004
42	Mo	0.02	92	U	0.0001
43	Tc	—			

* After Suess and Urey, *Rev. Mod. Phys.*, **28,** 53–74, 1956.

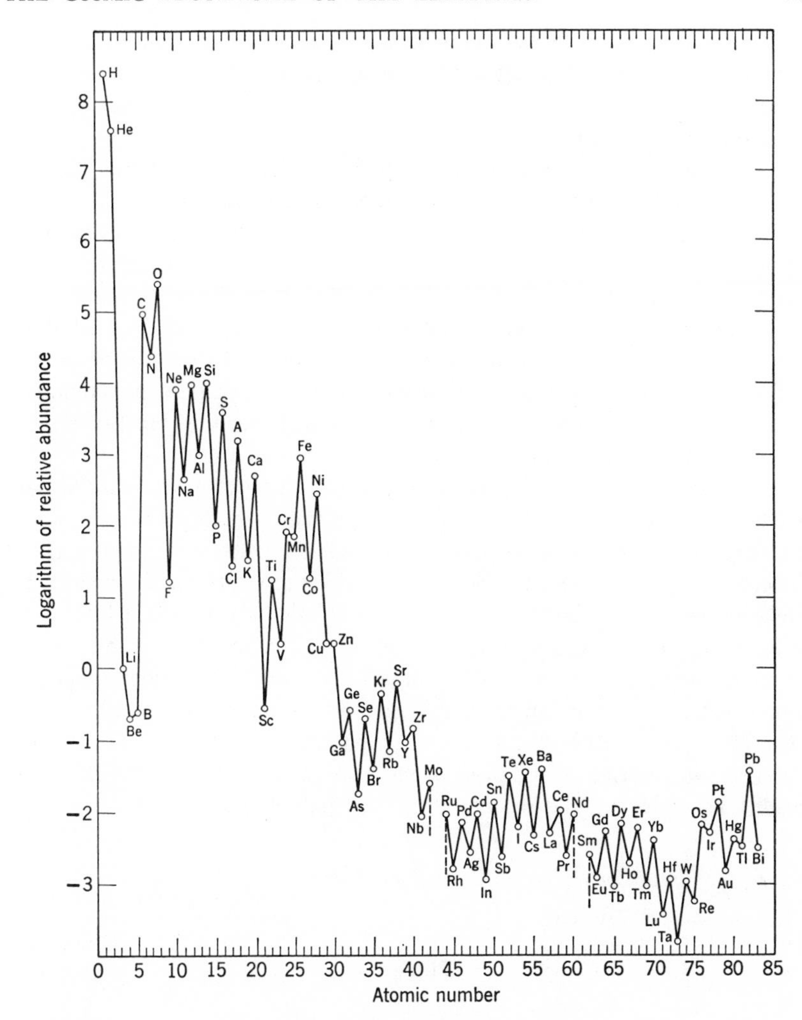

Figure 2.1 Relative abundances of the elements, referred to Si = 10,000, plotted against atomic number. (From Ahrens, *Distribution of the elements in our planet.* Copyright, 1965, McGraw-Hill Book Co. Used by permission.)

neighbor in the periodic table. Nevertheless, when the data are carefully analyzed numerous regularities are found. These may be summed up as follows:

1. The abundances show a rapid exponential decrease for elements of the lower atomic numbers (to about atomic number 30), followed by an almost constant value for the heavier elements.

2. Elements of even atomic number are more abundant than those of odd atomic number on either side. This regularity was first recognized independently by Oddo in 1914 and Harkins in 1917 and is sometimes referred to as the Oddo-Harkins rule.

3. The relative abundances for elements of higher atomic number than nickel vary less than those for elements of lower atomic number.

4. Only ten elements—H, He, C, N, O, Ne, Mg, Si, S, and Fe—all with atomic numbers less than 27, show appreciable abundance; of these, hydrogen and helium far outweigh the other eight.

The regularities displayed in Table 2.6 suggest that the absolute abundances of the elements depend on nuclear rather than chemical properties and are related to the inherent stability of the nuclei. An element is uniquely characterized by the number of protons (Z) in its nucleus, but the number of neutrons (N) associated with these protons can vary. As a result, an element may have several *isotopes* differing in mass number or atomic weight $A(A = N + Z)$ and stability but not appreciably in chemical properties. Similarly, there are *isobars*, which are different elements with the same A but different values of N and Z, and *isotones*, which are different elements with the same neutron number N but with different values of A and Z (Table 2.7).

Relatively few of the possible isotopes of any element are stable; of the thousand or more isotopes known to date, only about 270 are nonradioactive. The implication is that an isotope is abundant because the combination of protons and neutrons in its nucleus is a particularly stable one. On this basis the drop in relative abundances with increasing nuclear complexity can readily be explained; the absence from the earth of elements 43, 61, 85, and

Table 2.7. Illustration of Isotopes, Isobars, and Isotones

Isotopes $Z = 20$ (calcium)

N	A	Per Cent of Element
20	40	96.97
22	42	0.64
23	43	0.145
22	44	2.06
26	46	0.0033
28	48	0.185

Isotones $N = 20$

Element	Z	A	Per Cent of Element
Sulphur	16	36	0.0136
Chlorine	17	37	24.471
Argon	18	38	0.063
Potassium	19	39	93.10
Calcium	20	40	96.97

Isobars $A = 40$

Element	Z	N	Per Cent of Element
Argon	18	22	99.61
Potassium	19	21	0.0119
Calcium	20	20	96.97

87 is due to an almost complete instability of any nuclear arrangement for these atomic numbers (note that all four are odd-numbered). Then, too, it must be more than a coincidence that the nuclei of lithium, beryllium, and boron, which are exceptionally rare among low-numbered elements, are just those nuclei which are most readily disintegrated by bombardment with protons, alpha particles, and neutrons. As might be expected, the breakdown of relative abundance data for the elements into isotopic abundances has led to some significant results. Nuclei of the even N (neutron number)-even Z (proton number) type are both more numerous and more abundant than any of the other types. Nuclei of the even N-odd Z and odd N-even Z are about equally numerous and abundant. Nuclei of the odd N-odd Z type are few in number (only four nonradioactive ones are known—H^2, Li^6, B^{10}, N^{14}) and low in abundance. These features evidently reflect the nuclear binding energy, which is greatest for even N-even Z nuclei.

THE ORIGIN OF THE ELEMENTS

The structure of the nuclei of the elements as aggregates of protons and neutrons has resulted in theories to explain their origin and their relative abundances by a synthesis, or buildup, starting with either or both of the basic building blocks.

Several theories as to the mode of formation of the chemical elements have been proposed. One, which may be termed the equilibrium theory, proposes that the relative abundances of the elements are the result of a "frozen" thermodynamic equilibrium between atomic nuclei at some high temperature and density. By suitable assumptions as to the temperature, pressure, and density, good agreement with observed abundances is obtained for elements of atomic number up to 40. For elements of higher atomic number, however, these assumptions lead to impossibly low abundances. On this account, theories have been proposed which consider the relative abundances of the elements as resulting from nonequilibrium processes; on this basis the light nuclei were built up by thermonuclear processes and the remaining nuclei by successive neutron capture, with intervening β-disintegrations. This theory predicts the general trend of the observed data but fails to explain some of the detailed features, particularly bridging of the gap caused by the nonexistence of nuclei of atomic weights 5 and 8. The difficulty can be overcome by postulating the fusion of three He^4 nuclei to give C^{12}, thereby skipping the intermediate elements lithium, beryllium, and boron; these elements would be derived by secondary processes.

However, the complexities of the abundance data as established by Suess and Urey showed that no single process can satisfactorily account for these complexities. The problem was elucidated by Burbidge, Burbidge, Fowler, and Hoyle in 1957, when they showed the feasibility of the following eight processes for the synthesis of the elements and their role in accounting for the observed abundances.

1. Hydrogen "burning" to produce helium.
2. Helium "burning" to produce C^{12}, O^{16}, Ne^{20}, and perhaps Mg^{24}.
3. Alpha-particle processes, in which Mg^{24}, Si^{28}, S^{32}, Ar^{36}, and Ca^{40} are produced by successive additions of alpha-particles to O^{16} and Ne^{20}.
4. The equilibrium *e*-process, a statistical equilibrium between nuclei, protons, and neutrons, accounting for the abundance peak at iron.
5. The *s*-process, in which neutrons are captured at a relatively slow rate, producing elements up to and including Bi^{209}.
6. The *r*-process, in which neutrons are captured at a fast rate producing elements up to Cf^{254} (californium).
7. The *p*-process, in which proton-rich isobars are produced.
8. The *x*-process, responsible for the production of Li, Be, and B.

These processes are correlated with the observed features of stellar evolution. All stars convert hydrogen into helium, but only the most massive stars produce the elements in the upper part of the periodic table. Certain heavy nuclides appear to be formed only under catastrophic conditions, such as the development of a supernova. A supernova is essentially a stellar explosion, the catastrophic disintegration of a star. The explosion produces luminosity of the order of 10^8 that of the sun, and the luminosity falls off exponentially with a half-life of about 56 days. It can hardly be a coincidence that Cf^{254} decays by spontaneous fission with a half-life of 56 days. Evidently a supernova is triggered by the *r*-process. Man has reproduced the *r*-process on a comparatively modest scale; substantial quantities of californium are produced in H-bomb explosions when the U^{238} in the bomb is exposed to an intense neutron flux during the explosion.

SELECTED REFERENCES

Aller, L. H. (1961). *The abundance of the elements.* 283 pp. Interscience Publishers, New York. A comprehensive account of elemental abundances, with special reference to solar and stellar abundances and an excellent discussion of theories of element synthesis.

Burbridge, E. M., G. R. Burbridge, W. A. Fowler, and F. Hoyle (1957). Synthesis of the elements in stars. *Rev. Mod. Phys.* **29,** 547–650. A detailed account of the nuclear reactions required to account for the origin and relative abundances of the elements.

Faul, H. (ed.) (1954). *Nuclear geology.* 414 pp. John Wiley and Sons, New York. A useful volume on the applications of nuclear physics to geological problems, with good accounts of the properties of nuclear species, and on age determinations.

Fowler, W. A. (1964). The origin of the elements. *Proc. Nat. Acad. Sci.* **52,** 524–548. A careful survey of the theories of element formation, with special reference to processes of synthesis in stars and supernovae.

Goldschmidt, V. M. (1954). *Geochemistry.* Chapter 5.

Jastrow, R., and A. W. G. Cameron (eds.) (1963). *Origin of the solar system.* 176 pp. Academic Press, New York. A series of articles providing a comprehensive account of theories of origin of the solar system.

Kuiper, G. P. (ed.) (1953–1963). *The solar system.* 4 vols. University of Chicago Press, Chicago. A comprehensive reference work by many contributors; Vol. 1 deals with the sun, Vol. 2, with the earth, Vol. 3, with planets and satellites, and Vol. 4, with the moon, meteorites, and comets.

Levin, B. Y. (1958). *The origin of the earth and the planets.* 167 pp. Foreign Languages Publishing House, Moscow. An interesting account of theories of the origin of the solar system, with special reference to the work of O. J. Schmidt and his associates in the U.S.S.R.

Mason, B. (1962). *Meteorites.* 274 pp. John Wiley and Sons, New York. A general account of meteorites and tektites, with special emphasis on their chemical and mineralogical composition.

O'Keefe, J. A. (ed.) (1963). *Tektites.* 228 pp. University of Chicago Press, Chicago. A comprehensive account by a number of authors on the form, composition, distribution, and theories of origin of these enigmatic bodies.

Rankama, K., and Th. G. Sahama (1950). *Geochemistry.* Chapters 1, 2, and 9.

Ringwood, A. E. (1966). Chemical evolution of the terrestrial planets. *Geochim. Cosmochim. Acta* **30,** 41–104. A closely reasoned account of the formation of the earth from meteoritic material; extensive bibliography.

Tilton, G. R., and S. R. Hart (1963). Geochronology. *Science* **140,** 357–366. A useful summary of the possibilities and limitations of geological dating by means of radioactive elements.

Urey, H. C. (1952). *The planets.* 245 pp. Yale University Press, New Haven. A detailed discussion of the origin and development of the solar system, with special reference to the chemical processes which may have taken place during the formation of the planets.

Watson, F. G. (1956). *Between the planets.* 188 pp. Harvard University Press, Cambridge, Mass. An interesting account of the asteroids, meteorites, and comets.

Weizsäcker, C. F. von, (1949). *The history of nature.* 192 pp. University of Chicago Press, Chicago. A survey of the universe and the position of the earth and man therein, with a discussion of the philosophical background of current theories and ideas.

Whipple, F. L. (1963). *Earth, moon, and planets.* 278 pp. Harvard University Press, Cambridge, Mass. A comprehensive account of the solar system.

CHAPTER THREE

THE STRUCTURE AND COMPOSITION OF THE EARTH

INTRODUCTION

A knowledge of the composition and state of the earth's interior is prerequisite to an understanding of geochemistry. This problem is obviously one that cannot be solved by direct observation. The deepest boring yet made is one of about 7 km; mines are still shallower; and apart from borings and mines the only accessible parts of the earth are those exposed on the surface itself. Some material is brought to the surface by igneous activity, but as we know only vaguely the depth from which it comes few deductions can be made from this evidence. Thus to obtain some understanding of the earth's internal structure we must use an indirect approach. Here we turn to geophysics, the application to the properties of the earth and its several parts of physical laws dealing with gravitation, wave transmission, heat conduction, and other phenomena. The principal sources of information are (*a*) the acceleration of gravity at the earth's surface and the gravitational constant, from which the mean density of the earth can be determined; (*b*) the constant of precession of the equinoxes, from which the earth's moment of inertia can be calculated, thereby allowing important inferences to be drawn regarding density distribution within the earth; (*c*) seismological data, which indicate the presence of discontinuities within the earth and from which information can be derived on the elastic constants of the materials in the interior; (*d*) heat flow data, which reflect the abundance and distribution of radioactive elements in the crust and mantle. These facts, together with laboratory determinations of the elastic constants of various rocks, information on the probable abundances of the elements, and similar data, provide the basis for theories regarding the earth's internal structure and composition. To be acceptable any theory must be consistent with the data available.

However, the data may permit of several interpretations, none of which will formally conflict with what is known about the earth. Current ideas on the earth's internal constitution are certain to be modified by the discovery of new facts and the improvement of existing knowledge.

SEISMIC DATA ON THE EARTH'S INTERIOR

Much information about the earth's interior is derived from the analysis of earthquake waves. An earthquake generates waves of various kinds, of which the two types that pass through the body of the earth are the most important for our present purpose. These two types of waves travel with unequal velocity, even in the same medium. The faster are those transmitted by vibrations in the direction of propagation (analogous to sound waves in air). They are the first to be recorded by seismographs at an appreciable distance from the focus of the earthquake and are called primary or *P* waves. The slower waves are transmitted by vibrations at right angles to the direction of propagation (analogous to light waves) and are known as secondary or *S* waves. The velocities of *P* and *S* waves vary with the density and elastic constants of the material through which they pass, and they are subject to reflection and refraction at surfaces of discontinuity. By comparing the times at which *P* and *S* waves from the same shock arrive at different stations, travel-time tables can be drawn up from which the velocity of these waves as a function of depth can be calculated. Figure 3.1 depicts graphically the

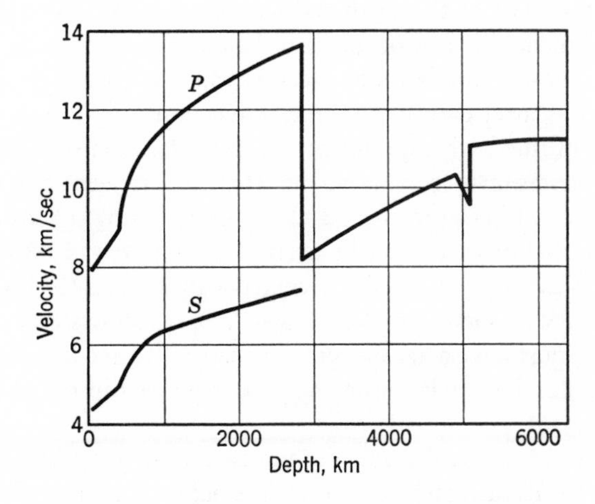

Figure 3.1 Velocity of P and S waves within the earth. (From Bullen, *An introduction to the theory of seismology*. Courtesy of Cambridge University Press)

data thus obtained. It shows that the interior of the earth is clearly heterogeneous, in the sense that at different depths the material has different elastic properties. This heterogeneity is not random but is present in zones separated from each other by discontinuities of greater or lesser sharpness. Two major or first-order discontinuities (a first-order discontinuity is one producing an abrupt break in the velocity-depth curve) have been recognized by all geophysicists. The earth is thus divided into three parts: the crust, from the surface down to the first discontinuity (the Mohorovičić discontinuity); the mantle, from the base of the crust to the second discontinuity (the Wiechert-Gutenberg discontinuity); and the core, from the Wiechert-Gutenberg discontinuity to the center of the earth. The extinction of the *S* waves at the base of the mantle is particularly significant, suggesting that the material of the underlying core lacks rigidity and behaves as a liquid. Second-order discontinuities are recognized within the crust, the mantle, and the core (a second-order discontinuity is marked by a sudden change in the *rate* at which wave velocity increases or decreases and produces a change in slope of the velocity-depth curve). There is no general agreement on the number and positions of these discontinuities, and their interpretation is much less certain.

DENSITY WITHIN THE EARTH

It may be said that Sir Isaac Newton was the founder of geophysics because his formulation of the law of gravitation provided the means for determining the mass of the earth and its mean density. Newton made the prescient statement: "It is probable that the quantity of the whole matter of the earth may be five or six times greater than if it consisted of water." It was not until the latter part of the eighteenth century that Newton's brilliant guess was confirmed by experiment. In 1798 Cavendish determined the constant of gravitation by comparing the attraction between two lead spheres with the attraction between them and the earth, using a sensitive torsion balance; from this measurement he derived the figure of 5.48 for the mean density of the earth. The present accepted figure is 5.517 ± 0.004. Since the mean density of surface rocks is about 2.8, it follows that at least part of the interior must have a density greater than 5.5 to account for that of the earth as a whole. The high density may be explained in two ways:

1. A change in physical state, the increase in density being due to the contraction of crustal material into much smaller volume under enormous pressure.
2. A change in chemical composition, the increase in density then being due to the presence of some intrinsically heavier substance, such as a heavy metal.

The next step is the determination of the density distribution within the earth. It was mentioned in the previous section that the velocities of P and S waves vary with the density and elastic constants of the material through which they pass. The relevant equations are

$$V_p^2 = \left(\frac{1}{\rho}\right)\left(k + \frac{4\mu}{3}\right)$$

$$V_s^2 = \frac{\mu}{\rho}$$

(V_p and V_s are the velocities of P and S waves, respectively; ρ is the density of the material, k the bulk modulus, μ the rigidity.) Of the variables in these equations only the velocities are accurately known for conditions in the earth's interior, and thus no unique solution for the other factors is possible from seismic data alone. However, the density distribution within the earth must comply with two stringent conditions: the integrated density must agree with the known density of the earth as a whole, and it must also give the correct moment of inertia as determined from the precession of the equinoxes. By making some plausible assumptions in the interpretation of the seismic data, Bullen computed figures for the density distribution within the earth which are consistent with the independent controls of bulk density and moment of inertia (Figure 3.2). From these results he also calculated the pressure distribution within the earth (Figure 3.3). The pressure at the

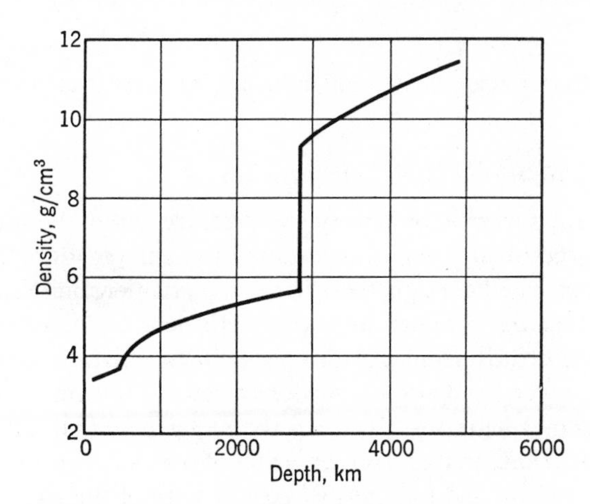

Figure 3.2 Density variation in the earth's interior. (From Bullen, *An introduction to the theory of seismology*. Courtesy of Cambridge University Press)

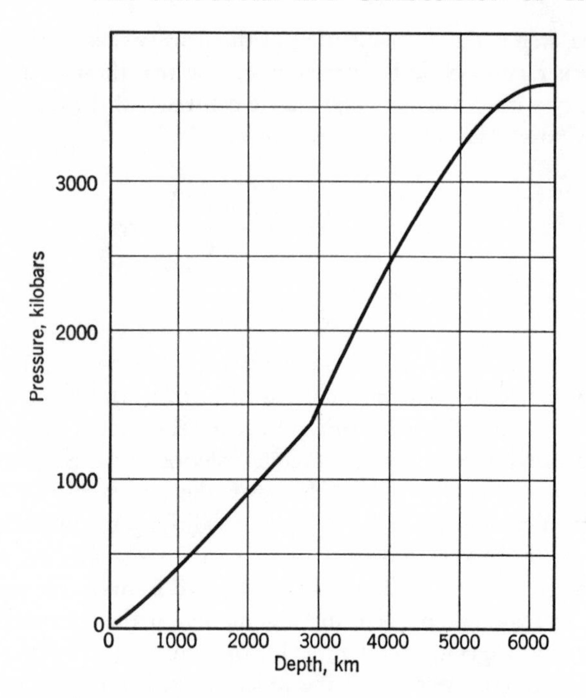

Figure 3.3 Pressure distribution in the earth's interior. (From Bullen, *An introduction to the theory of seismology*. Courtesy of Cambridge University Press)

earth's center is found to be 3640 kilobars, or somewhat over 3,000,000 atmospheres.

TEMPERATURES WITHIN THE EARTH

We know the variation of density and pressure within the earth to a considerable degree of precision, but estimates of temperatures at depth are little more than intelligent guesses. Direct observations in mines and boreholes show that temperature increases with depth, although the rate of increase varies greatly from place to place. This variation is characteristic of the crustal layers, and at quite moderate depths the temperature gradient probably becomes uniform. Measured thermal gradients in the crust range from 10 to 50°/km, and an average value of 30°/km is often used. From the thermal gradient and the thermal conductivity of the rocks the flow of heat toward the surface can be calculated. The average value of this heat flow is 1.4×10^{-6} cal/cm²/sec. This is about 50 cal/cm² annually, sufficient to melt a sheet of ice 6 mm thick (latent heat of fusion of ice is 80 cal/g).

This shows how little the earth's internal heat can influence climatic conditions; 6 mm of ice can easily be melted away by a few hours' sunshine or be formed in one frosty night. However, when computed for the area of the earth it is nevertheless much larger than the amount of heat brought to the surface by the spectacular activity of volcanoes.

A significant fact is that the average oceanic heat flow is approximately the same as the average continental heat flow. This was a surprising discovery, because a large fraction of the heat flow from the continents is accounted for by the radioactive elements in the continental crust, but the oceanic crust is thin, less radioactive, and incapable of providing more than about 10% of the oceanic heat flow. Hence the oceanic heat flow must come mainly from the mantle, whereas the continental heat flow comes mainly from the crust. Therefore, the upper mantle below the oceans must be chemically different than that below the continents, at least with respect to the radioactive elements. The upper mantle below the continents is evidently impoverished in these heat-producing elements, and the temperature distribution in the upper mantle is quite unsymmetrical, as indicated by Figure 3.4. The thermal gradient decreases quite rapidly with depth. Below the

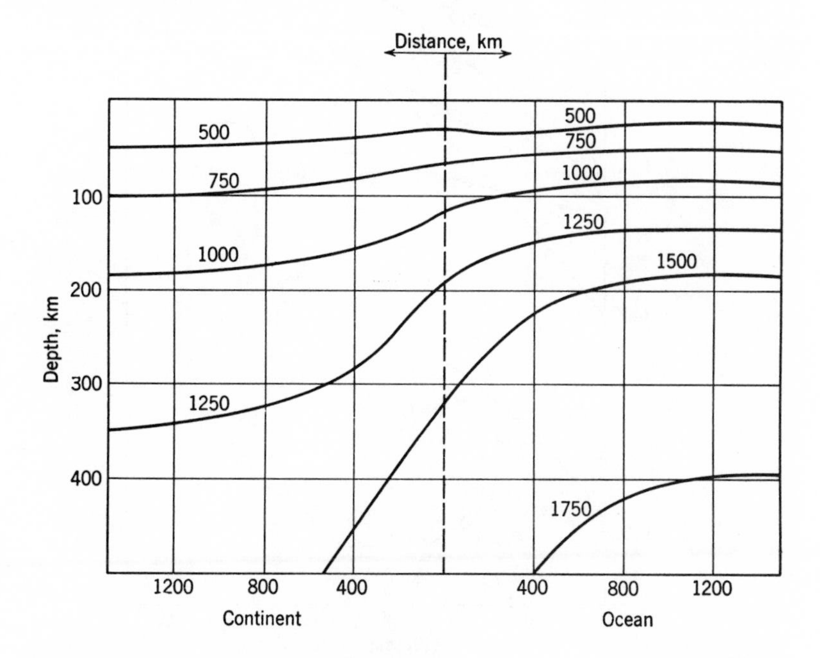

Figure 3.4 The distribution of temperatures under oceans and continents for a mantle having an average uranium concentration of 0.033 parts per million. (MacDonald, 1964)

upper mantle it must be quite low, for otherwise the flow of heat to the surface would be greater than the observed amount. Seismic evidence shows that the mantle is solid throughout, and therefore the temperature at any depth within it cannot exceed the melting range of the material at that depth. Data on the variation of melting point with pressure for silicates are scanty, but for diopside (Figure 3.5) about 10°/kilobar is indicated, or 3°/km. On this basis the temperature at the bottom of the mantle cannot be greater than about 10,000°. However, the melting point curves for the silicates in Figure 3.5 have a pronounced curvature, and their slopes decrease markedly with increasing pressure, so that the melting points at great depths will be considerably lower than given by the preceding extrapolation. The melting-

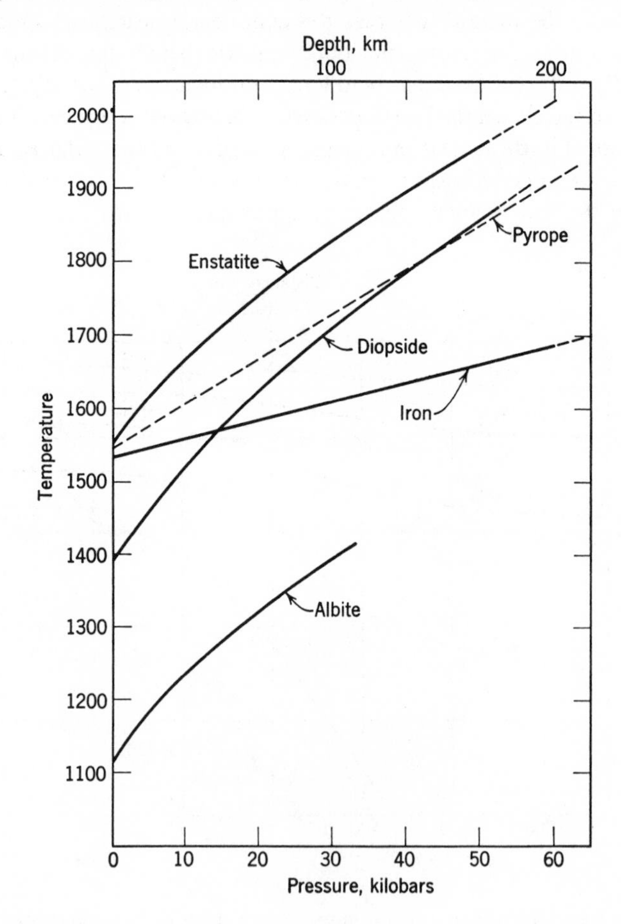

Figure 3.5 Melting temperatures versus pressure. (Boyd, 1964)

point of the material at the base of the mantle has been estimated at 7500°, with an uncertainty of the order of 2000°. This is consistent with a molten iron core, because the melting temperature of iron at the pressure of the core-mantle boundary is believed to be about 5500°.

THE INTERNAL STRUCTURE OF THE EARTH

The interpretation of seismic data provides a primary threefold division of the earth into crust, mantle, and core (Figure 3.6). Together with other geophysical evidence, these data also give some indication of the physical properties of the material making up the three parts. It now remains to make plausible deductions as to the actual constitution of the earth's interior, using both the above-mentioned evidence and other significant information, such as the relative abundances of the elements and the composition of meteorites.

The crust is directly accessible to our observations, in the upper part at least, and general agreement exists as to its major features. It is heterogeneous and varies in thickness from place to place. Marked differences

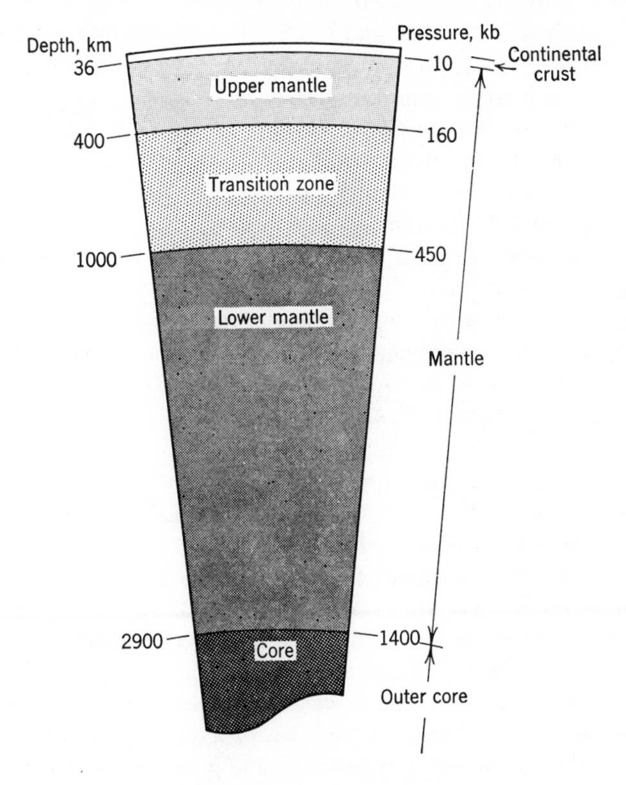

Figure 3.6 The internal structure of the earth.

exist, especially between the continents and the deep ocean basins. The Mohorovičić discontinuity is a datum which is at different depths in different geological environments. Under all explored ocean basins it is between 10 and 13 km below sea level; under the continents it is usually about 35 km below sea level, descending to greater depths (as much as 60 km) below active mountain belts. The evidence indicates than in the ocean basins we have a depth of about 4 km of sea water, underlain by ½–2 km of unconsolidated sediments, followed by material of basaltic composition (5–8 km) to the Mohorovičić discontinuity.

It has been customary to interpret the seismic evidence on the continental crust as indicating two principal layers, an upper one of granitic or granodioritic composition and a lower one of basaltic composition. These two layers correspond to the sial (i.e., material rich in Si and Al) and the sima (rich in Si and Mg), respectively. More refined seismic work, coupled with the geological heterogeneity of exposed continental areas, has largely refuted this concept of crustal layering. The continental crust is a mosaic of sediments, metamorphosed sediments, igneous intrusions of different kinds, and volcanics, faulted and broken into blocks of various shapes and sizes; however, there is probably a gradual change in average composition of the material, from granitic near the surface to gabbroic farther down. The earth's crust is in isostatic adjustment, the irregularities of the surface being compensated by the distribution of materials of different density within the crust, and possibly within the upper mantle.

The Mohorovičić discontinuity separates the heterogeneous crust from the more homogeneous mantle and is marked by a sudden increase in the velocities of seismic waves. The geophysical data indicate that the mantle has a layered structure; the upper mantle, to a depth of about 400 km, is separated from the lower mantle by a transition zone about 600 km thick. The nature of the upper mantle and the transition zone has been the subject of intensive study and extensive speculation in recent years. It is now realized that the key to many geological and geochemical problems—such as the origin of magmas, the triggering of deep-focus earthquakes, the possibility of continental drift—lies in the upper mantle and the transition zone. Our understanding of the nature of the upper mantle has been greatly increased by improved equipment and techniques for studying minerals and rocks at high temperatures and pressures. It is now possible to reproduce in the laboratory the physical conditions corresponding to depths of 400 km or more within the earth. The identification of the material making up the mantle is largely based on samples of possible mantle material brought up as inclusions in volcanic pipes, laboratory experiments on the behavior of minerals and rocks at high temperatures and pressures, and our knowledge of elemental abundances. Experiments indicate that only three rock types—

dunite (olivine), peridotite (olivine and pyroxene), and eclogite (garnet and pyroxene)—have elastic properties of the right order to give the observed wave velocities in the upper mantle. This would also include material of chondritic meteorite composition, essentially a peridotite with a small content of plagioclase feldspar. These rocks are all made up largely of magnesium-iron silicates. Inclusions of dunite and peridotite are sometimes abundant in volcanic rocks (basalts and the diamond-bearing kimberlites); eclogite inclusions are less common. Basalt magmas on Hawaii appear to originate at depths of about 60 km, whereas the presence of diamonds in kimberlite indicates a greater depth of origin, over 100 km.

The nature of the Mohorovičić discontinuity has been the subject of considerable controversy. One school of thought considers it a physical discontinuity, the result of a phase change from lower crustal rocks of gabbroic composition to eclogite, which is higher-density material of essentially the same composition. The alternative view is that the discontinuity is a chemical one, the upper mantle having ultrabasic composition (dunite or peridotite). The problem has been reviewed by Clark and Ringwood (1964), who endeavor to correlate the geochemical, geophysical, and petrological information. They prefer an ultrabasic model for the upper mantle, with an overall composition corresponding to a mixture of one part of basalt to three parts of dunite, which they call pyrolite (pyroxene-olivine rock). Fractional melting of this material would provide the basaltic magma, which has been copiously injected into and through the crust throughout geological time, and leave a residual dunite or peridotite. The mineralogy of pyrolite varies as a function of temperature and pressure, and Ringwood has shown that material of this composition could crystallize in four distinct assemblages, as follows:

(a) Olivine and amphibole	Ampholite
(b) Olivine + Al-poor pyroxene + plagioclase	Plagioclase pyrolite
(c) Olivine + Al-rich pyroxene + spinel	Pyroxene pyrolite
(d) Olivine + Al-poor pyroxene + garnet	Garnet pyrolite

Because the geotherms in the upper mantle differ considerably below continents and oceans (Figure 3.4), there are regional differences in the mineralogical composition of the pyrolite. The inferred composition of the upper mantle below Precambrian shield areas, average continental areas, and oceanic areas is illustrated in Figure 3.7. Under the Precambrian shields the mantle consists, to a considerable depth, of dunite or peridotite with minor segregations of eclogite. This layer is not so thick under the average continental areas and is thin or replaced by ampholite under the oceanic areas.

An important development in our understanding of the transition zone in

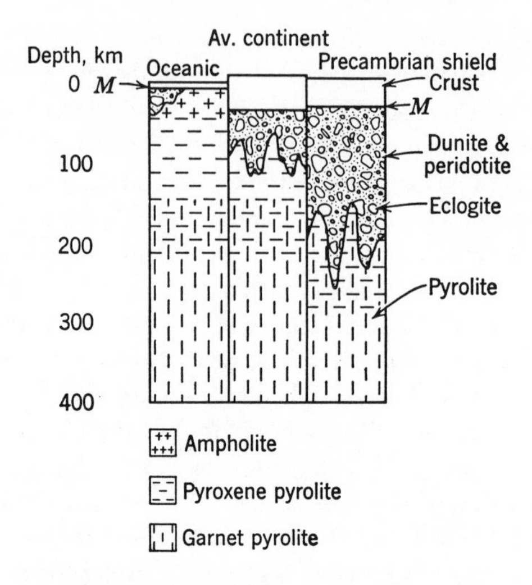

Figure 3.7 Petrological model for the upper mantle, as suggested by Clark and Ringwood (1964); *M* = Mohorovičić discontinuity.

the mantle has been the realization of the significance of polymorphic changes taking place at high temperatures and pressures. This idea was originally due to Bernal, who in 1936 suggested that the discontinuity between the upper mantle and the transition zone results from the formation of a high-pressure polymorph of olivine. He pointed out that the analogous compound Mg_2GeO_4 exists in two modifications, one with the structure of olivine and the other an isometric phase with the spinel structure and 9% higher density. Birch (1952) suggested other polymorphic transformations that might be expected within the mantle, such as the change of $MgSiO_3$ from a pyroxene to an ilmenite structure, and of SiO_2 to a rutile structure. Some of these predicted polymorphic changes have been confirmed by recent laboratory work. The rutile form of SiO_2 was made by Stishov and Popova in 1961 and later named stishovite; it is stable at about 130 kb and 1600°. Extrapolation of experimental results on $(Mg,Fe)_2SiO_4$ solid solutions shows that an olivine-spinel transition will occur in Mg_2SiO_4 at about 130 kb and 600°. Some substances with pyroxene structure are converted into ilmenite structures at high pressures, but laboratory experiments indicate that $MgSiO_3$ is probably converted into Mg_2SiO_4 and SiO_2 (stishovite) at about 120 kb in the temperature range 500°–2000°. All these transformations can be expected in the transition zone.

Clark and Ringwood have applied these data to the elucidation of the transition zone. As a first approximation they assume it is composed essentially of MgO and SiO_2, corresponding to a mixture of olivine, Mg_2SiO_4, and pyroxene, $MgSiO_3$. These phases would be stable to a depth of about 400 km. Below that depth a series of transformations into closer-packed phases would take place. In the light of available data, they postulate the following transformations with increasing depth:

$$2MgSiO_3 \text{ (pyroxene)} = Mg_2SiO_4 \text{ (olivine)} + SiO_2 \text{ (stishovite)}$$

$$Mg_2SiO_4 \text{ (olivine)} = Mg_2SiO_4 \text{ (spinel)}$$

$$Mg_2SiO_4 \text{ (spinel)} + SiO_2 \text{ (stishovite)} = 2MgSiO_3 \text{ (ilmenite)}$$

$$Mg_2SiO_4 \text{ (spinel)} = MgSiO_3 \text{ (ilmenite)} + MgO \text{ (periclase)}$$

This series of transformations would become complete around a depth of 1000 km. They result in a density increase from 3.2 to 3.9 (referred to zero pressure). It is significant that the ultimate attainment of a close-packed state involves successive transformations through several intermediate states, reflected in the considerable depth of the transition zone. In addition, the mantle contains additional elements, mainly iron, calcium, aluminum, and sodium, which are in atomic substitution in the principal phases. Their presence influences the preceding reactions to some degree, in effect smearing out the transitions and resulting in a continuous rather than a stepwise increase in density.

The lower mantle, between 1000 and 2900 km, appears to be homogeneous, and presumably consists of a mixture of $(Mg, Fe)SiO_3$ with the ilmenite structure, and $(Mg, Fe)O$, periclase. The $FeO/FeO + MgO$ ratio (molecular) has been estimated to be between 0.1 and 0.2. Ringwood has suggested the possibility of even more closely packed structures in the lower mantle, the ilmenite structure transforming to the perovskite ($CaTiO_3$) structure and the periclase structure to the CsCl structure; however, experimental evidence for these transformations is lacking.

The belief that the earth has an iron core predates the seismic evidence for its existence. The idea apparently originated with Daubrée in 1866 and was based on the composition of meteorites. The concept of the iron core has become thoroughly entrenched in geophysical and geochemical thought, although Kuhn and Rittmann in 1941 and Ramsey in 1948 suggested that the core consists of the same material as the mantle but in a highly compressed form. However, the chemical difference between mantle and core has been confirmed by experiments with explosively generated shock pressures equal to those of the core. The results show that it is not possible to make a satisfactory core of light metals or their oxygen compounds. The physical properties of the core require elements of the transition group, and only iron is sufficiently abundant. The properties of iron are close to those required and can be adjusted with small amounts of lighter elements.

The seismic discontinuity between the mantle and the core is a sharp one, and its position is known with considerable accuracy (± 2.5 km, according to Jeffreys). The passage of *S* waves through the core has not been observed (a few reported instances are generally regarded as spurious), and this is interpreted as indicating that the core is liquid at least in its outer part. Additional evidence for the fluidity of the core is obtained from the analysis of the bodily tide of the earth. Bullen suggests that the second-order discontinuity at a depth of about 5000 km is due to a change in rigidity and that below this depth the core is solid.

A plausible origin for the magnetic field of the earth and for its secular variation can be deduced from the presence of a liquid iron core, as shown by Elsasser. He suggests that the earth's magnetic field is the result of electric currents flowing in the interior and that such currents are to be expected in the highly conducting metal core rather than in the silicate mantle. The secular variation of the geomagnetic field is then interpreted as the expression of thermal convection currents in the fluid core. Bullen points out, however, that if the inner core is solid it may conceivably be ferromagnetic and would thus provide a possible explanation of the earth's magnetic field. Nevertheless, since iron loses its ferromagnetism at a comparatively low temperature, the Curie point, and since the Curie point is hardly affected by pressure, serious objections can be raised to the idea of a ferromagnetic core.

THE ZONAL STRUCTURE OF THE EARTH

The earth may therefore be considered as made up of an iron core, a fairly homogeneous silicate mantle, and a heterogeneous silicate crust. This picture of its internal structure and composition is consistent with its mass, its moment of inertia, and the presence of discontinuities indicated by seismic waves. The decreasing silica content in the succession of materials from the crust into the mantle agrees with petrological experience. The probable composition of an earth of this kind is also consistent with the relative abundances of the elements. No one of these points, nor all of them together, proves that this picture is necessarily a true one, but at least it agrees with the available data. To complete the picture, we must add to the crust, mantle, and core three further zones: the atmosphere, the hydrosphere, and the biosphere. The atmosphere is the gaseous envelope that surrounds the earth. The hydrosphere is the discontinuous shell of water, fresh and salt, making up the oceans, lakes, and rivers. The biosphere is the totality of organic matter distributed through the hydrosphere, the atmosphere, and on the

Table 3.1. The Structure of the Earth

Name	Important Chemical Characters	Important Physical Characters
Atmosphere	N_2, O_2, H_2O, CO_2, inert gases	Gas
Biosphere	H_2O, organic substances, and skeletal matter	Solid and liquid, often colloidal
Hydrosphere	Salt and fresh water, snow, and ice	Liquid (in part solid)
Crust	Normal silicate rocks	Solid
Mantle	Silicate material, probably largely olivine and pyroxene or their high-pressure equivalents	Solid
Core or siderosphere	Iron-nickel alloy	Upper part liquid, lower part possibly solid

*Table 3.2. Volume and Masses of Earth Shells**

	Thickness (km)	Volume ($\times 10^{27}$ cm^3)	Mean Density (g/cm^3)	Mass ($\times 10^{27}$ g)	Mass (per cent)
Atmosphere	—	—	—	0.000005	0.00009
Hydrosphere	3.80 (mean)	0.00137	1.03	0.00141	0.024
Crust	17 "	0.008	2.8	0.024	0.4
Mantle	2883	0.899	4.5	4.016	67.2
Core	3471	0.175	11.0	1.936	32.4
Whole earth	6371	1.083	5.52	5.976	100.00

* Data for the biosphere is not included on account of its relatively small mass and the lack of precise figures.

surface of the crust. Table 3.1 gives the important features of these zones, and Table 3.2 gives data on thickness, volume, mean density, and mass of each.

The atmosphere, the hydrosphere, and the biosphere, although geochemically important, contribute less than 0.03% of the total mass of the earth. Hence, in arriving at an average composition for the earth, these zones may be ignored. Even the crust makes up less than 1% of the whole. Thus the bulk composition of the earth is essentially determined by that of the mantle and the core. However, in view of the significance of the crust in geology and geochemistry its composition is discussed in detail at this point.

THE COMPOSITION OF THE CRUST

The average composition of the crust is in effect that of igneous rocks, since the total amount of sedimentary and metamorphic rocks is insignificant in comparison to the bulk of the igneous rocks, and in any case their average composition is not greatly different. Clarke and Washington (1924) estimate that the upper 10 miles of the crust consist of 95% igneous rocks, 4% shale, 0.75% sandstone, and 0.25% limestone. Where sedimentary rocks are present they form a relatively thin veneer on an igneous basement, except where locally thickened in orogenic belts.

Clarke and Washington have made an exhaustive study of the data available for computing an average composition of igneous rocks. The basis for their computations was Washington's compilation of 5159 "superior" analyses. The analyses were grouped geographically, and the averages of these groups agreed fairly well with one another. In other words, the composition of the earth's crust is approximately the same in different regions, provided these are large enough to eliminate local variations. However, the SiO_2 percentage is markedly lower for rocks from oceanic areas, such as the islands of the Atlantic and Pacific Oceans, further evidence for the belief that the sial shell is thin or absent below the ocean basins. Clarke and Washington's estimate of the average composition of igneous rocks, which is sometimes used as the average composition of the crust, is, in effect, the composition of continental areas rather than the crust as this term has been used earlier in this chapter.

The overall average of the 5159 analyses, recalculated to 100 with the elimination of H_2O and minor constituents, is as follows:

SiO_2	Al_2O_3	Fe_2O_3	FeO	MgO	CaO	Na_2O	K_2O	TiO_2	P_2O_5
60.18	15.61	3.14	3.88	3.56	5.17	3.91	3.19	1.06	0.30

This composition does not correspond to any common igneous rock but is intermediate between that of granite and basalt, which incidentally make up the bulk of all igneous rocks.

Numerous objections have been raised to the method of arriving at an average composition of igneous rocks by averaging analyses, but a more satisfactory procedure has yet to be proposed. The principal objections are

1. The uneven geographical distribution of analyses.
2. Their non-statistical distribution over the different rock types.
3. The lack of allowance for the actual amounts of the rocks represented by the analyses.

The basis for the first objection is, of course, that Europe and North America have been more adequately investigated and are represented by far more analyses per unit-area than other parts of the earth. However, Clarke showed that the averages for the individual continental areas are in marked agreement in spite of the widely different coverage, a fact which suggests that the results can probably be accepted as a reasonable approximation.

The second objection is valid in that an average of published rock analyses must inevitably give undue weight to the rare and unusual rock types, and insufficient weight to the abundant and uniform types such as granites and basalts. This objection is probably not so serious as has been asserted, since in a large number of analyses the unusual types will be drawn from the whole range of rock compositions and will tend to give a true average. Also, as Clarke and Washington point out, their 1924 average was made from analyses of fresh, unaltered rocks only, thus eliminating many analyses of unusual rocks.

The third objection, that all analyses are given equal weight regardless of the areas occupied by the rocks and therefore of their relative amounts, can be countered if it can be shown that in the average the variations due to this cause offset one another. It is true that one rock, say a basalt, is exceedingly abundant, whereas another may be merely a narrow dike. Such differences may not affect the mean appreciably, since the relatively insignificant rocks range in composition from persilicic to subsilicic just as the more abundant rocks do. Furthermore, the surface exposure of a rock is no certain measure of its real volume and mass, for a small exposure may be merely the peak or crest of a large subterranean body and a large exposure may represent only a thin layer.

An interesting confirmation of the general reliability of the figures of Clarke and Washington was provided by Goldschmidt. He suggested that if it were possible to obtain an average sample of a large part of the earth's crust consisting mainly of crystalline rocks, its analysis would give a reliable picture of the composition of the crust as a whole. Such an average sample was provided, he pointed out, by the glacial clay widely distributed in southern Norway. This clay represents the finest rock flour deposited by melt water from the Fennoscandian ice sheet. From 77 analyses of different samples of this material he calculated the following average analysis:

SiO_2	Al_2O_3	Fe_2O_3 + FeO	MgO	CaO	Na_2O	K_2O	H_2O	TiO_2	P_2O_5
59.12	15.82	6.99	3.30	3.07	2.05	3.93	3.02	0.79	0.22

These figures agree remarkably with those of Clarke and Washington, especially when the effects of hydration and solution, which result in the leaching of sodium and calcium, are taken into account.

A different approach to this problem is that of Poldervaart (1955). He has analysed the composition of the crust in terms of four major geological divisions—the deep oceanic region, the continental shields, the young folded belts, and the continental platform and slopes. For each of these divisions he has determined an average composition in terms of the estimated compositions and amounts of the major rock types. From these data he finally arrived at an average composition for the crust, that is, the rock material above the Mohorovičić discontinuity. This average, reduced to a water-free basis, is as follows:

SiO_2	Al_2O_3	Fe_2O_3	FeO	MnO	MgO	CaO	Na_2O	K_2O	TiO_2	P_2O_5
55.2	15.3	2.8	5.8	0.2	5.2	8.8	2.9	1.9	1.6	0.3

This average reflects more truly than the figures of Clarke and Washington the composition of the crust as a whole, since it gives adequate consideration to the submarine region. Comparison with the average made by Clarke and Washington shows somewhat lower silica and alkalies and higher iron, magnesium, and calcium, reflecting the ferromagnesian nature of the submarine crust.

So far we have considered only the major elements, those which are determined as a matter of course in a rock analysis, and have omitted those present in lesser amounts. Obviously, the determination of the average amounts of the minor elements is a more difficult problem. However, in the last 30 years our knowledge of the relative and absolute abundances of the less common elements has greatly increased, mainly as a result of the introduction of methods of quantitative spectrographic analysis, colorimetric determination, neutron activation, and isotope dilution. Three distinct procedures have been used in arriving at abundance figures for minor and trace elements: (*a*) the averages of many individual analyses; (*b*) the analysis of mixtures of many different rock types; and (*c*) the determination of the proportion of the trace element to some more common element with which it is geochemically associated (e.g., the abundance of rubidium can be estimated from the abundance of potassium and the average Rb:K). Table 3.3 presents the data on the average abundances of the elements in the earth's crust (illustrated graphically in Figure 3.8), and the abundances in a standard granite and a standard diabase. The data on the average crustal abundances are those for the continental crust (the suboceanic crust is probably close in composition to average basalt); data for those elements more abundant than 1000 g/ton are taken from Clarke and Washington; for the minor and trace elements most of the figures are from a critical compilation by Taylor (1964), who combined averages for granites and basalts in the proportions of 1:1. The standard granite G-1 and diabase W-1 were originally prepared about 1948 from a granite from Westerly, Rhode Island, and a diabase from Centerville, Virginia, and have the distinction of being the most extensively analysed rocks in the world. These rocks were prepared to serve as geochemical standards; they were analyzed for major constituents by many laboratories in all parts of the world, and practically all the minor and trace elements have been determined in them by a wide variety of techniques. The figures given for the individual elements are considered the

Table 3.3. The Average Amounts of the Elements in Crustal Rocks in Grams per Ton or Parts per Million

(omitting the rare gases and the short-lived radioactive elements)

Atomic Number	Element	Crustal Average	Granite (G-1)	Diabase (W-1)
1	H	1,400	400	600
3	Li	20	24	12
4	Be	2.8	3	0.8
5	B	10	2	17
6	C	200	200	100
7	N	20	8	14
8	O	466,000	485,000	449,000
9	F	625	700	250
11	Na	28,300	24,600	15,400
12	Mg	20,900	2,400	39,900
13	Al	81,300	74,300	78,600
14	Si	277,200	339,600	246,100
15	P	1,050	390	650
16	S	260	175	135
17	Cl	130	50	
19	K	25,900	45,100	5,300
20	Ca	36,300	9,900	78,300
21	Sc	22	3	34
22	Ti	4,400	1,500	6,400
23	V	135	16	240
24	Cr	100	22	120
25	Mn	950	230	1,320
26	Fe	50,000	13,700	77,600
27	Co	25	2.4	50
28	Ni	75	2	78
29	Cu	55	13	110
30	Zn	70	45	82
31	Ga	15	18	16
32	Ge	1.5	1.0	1.6
33	As	1.8	0.8	2.2
34	Se	0.05		
35	Br	2.5	0.5	0.5
37	Rb	90	220	22
38	Sr	375	250	180
39	Y	33	13	25
40	Zr	165	210	100
41	Nb	20	20	10
42	Mo	1.5	7	0.05
44	Ru	0.01		

Table 3.3. The Average Amounts of the Elements in Crustal Rocks in Grams per Ton or Parts per Million (Continued)

Atomic Number	Element	Crustal Average	Granite (G-1)	Diabase (W-1)
45	Rh	0.005		
46	Pd	0.01	0.01	0.02
47	Ag	0.07	0.04	0.06
48	Cd	0.2	0.06	0.3
49	In	0.1	0.03	0.08
50	Sn	2	4	3
51	Sb	0.2	0.4	1.1
52	Te	0.01		
53	I	0.5		
55	Cs	3	1.5	1.1
56	Ba	425	1,220	180
57	La	30	120	30
58	Ce	60	230	30
59	Pr	8.2	20	2
60	Nd	28	55	15
62	Sm	6.0	11	5
63	Eu	1.2	1.0	1.1
64	Gd	5.4	5	4
65	Tb	0.9	1.1	0.6
66	Dy	3.0	2	4
67	Ho	1.2	0.5	1.3
68	Er	2.8	2	3
69	Tm	0.5	0.2	0.3
70	Yb	3.4	1	3
71	Lu	0.5	0.1	0.3
72	Hf	3	5.2	1.5
73	Ta	2	1.6	0.7
74	W	1.5	0.4	0.45
75	Re	0.001	0.0006	0.0004
76	Os	0.005	0.0001	0.0004
77	Ir	0.001	0.006	
78	Pt	0.01	0.008	0.009
79	Au	0.004	0.002	0.005
80	Hg	0.08	0.2	0.2
81	Tl	0.5	1.3	0.13
82	Pb	13	49	8
83	Bi	0.2	0.1	0.2
90	Th	7.2	52	2.4
92	U	1.8	3.7	0.52

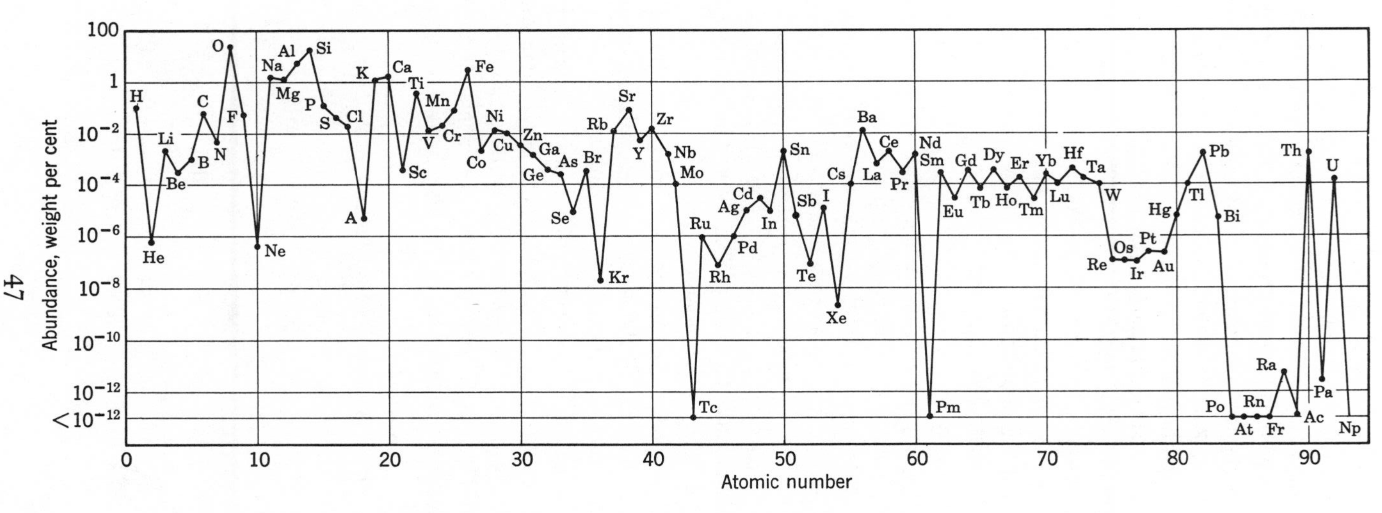

Figure 3.8 Crustal abundances of elements of atomic numbers 1 to 93.

best choice from the available data, which not infrequently vary considerably. The data on these two rocks provide a meaningful comparison with those for the crustal average, based as that is on equal quantities of granite and basalt (the volcanic equivalent of diabase). Comparisons show that for many elements the figures are consistent throughout. For a few there are notable discrepancies, some of which can readily be explained—for example, G-1 is unusually high in thorium, a peculiar characteristic of the Westerly granite. Thanks in considerable part to the availability of G-1 and W-1, the quality and extent of minor and trace element data on rocks has improved markedly in recent years. It can fairly be said that the figures in Table 3.3 provide reliable norms for most elements, and thus make it possible to recognize abnormal enrichment or depletion of an element in any rock. The figures for G-1 and W-1 also show that some elements (such as Be, Rb, Ba) are markedly enriched in the granite, others (such as B, Sc, Ni) are enriched in the diabase, whereas some (such as Zn, Ga, Ge) show a rather uniform abundance in these contrasted rock types.

Some interesting features of Table 3.3 may be noted. Eight elements—O, Si, Al, Fe, Ca, Na, K, Mg—make up nearly 99% of the total. Of these oxygen is absolutely predominant. As Goldschmidt first pointed out, this predominance is even more marked when the figures are recalculated to atom per cent and volume per cent (Table 3.4). The earth's crust consists almost entirely of oxygen compounds, especially silicates of aluminum, calcium, magnesium, sodium, potassium, and iron. In terms of numbers of atoms oxygen exceeds 60%. If the volume of the different atoms, or rather ions, is calculated, oxygen makes up more than 90% of the total volume occupied by the elements. Thus the crust of the earth is essentially a packing of oxygen anions, bonded by silicon and the ions of the common metals. As Goldschmidt remarked, the lithosphere may well be called the oxysphere.

Table 3.3 shows that some of the elements that play a most important part in our economic life and that have long been known to and used by man

Table 3.4. The Commoner Chemical Elements in the Earth's Crust

	Weight Per Cent	Atom Per Cent	Radius (A)	Volume Per Cent
O	46.60	62.55	1.40	93.77
Si	27.72	21.22	0.42	0.86
Al	8.13	6.47	0.51	0.47
Fe	5.00	1.92	0.74	0.43
Mg	2.09	1.84	0.66	0.29
Ca	3.63	1.94	0.99	1.03
Na	2.83	2.64	0.97	1.32
K	2.59	1.42	1.33	1.83

are actually quite rare. For example, copper is less abundant than zirconium; lead is comparable in abundance with gallium; mercury is rarer than any of the so-called "rare earths." Complementary to this is the relative abundance of many unfamiliar elements: rubidium is present in amounts comparable with nickel, vanadium is much more abundant than tin, scandium more abundant than arsenic, and hafnium, one of the last elements to be discovered, is more abundant than iodine. Evidently we must draw a clear distinction between the abundance of an element and its availability. Some elements, although present in the crust in considerable amounts, are systematically dispersed throughout common minerals and never occur in any concentration. Vernadsky called these *dispersed elements;* examples are rubidium, dispersed in potassium minerals, and gallium, in aluminum minerals. Other elements, such as titanium and zirconium, form specific minerals which, in turn, are widely dispersed in small amounts through some of the commonest rocks. The longest known and most familiar elements are those forming the major constituents of easily recognized minerals, minerals that are readily converted into useful industrial materials.

Fersman introduced the term *clarke*, defining it as the average percentage of an element in the earth's crust. Thus the clarke of oxygen is 46.60, of silicon, 27.72. In discussing dispersion and concentration of the elements, Vernadsky introduced a further term, the *clarke of concentration*, which is a factor showing the concentration of an element within a particular deposit or even a particular mineral. Thus, if the clarke of manganese is 0.1, the clarke of concentration of manganese is pyrolusite is 632, in rhodonite, 419, in a psilomelane with 50% Mn, 500. This factor is useful in the consideration of the migration and deposition of the elements and in the discussion of ore deposits. An ore is simply a deposit in which the concentration clarke of the element sought reaches a figure sufficient to make extraction profitable. Table 3.5 is an evaluation of the concentration clarkes necessary to form ore bodies of the commoner metals.

The availability of an element depends in large part on its ability to form individual minerals in which it is a major constituent; the most unavailable elements are those which form no minerals of their own but occur in amounts generally much less than 1% in minerals of other elements—indium, rubidium, gallium, hafnium, rhenium, etc. Even for the "common" elements, however, the dispersed amounts exceed, by a vast factor, the amount available in so-called "ore deposits"; for example, clay minerals are a far more readily available source of aluminum than is bauxite, but the problem with the clay minerals is the technical one of extraction. Magnesium is extracted from sea water, in which it is present to the extent of 0.13%, although there are vast deposits of olivine that contain about 30%. The technical availability of a number of the rarer elements is conditioned in part by their

Table 3.5. Concentration Clarkes for Ore Bodies of the Commoner Metals

Metal	Clarke	Minimum Per Cent Profitably Extracted	Concentration Clarke Necessary for an Ore Body
Al	8.13	30	4
Fe	5.00	30	6
Mn	0.10	35	350
Cr	0.02	30	1500
Cu	0.007	1	140
Ni	0.008	1.5	175
Zn	0.013	4	300
Sn	0.004	1	250
Pb	0.0016	4	2500
U	0.0002	0.1	500

inherently useful properties and in part by their being obtainable as byproducts from the extraction of more abundant elements. An example of the first kind is beryllium, which, although absolutely and relatively rare (its sole industrial source is as the mineral beryl, irregularly dispersed in a few granite pegmatites), has such valuable properties as an alloying element that it is an important industrial material. On the other hand, a number of exceedingly rare elements could be produced relatively easily as byproducts from extraction processes. For example, the electrolytic refining of copper can provide tellurium in sufficient quantities to encourage the development of industrial uses of this element. Similarly, Goldschmidt pointed out that if any demand for gallium or germanium existed large amounts of these elements (1000 tons or so annually) could be extracted from the ashes of certain coals. The demand for germanium for industrial purposes has resulted in the application of Goldschmidt's suggestion to the production of commercial amounts of this element. The annual world consumption of the elements and their compounds is summarized in Table 3.6, and details of sources and price in Appendix III.

Table 3.6. Annual World Consumption of the Elements (and their compounds) in Tons, in Powers of 10

10^9–10^{10}:	C
10^8–10^9:	Na, Fe
10^7–10^8:	N, O, S, K, Ca
10^6–10^7:	H, F, Mg, Al, P, Cl, Cr, Mn, Cu, Zn, Ba, Pb
10^5–10^6:	B, Ti, Ni, Zr, Sn
10^4–10^5:	Ar, Co, As, Mo, Sb, W, U
10^3–10^4:	Li, V, Se, Sr, Nb, Ag, Cd, I, Rare Earths, Au, Hg, Bi
10^2–10^3:	He, Be, Te, Ta

THE COMPOSITION OF THE EARTH AS A WHOLE

The bulk composition of the earth is essentially determined by the composition and relative amounts of the mantle and the core, since they make up over 99% of its mass. By using the deductions regarding the earth's interior discussed above, an average composition for the earth as a whole can be calculated.

For the purpose of this calculation the following assumptions about the composition of the core and the mantle are made:

1. The iron alloy of the core has the composition of the average for nickel-iron in chondrites, and includes the average amount (5.3%) of FeS in these meteorites.
2. The composition of mantle plus crust is the same as the oxidic material (silicates plus small amounts of phosphates and oxides) of the average chondrite.

These assumptions are certainly oversimplifications. For example, all Ni is assumed to be in the core, whereas terrestrial olivine always contains nickel, that from peridotite inclusions in basalt usually having 0.2–0.3% Ni. Some nickel is certainly combined in mantle silicates. Again, the assumption that all iron sulfide is incorporated in the core is clearly an approximation at best, for some sulfide is certainly present in mantle and crust. However, the incorporation of iron sulfide in the core rather neatly meets the geophysical requirement that the density of the outer core is less than that of iron by about 10%. If 27.1% nickel-iron, density 7.90, is mixed with 5.3% troilite, density 4.80, the resulting mixture will have a density, at zero pressure, of 7.15, thereby meeting this requirement quite closely. This obviates the necessity of incorporating Si in the core, as has been suggested by Ringwood and others. A nickel-iron core with free Si is also incompatible with a mantle containing appreciable amounts of FeO, if mantle and core are in chemical equilibrium.

Mass of core equals 32.4% mass of earth and mass of mantle plus crust equals 67.6% mass of earth; the core is assumed to consist of nickel-iron of composition of average chondrite plus 5.3% troilite, and mantle and crust are assumed to have a composition of silicate material of average chondrite.

The results of this calculation are given in Table 3.7. The figures obtained (rounded to two significant figures) are compared in Table 3.8 with the published estimates of Washington (1925) and Niggli (1928). Earlier estimates, such as those of Farrington, Linck, and Clarke, are based on erroneous earth models and give values too high for nickel and iron, as a result of assuming an iron core considerably larger than that now accepted. Washington based his calculation on the following structure of the earth:

1. Central core, composition average of iron meteorites (27.30%).
2. Lithosporic shell, composition average of equal weights of meteoritic nickel-iron and pallasite olivine (8.51%).
3. Ferrosporic shell, composition average of stony meteorites (22.55%).
4. Peridotitic shell, composition average of achondritic meteorites (40.08%).
5. Basaltic shell, composition Daly's average of basalts (1.08%).
6. Granitic shell, composition that of "average igneous rock" (Clarke and Washington) (0.48%).

Niggli assumed an eclogite shell within the earth, which served to raise the figure for aluminum somewhat; otherwise his earth model corresponds quite closely with that of Washington. As can be seen, the three estimates are in reasonable accord for the major elements.

In view of the uncertainties and assumptions involved in calculations of this kind too much weight should not be given to the numerical results obtained. Nevertheless, some significant deductions of a semiquantitative order are probably justified. The results indicate that about 90% of the earth is made up of four elements, Fe, O, Si, and Mg. The only other elements that may be present in amounts greater than 1% are Ni, Ca, Al, and S. Seven elements, Na, K, Cr, Co, P, Mn, and Ti may occur in amounts from 0.1 to 1%. Thus the earth is made up almost entirely of 15 elements, and the percentage of all the others is negligible, probably 0.1% or less of the whole.

Table 3.7. Calculation of the Bulk Composition of the Earth

	Metal	Troilite	Silicate	Total
Fe	24.58	3.37	6.68	34.63
Ni	2.39			2.39
Co	0.13			0.13
S		1.93		1.93
O			29.53	29.53
Si			15.20	15.20
Mg			12.70	12.70
Ca			1.13	1.13
Al			1.09	1.09
Na			0.57	0.57
Cr			0.26	0.26
Mn			0.22	0.22
P			0.10	0.10
K			0.07	0.07
Ti			0.05	0.05
	27.10	5.30	67.60	100.00

Table 3.8. The Chemical Composition of the Earth

Element	Atomic Number	Table 12	Washington's Estimate	Niggli's Estimate†
Fe	26	35	39.76	36.9
O	8	30	27.71	29.3
Si	14	15	14.53	14.9
Mg	12	13	8.69	6.73
Ni	28	2.4	3.16	2.94
S	16	1.9	0.64	0.73
Ca	20	1.1	2.52	2.99
Al	13	1.1	1.79	3.01
Na	11	0.57	0.39	0.90
Cr	24	0.26	0.20	0.13
Co	27	0.13	0.23	0.18
P	15	0.10	0.11	0.15
K	19	0.07	0.14	0.29
Ti	22	0.05	0.02	0.54
Mn	25	0.22	0.07	0.14

* *Am. J. Sci.* (5) **9,** 361, 1925.
† *Fennia* **50,** No. 6, 1928.

Let us now compare the order of abundance of the major elements in the earth with their relative abundances in other parts of the solar system. Table 3.9 gives these data for the earth as a whole, for the earth's crust, for the average composition of meteorites, and for the sun. The most striking feature is the relative uniformity throughout; the same elements appear in all columns, albeit in different order. The rarity of hydrogen and helium in the earth and meteorites is, of course, easily understood. When we come to consider the nonvolatile elements, it is seen that iron, silicon, and magnesium in general head the lists and are followed by nickel, sodium, calcium, and aluminum. Note particularly that the most abundant elements are all of low atomic number; none with atomic number greater than 30 appears in any of the lists, except for Ba in the crust.

The relative abundances of the elements in the crust are included in Table 3.9 for comparison with the bulk composition of the earth. The main differences are the lesser abundance of iron and magnesium in the crust, the nonappearance of nickel and sulfur, and the increased significance of aluminum, potassium, and sodium. This suggests that the differentiation of the earth has led to a concentration of relatively light, easily fusible alkali-aluminum silicates at the surface.

A comparison of the elemental composition of the mantle plus crust (the silicate material of Table 3.7) with that of the crust (Table 3.3) reveals some interesting fractionations. The mass of the crust (the solid material above

Table 3.9. Relative Abundances (by Weight) of the Elements

Crust*	Whole Earth†	Meteorites‡	Sun§
O	Fe	O	H
Si	O	Fe	He
Al	Si	Si	O
Fe	Mg	Mg	C
Ca	Ni	S	N
Na	S	Ni	Si
K	Ca	Ca	Mg
Mg	Al	Al	S
Ti	Na	Na	Fe
H	Cr	Cr	Ca
P	Mn	Mn	Ni
Mn	Co	P	Na
F	P	Co	Al
Ba	K	K	Cu

* From Table 3.9.
† From Table 3.13.
‡ From Table 3.4.
§ From Table 3.3.

the Mohorovičić discontinuity) is 0.024×10^{24} g, that of the mantle 4.016×10^{27} g, so the mass of the crust is only 0.6% that of the mantle. The major elements O and Si and the minor elements Mn and P show no marked fractionation between mantle and crust. In the crust relative to the mantle, Mg is much diminished; Cr is also strongly diminished, with the average in the crust about 0.02%; Fe, on the other hand, is only slightly less concentrated in the crust than in the mantle. Elements Al and K are strongly enriched in the crust, Na and Ca to a lesser extent—in terms of mineralogical composition the differentiation of crust from mantle has largely been the accumulation of feldspar in the crust.

THE PRIMARY DIFFERENTIATION OF THE ELEMENTS

It seems probable that the earth was formed by the gradual accretion of solid planetesimals—indeed, the chondritic meteorites may be planetesimals left over from the preplanetary stage of the solar system. The chondritic meteorites are made up of three different phases or groups of phases: nickel-iron, iron sulfide, and silicate minerals (mainly olivine and pyroxene). The elements are distributed between these phases according to their relative affinity for metal, for sulfide, or for silicate. The composition of these phases in meteorites has been essentially determined by equilibrium in the Fe—

Mg—Si—O—S system, in which oxygen greatly exceeded sulfur and the sum of the two was insufficient to combine completely with the electropositive elements. Since Fe was more abundant than Mg and Si, most easily reduced to the metal, and with the greatest affinity for sulfur, a system of three essentially immiscible phases—iron-magnesium silicate, iron sulfide, and free iron—resulted, and the amount of combined iron was fixed by the amount of oxygen plus sulfur. The distribution of the remaining electropositive elements was governed by reactions of the type

$$\text{M} + \text{Fe silicate} \rightleftharpoons \text{M silicate} + \text{Fe}$$

$$\text{M} + \text{Fe sulfide} \rightleftharpoons \text{M sulfide} + \text{Fe}$$

that is, by the free energies of the corresponding silicates and sulfides in relation to those of iron sulfide and iron silicate.

In broad outline, we can perceive the factors that determine the distribution of the elements in a system of this kind. Iron, because of its preponderant abundance, was common to all the condensed phases—the metallic phase rich in free electrons, the ionic silicate phase, and the semimetallic, probably covalent, sulfide phase. Elements more electropositive than iron displaced iron from the silicates, in which they were accordingly concentrated. Conversely, less electropositive elements concentrated in the metal, being displaced by iron from ionic compounds. The sulfide phase attracted those elements that form essentially homopolar compounds with sulfur and the metalloids and that cannot exist in an ionic environment with appreciable concentrations of metalloid ions; such elements are in the main the metals of the sulfide group of analytical chemistry.

It is highly significant that the distribution of elements in a gravitational field, such as that of the earth, is controlled not by their densities or atomic weights, as might perhaps be expected, but by their affinities for the major phases that can be formed. This is, in turn, controlled by the electronic configurations of their atoms. For example, uranium and thorium, although of high density, are strongly electropositive elements and have concentrated in the earth's crust as oxides or silicates. Gold and platinum metals, on the other hand, have no tendency to form oxides or silicates, and alloy readily with iron; hence they are presumably concentrated in the earth's core. The distribution of the elements is not directly controlled by gravity, which merely controls the relative positions of the phases; the distribution of the elements within these phases depends upon chemical potentials.

THE GEOCHEMICAL CLASSIFICATION OF THE ELEMENTS

Goldschmidt was the first to point out the importance of this primary geochemical differentiation of the elements. He coined the terms siderophile, chalcophile, lithophile, and atmophile to describe elements with affinity for metallic iron, for sulfide, for silicate, and for the atmosphere, respectively. When he put forward this concept in 1923, few quantitative data were available on which to base his ideas. The geochemical nature of an element could, of course, be established by measuring its distribution between three liquid phases of metal, sulfide, and silicate. He recognized the difficulty of carrying out these measurements in the laboratory, but remarked that meteorites provide us with such an experiment in a fossilized condition. Many meteorites consist of nickel-iron, troilite, and silicate, all of which have presumably solidified from a liquid state. The distribution of a particular element between those three phases would be established when the system was in a liquid condition and can be determined by mechanically separating these phases and analyzing them individually. From such analyses the partition of the element between metal, silicate, and sulfide can easily be calculated.

In later years Goldschmidt and his coworkers made many measurements of the content of various elements in nickel-iron, troilite, and silicate of meteorites. Much work along these lines was also carried out by I. and W. Noddack in Berlin. These measurements have been greatly refined and extended in recent years by the application of new techniques, especially neutron activation analysis. The information derived from meteorites has been supplemented by that obtained by a study of smelting processes, such as the distribution of elements during the smelting of the Mansfeld copper slate in Germany. The smelting of this slate gives a silicate slag, a matte rich in iron and copper sulfide, and metallic iron. Spectrographic measurements of the concentration of many minor elements in the different phases have given distribution coefficients agreeing on the whole with those determined from examination of meteorites.

On the basis of these results an element may be classified according to its geochemical affinity into one of four groups: siderophile, chalcophile, lithophile, and atmophile (Table 3.10). Some elements show affinity for more than one group, because the distribution of any element is dependent to some extent on temperature, pressure, and the chemical environment of the system as a whole. For instance, chromium is a strongly lithophile element in the earth's crust, but if oxygen is deficient, as in some meteorites, chromium is decidedly chalcophile, entering almost exclusively into the sulfo-spinel daubréelite, $FeCr_2S_4$. Similarly, under strongly reducing conditions carbon and phosphorus are siderophile. The mineralogy of an element, although a general guide, may not be altogether indicative of its geochemical character. For example, although all thallium minerals are sulfides, the greater part of the thallium in the earth's crust is contained in potassium minerals, in which the Tl^+ ion proxies for the K^+ ion. In general, the classification of an element as lithophile, chalcophile, or siderophile refers to its behavior in liquid-liquid

Table 3.10. Geochemical Classification of the Elements

(based on distribution in meteorites)

Siderophile	Chalcophile	Lithophile	Atmophile
Fe* Co* Ni*	(Cu) Ag	Li Na K Rb Cs	(H) N (O)
Ru Rh Pd	Zn Cd Hg	Be Mg Ca Sr Ba	He Ne Ar Kr Xe
Os Ir Pt	Ga In Tl	B Al Sc Y La-Lu	
Au Re† Mo†	(Ge) (Sn) Pb	Si Ti Zr Hf Th	
Ge* Sn* W†	(As) (Sb) Bi	P V Nb Ta	
C‡ Cu* Ga*	S Se Te	O Cr U	
Ge* As† Sb†	(Fe) Mo (Os)	H F Cl Br I	
	(Ru) (Rh) (Pd)	(Fe) Mn (Zn) (Ga)	

* Chalcophile and lithophile in the earth's crust.
† Chalcophile in the earth's crust.
‡ Lithophile in the earth's crust.

equilibria between melts. When an element shows affinity for more than one group, it is given in parentheses under the group or groups of secondary affinity.

The geochemical character of an element is largely governed by the electronic configuration of its atoms and hence is closely related to its systematic position in the periodic table (Table 3.11). Lithophile elements are those that readily form ions with an outermost 8-electron shell; the chalcophile elements are those of the B subgroups, whose ions have 18 electrons in the outer shells; the siderophile elements are those of Group VIII and some neighboring elements, whose outermost shells of electrons are for the most part incompletely filled. These factors are reflected by other properties also. Goldschmidt pointed out the marked correlation between geochemical character and atomic volume. If the atomic volume of the elements is plotted against atomic number, the resulting curve shows maxima and minima. All siderophile elements are near the minima; the chalcophile elements are on sections in which the atomic volume increases with the atomic number; they are followed by the atmophile elements, whereas the lithophile elements are near the maxima and on the declining sections of the curve.

Brown and Patterson have shown that if the heat of formation of an oxide is greater than that of FeO the element is lithophile; the difference between the two heats of formation is a measure of the intensity of the lithophile character. Similarly, those elements having oxides with heats of formation lower than FeO are chalcophile or siderophile. Semiquantitative measure of lithophile, siderophile, or chalcophile character is also provided by the electrode potential. Elements with high positive potentials (1–3 volts), such as the alkali and alkaline earth metals, are lithophile; the noble metals, with

Table 3.11. The Geochemical Classification of the Elements in Relation to the Periodic System

Atmophile: N
Lithophile: Na
Chalcophile: Zn
Siderophile Fe

H																	He
Li	Be											B	C	N	O	F	Ne
Na	Mg											Al	Si	P	S	Cl	A
K	Ca	Sc	Ti	V	Cr	Mn	Fe	Co	Ni	Cu	Zn	Ga	Ge	As	Se	Br	Kr
Rb	Sr	Y	Y	Nb	Mo		Ru	Rh	Pd	Ag	Cd	In	Sn	Sb	Te	I	Xe
Cs	Ba	La-Lu	Hf	Ta	W	Re	Os	Ir	Pt	Au	Hg	Tl	Pb	Bi			
			Th		U												

high negative potentials, are siderophile; elements falling in the intermediate range are generally chalcophile.

THE PREGEOLOGICAL HISTORY OF THE EARTH

The pregeological history of the earth comprises the sequence of events through which it passed before the time when the physical condition of the surface became much as it is today—a surface partly of rocks, partly of water, with a mean temperature essentially determined by solar radiation. This is the zero datum for geological time, a datum following which the earth's surface has been subject to the normal processes of weathering and erosion. The pregeological history began when the earth originated as an individual body within the universe. We do not know the length of time covered by this period with any degree of precision; the earth's crust has existed for over 3000 million years, and the age of the earth as an individual body is probably of the order of 4500 to 5000 million years.

As discussed in Chapter 2, the earth is believed to have formed from the same material that gave rise to the sun and the other planets. Hypotheses differ as to the mode of aggregation; one considers that the earth condensed from incandescent gas, the other that it grew by the gradual accretion of solid particles in a cosmic dust cloud. With these things in mind it is interesting to compare the composition of the earth with that of the sun. The major differences are the great abundance of hydrogen and helium and to a lesser extent carbon and nitrogen in the sun. These differences can be adequately explained on either hypothesis, since these elements are either gaseous or form stable gaseous compounds. If the earth grew by the accretion of solid particles, it is readily seen that these elements have always been of low abundance on the earth; if the earth was formed by the condensation of a mass of incandescent material, the light gases would have tended to escape from the earth's gravitational field during cooling from the high temperatures. One piece of evidence, however, favors the accretion hypothesis; not only are the light gases of low abundance on the earth but the heavy gases are also. Krypton and xenon are about a million times less abundant than their immediate neighbors in the periodic table, a definite deficiency compared to their probable abundance throughout the universe. If the earth condensed from incandescent material, these gases should be more abundant than they are; if the earth was formed by the accretion of solid particles, then they were never present in any amount. On the accretion hypothesis the earth's atmosphere was formed by the release of gases occluded and chemically combined in the solid particles.

Whichever theory of origin we accept, the internal structure of the earth with its marked density stratification seems to demand that at an early period in its history it was sufficiently hot for metallic iron to liquefy and undergo gravitational accumulation to form the core. If, as appears probable, the earth formed by the accretion of planetesimals resembling chondritic meteorites, the initial state of our planet was an intimate mixture of nickel-iron, troilite, and silicate minerals, broadly homogeneous throughout. The time of formation of the earth by this accretion of planetesimals is usually placed 4.5×10^9 years ago, the figure obtained for the time when certain stony meteorites were enriched in and certain iron meteorites were depleted in uranium and thorium. However, it is not an obvious fact that the latter event and the formation of the earth were necessarily contemporaneous, and there has been a tendency to push back the time of accretion of the earth to about 5×10^9 years.

The next stage after accretion of the earth was one of heating, mainly by radioactivity. Here it is essential to bear in mind the greater effect of the shorter-lived radioactive nuclides as we go back in time. This is graphically illustrated in Figure 3.9, which is based on the known heat production by radioactive nuclides and their estimated abundances in the earth. This figure shows that at 5×10^9 years ago radiogenic heat production in the earth would be approximately six times greater than today, and most of this heat was contributed by K^{40} and U^{235}, whereas today these nuclides are greatly diminished in amount and the principal sources of radiogenic heat are U^{238} and Th^{232}.

Under these circumstances the roughly homogeneous undifferentiated earth would heat up comparatively rapidly. With the chondritic values for the radioactive elements, the temperatures within the primitive earth would increase as shown in Figure 3.10. On this basis, the melting temperature of iron was reached in about 600 million years at a depth of a few hundred kilometers; at this stage a change in regime sets in, and the curves for later times are not realized.

Elsasser (1963) has cogently linked this thermal evolution with the formation of the iron core. He points out that the thermal evolution depicted in Figure 3.10 results in the melting of the free iron in a zone at a depth of several hundred kilometers and the accumulation of this metal into a layer of molten iron. A coherent layer of molten iron within a predominantly silicate mantle is clearly unstable, because material of higher density overlies that of lower density. Elsasser shows that this results in the development of a large "drop" (Figure 3.11), which then sinks towards the center of the earth, displacing the lighter silicates. It is essentially the mechanism of salt dome or igneous intrusion, on a giant scale and in the reverse direction. The sinking of this large mass of molten iron to the center would transform a considerable amount of gravitational energy into heat, sufficient to raise the temperature of the interior of the earth by some 2000°. Thus the process of core formation was strongly exothermic and self-accelerating, and Elsasser

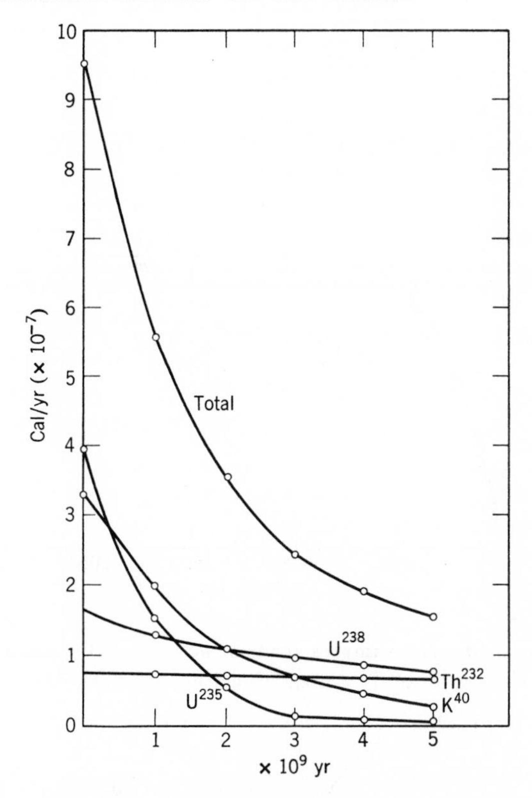

Figure 3.9 Radiogenic heat formed in the Earth by U, Th and ^{40}K separately and together plotted against time. (After Vinogradov)

believes that the time required for such a "drop" to form and fall was of the order of 100,000 years.

With the movement of the free iron, approximately one third of the total mass of the earth, towards the center, the whole planet was profoundly reorganized. While the iron was unsymmetrically sinking the lighter fractions of the silicate material were unsymmetrically rising, undergoing partial fusion, mixing, and unmixing. The original surface was probably engulfed and digested. Under these circumstances it is not surprising that crustal rocks older than 3.5×10^9 years have not been found. The development of thick, stable crustal blocks had to await the decay of radioactivity and the solidification of most of the mantle.

The development of the crust is a further problem and one of special importance to geochemistry. Considerable diversity of opinion exists. One

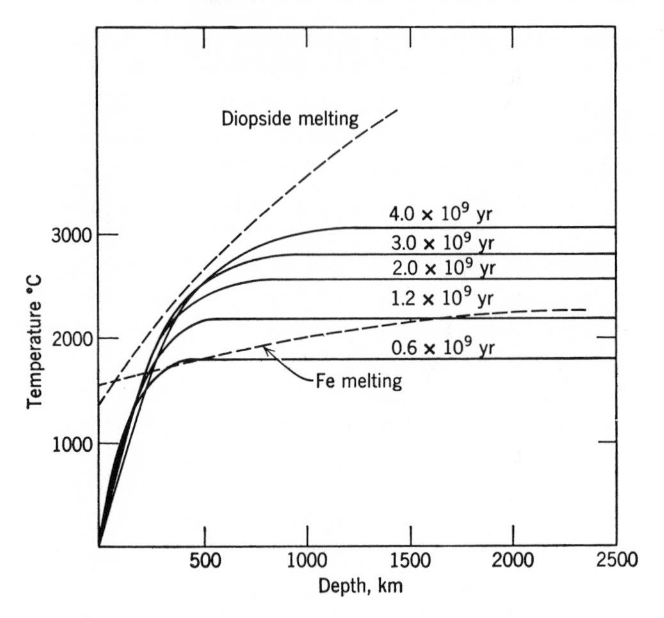

Figure 3.10 Temperatures in a homogeneous Earth heated by the radioactivity of the average chondrite. (MacDonald)

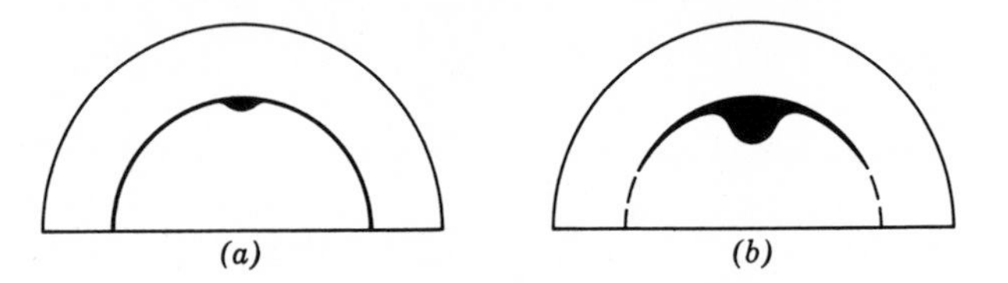

Figure 3.11 Stages of drop formation from a heavy liquid layer. (Elsasser, 1963)

school of thought pictures the entire crust of the earth solidifying as a basalt. No differentiation into continents and ocean basins would then exist. Localized fractures in this initially thin basaltic crust would channel lava, hot gases, and solutions from the hotter subcrustal layers, and these fractures would provide nuclei (something like present-day island arcs) for the nascent continents. The action of solutions carrying silica and alkalies would produce less basic crustal rocks. Erosion and sedimentation, at first on a small scale, would accelerate the chemical differentiation of crustal material. This concept pictures the continents growing through pregeological (and geological) time by the successive welding of sedimentary deposits and new island arcs to the initial continental nuclei.

The other school of thought pictures the initial continents as the final product of the solidification of the crust. It carries the differentiation by crystallization, already exemplified by the separation of basaltic crust from peridotitic mantle, a step further and considers the continental nuclei as being raft-like masses essentially of quartz and feldspar (i.e., of granitic or granodioritic composition). Elsasser suggest that the asymmetrical distribution of continents and oceans, the land areas being concentrated in the Northern Hemisphere, may have been produced by the asymmetrical production of the large iron "drops." The first protocontinent would form in the region above the first large falling drop, and the antipodal region would remain oceanic.

Present evidence suggests that the development of a stable crust was not a rapid, once-for-all process. The interval between the age of the earth and the age of the oldest dated rocks is over a billion years. Great crustal instability during this interval is indicated. The original continental nuclei may have been partly remelted and rebuilt many times before they grew to a sufficient size and thickness to resist further engulfment. However, by about 3.5×10^9 years ago the pattern of continental nuclei and ocean basins was probably established, since rocks approaching this age are known in most of the continents.

The origin of the atmosphere and hydrosphere is discussed in detail in later chapters. The growth of the earth by accretion in a solar dust cloud suggests an initial atmosphere rich in hydrogen and lacking free oxygen—it probably consisted largely of hydrogen, helium, nitrogen, water vapor, methane, and ammonia (methane and ammonia are present in the atmospheres of the larger planets). The gravitational attraction of the earth is insufficient to retain hydrogen and helium, and they would gradually diffuse into outer space. The evolution of the primeval atmosphere into the present one, which consists essentially of nitrogen and oxygen, is pictured as beginning with the photochemical dissociation of water vapor by solar radiation, thereby producing free oxygen. Methane would then oxidize to carbon dioxide and water. Other photochemical reactions would produce more complex organic compounds, ultimately leading to self-reproductory systems, in effect living matter. The probability of such reactions has been demonstrated in the laboratory by Miller (1957), who passed electrical discharges through a mixture of methane, ammonia, water vapor, and hydrogen and was able to demonstrate the presence of several carboxylic and amino acids in the products.

The various suggestions as to how life originated are summarized by Oparin (1957) and Barghoorn (1957). They show that the initial development of living matter was dependent on the preexistence of organic molecules such as amino acids. However, once cells of a kind were established photosynthesis became possible. Photosynthesis was a revolutionary event in earth history: it enabled organisms to collect and use solar radiation directly and thereby utilize carbon dioxide to synthesize more complex compounds; it was also responsible for the gradual release of oxygen into the atmosphere. Photosynthesis may have originated through the development of certain colored compounds capable of acting as catalysts in photochemical reactions. These compounds were probably porphyrins, which are easily synthesized and are extremely stable for organic substances. Porphyrins are formed by the condensation of four pyrrole derivatives around a metal atom [pyrrole is a five-membered ring compound with the empirical formula $(CH)_4NH$]. It is extremely suggestive that porphyrins act as light receivers in the chlorophyll of plants, as oxygen carriers in many animals, and as respiratory pigments in cells. Once photosynthesis was established, a vastly greater supply of energy was available for organisms, whether they used light themselves or ate others that did so.

The length of the period between the establishment of a stable crust and the appearance of life is difficult to evaluate. The oldest well-dated fossils are the microflora in the Gunflint chert, a Precambrian formation on the north shore of Lake Superior, whose age is fairly well established at 1.9×10^9 years. Stromatolites, laminated calcareous structures interpreted as fossil algae, are known from the Bulawayan System of Southern Rhodesia, believed to be older than 2.6×10^9 years, but neither the interpretation nor the age is firmly established. Most of the oldest sedimentary rocks are conglomerates, sandstones, and shales, showing few features indicating that conditions in the atmosphere and hydrosphere were very different from those today. However, Cloud (1965) and others have pointed out a number of suggestive features. The oldest formations show a notable dearth of limestones, which first become abundant in the younger Precambrian deposits, with ages less than 1.2×10^9 years; this indicates a rather high partial pressure of CO_2 in the atmosphere and hydrosphere, acting to keep $CaCO_3$ in solution. Red beds, colored by ferric oxide produced by oxidative weathering, appear about the same time, suggesting an increase in atmospheric oxygen over that present earlier. The most characteristic of Precambrian sediments, which are apparently unique to this system and are not repeated in later times, are banded-iron formations of the Lake Superior type, none of which is younger than about 1.7×10^9 years. These banded-iron formations occur on all the continents, and their formation implies a nonoxidizing, CO_2-rich atmosphere under which large quantities of iron could be transported to the sites of deposition in the soluble ferrous state. The combined evidence of geochemistry, paleontology, and stratigraphy indicates that photosynthesis by green plants was probably established at least 2.0×10^9 years ago, but that atmospheric oxygen was first available in relatively large quantities about 1.2 × 10^9 years ago. By the end of the Precambrian organic evolution had produced the metazoa (multicelled organisms), and probably by that time the atmosphere and hydrosphere were not notably different geochemically from their present state.

SELECTED REFERENCES

Barghoorn, E. S. (1957). Origin of life. *Geol. Soc. Amer. Mem.* **67,** 75–86. A useful summary of this problem, with a comprehensive bibliography.

Birch, F. (1952). Elasticity and constitution of the earth's interior. *J. Geophys. Research* **57,** 227–286. A comprehensive survey of the data and the deductions that can be made therefrom on the nature of the mantle and the core.

Birch, F. (1965). Speculations on the earth's thermal history. *Bull. Geol. Soc. Amer.* **76,** 133–154. A critical examination of the development of core, mantle, and crust as a result of radioactive heating in an originally cool accreted body.

Boyd, F. R. (1964). Geological aspects of high-pressure research. *Science* **145,** 13–20. An informative review article on the application of laboratory investigations to the nature of the earth's interior.

Bullen, K. E. 1963. *An introduction to the theory of seismology* (third edition). 381 pp. Cambridge University Press, Cambridge, England. Chapters 12 and 13 give the methods used and results obtained in deducing the nature of the interior of the earth from seismic data.

Clark, S. P., and A. E. Ringwood (1964). Density distribution and constitution of the mantle. *Rev. Geophys.* **2,** 35–88. A thorough review of this problem, utilizing geochemical, petrological, geothermal, and seismic data.

Clarke, F. W. (1924). *The data of geochemistry.* Chapter 1.

Clarke, F. W., and H. S. Washington (1924). The composition of the earth's crust. *U. S. Geol. Survey, Profess. Paper* 127. 117 pp. An exhaustive discussion and analysis of the data on the chemistry of the earth's crust.

Cloud, P. E. (1965). Significance of the Gunflint (Precambrian) microflora. *Science* **148,** 27–35. A carefully reasoned and well-documented account of the probable course of organic evolution in Precambrian time.

Day, F. H. (1963). *The chemical elements in nature.* 372 pp. Harrap & Co., London, England. A useful account of the geochemistry of the individual elements, with special reference to their mode of occurrence and extraction.

Donn, W. L., B. D. Donn, and W. G. Valentine (1965). On the early history of the earth. *Bull. Geol. Soc. Amer.* **76,** 287–306. A carefully reasoned discussion of the astrophysical and geological conditions during the first billion years of earth history.

Elsasser, W. M. (1963). Early history of the earth. *Earth Science and Meteoritics*, pp. 1–30. John Wiley & Sons, New York. This paper introduces a novel but convincing hypothesis for the formation of the core by partial melting in a homogeneous earth.

Fleischer, M. (1965). Summary of new data on rock samples G-1 and W-1. *Geochim. Cosmochim. Acta* **29,** 1263–1283. An evaluation of the data on minor and trace elements in these two standard rocks; for earlier work, see *U. S. Geol. Surv. Bull.* 1113 (1960) and 980 (1951) and *Geochim. Cosmochim. Acta* **26,** 525–543 (1962).

Goldschmidt, V. M. (1954). *Geochemistry.* Chapter 2.

Jeffreys, H. (1959). *The earth* (fourth edition). 420 pp. Cambridge University Press, Cambridge, England. A classic work on the nature and evolution of the earth.

Kuiper, G. P. (ed.) (1954). *The earth as a planet.* 751 pp. University of Chicago Press, Chicago. This volume contains a number of papers relevant to the present chapter, especially one by Bullard on the interior of the earth and one by Wilson on the development and structure of the crust.

MacDonald, G. J. F. (1964). Dependence of the surface heat flow on the radioactivity of the earth. *J. Geophys. Res.* **69,** 2933–2946. An examination of temperature variations within the earth as a result of radioactive disintegrations, and their geophysical and geochemical implications.

Miller, S. L. (1957). The formation of organic compounds on the primitive earth. *Ann. New York Acad. Sci.* **69,** 260–275. Describes experiments designed to demonstrate the production of complex organic compounds by photochemical reactions in simple gas mixtures.

Oparin, A. I. (1957). *The origin of life on the earth* (third edition). 516 pp. Oliver and Boyd, London. A new and revised edition of a classic work on this subject.

Poldervaart, A. (ed.) (1955). Crust of the earth. *Geol. Soc. Amer. Spec. Paper* 62. 762 pp. A symposium on the crust of the earth with contributions from many authorities; papers of special pertinence for this chapter are those by Poldervaart on the chemistry of the crust, by Birch on the physics of the crust, and by Rubey on the development of the hydrosphere and atmosphere.

Rankama, K., and Th. G. Sahama (1950). *Geochemistry*. Chapters 2, 3, 4, and 10.

Taylor, S. R. (1964). Abundance of chemical elements in the continental crust: a new table. *Geochim. Cosmochim. Acta* **28,** 1273–1285. A critical compilation and evaluation of the large amount of recent data on minor and trace elements in granites and basalts.

Vinogradov, A. P. (1961). The origin of the material of the earth's crust. *Geochemistry,* 1–32. A thorough discussion of the processes by which the earth's crust may have formed by the partial melting of meteoritic material.

CHAPTER FOUR

SOME THERMODYNAMICS AND CRYSTAL CHEMISTRY

INTRODUCTION

In the introductory chapter geochemistry was described as dealing with the abundance, distribution, and migration of the chemical elements in the earth. The basic units of geochemical investigation are thus the elements, either in the form of atoms or more often as charged particles or ions. Atoms and ions have a certain energy content which changes when they undergo a physical or chemical transformation. In redistribution and recombination of the chemical elements in minerals and rocks the atoms or ions lose part of their energy and yield more stable systems. Every rock exemplifies the laws conditioning the stability of crystal lattices, laws which follow the general principles of the structure of matter and of thermodynamics. Geochemical concepts are completely meaningful when they indicate the relationship between atoms, ions, and crystal lattices and the factors determining their equilibrium conditions. Such relations properly belong to the field of chemistry, but in view of their significance in geochemistry a brief summary is given here.

FUNDAMENTAL THERMODYNAMIC EQUATIONS

The ideas of thermodynamics are most readily expressed in the form of equations. Only a brief summary of these equations is presented here; for further information reference should be made to a textbook of chemical thermodynamics or to the discussions of Turner and Verhoogen (1960) and Ramberg (1955) on the application of thermodynamic principles to geological processes.

A thermodynamic system is characterized by certain fundamental properties, divisible into two types: (*a*) extensive or capacity properties, such as mass, volume, and entropy, which depend upon the quantity of matter in the system; (*b*) intensive properties, such as temperature, pressure, and chemical potential, which are independent of the amount of matter in the system. The total energy of all kinds contained within a system is called its internal energy (E). It depends only on the state of the system and cannot be determined in absolute values; it is the change in internal energy which the system undergoes in passing from one state to another that is significant.

The first law of thermodynamics states that energy can neither be created nor destroyed. If a system undergoes a change of state, and E_1 is the internal energy in the first state and E_2 the internal energy in the second state, then ΔE, the change in internal energy, is

$$\Delta E = E_2 - E_1$$

If in this change an amount of energy q is absorbed by the system in the form of heat and an amount of energy w leaves the system as mechanical work, then

$$\Delta E = q - w$$

For an infinitesimal change

$$dE = dq - dw$$

The mechanical work, dw, is usually measured by a change in volume dV acting against a hydrostatic pressure P, in which case

$$dw = P\,dV$$

so that

$$dE = dq - P\,dV$$

The second law of thermodynamics can be stated in the following form: "In any reversible process the change in entropy (dS) of a system is measured by the heat (dq) received by the system divided by the absolute temperature (T), that is, $dS = dq/T$; for any spontaneous irreversible process $dS > dq/T$." Thus, for a reversible process, the preceding equation can be restated in the form

$$dE = T\,dS - P\,dV$$

Because many processes take place at constant pressure with only heat energy and mechanical energy involved and because under these circumstances the heat energy absorbed by the system from its surroundings is equal to the increment in the $(E + PV)$ function of the system, it has been found convenient to define a function H, called the enthalpy, such that

$$H = E + PV$$

Hence for any infinitesimal transformation

$$dH = dE + P\,dV + V\,dP$$

If the transformation occurs at constant pressure (i.e., $dP = 0$), then

$$dH = dq$$

that is, the change in enthalpy in any process at constant pressure is measured by the heat received and for this reason is often referred to as the heat of a reaction.

The Helmholtz free energy (A) and the Gibbs free energy (G) are defined by the following equations:

$$A = E - TS$$

$$G = E - TS + PV$$

The Gibbs free energy is especially significant in connection with processes that take place at constant temperature and pressure. Under these conditions

$$dG = dE - T\,dS + P\,dV$$

If the reaction is reversible

$$dE = T\,dS - P\,dV$$

and

$$dG = 0$$

Now a reversible reaction is synonymous with a state of equilibrium, and so we have as a criterion of equilibrium at constant pressure that the Gibbs free energy (hereafter referred to simply as the free energy) of the reactants must be equal to that of the products.

All geochemical processes may be regarded as striving towards equilibrium, which may be approached rather closely when composition, temperature, and pressure remain approximately constant for a long time. However, equilibrium if attained is seldom preserved, owing to changes in physical conditions. For example, if a thermally metamorphosed shale reached equilibrium under the conditions of metamorphism it is probably no longer in equilibrium at ordinary temperatures. However, the study of equilibria in laboratory experiments and by thermodynamic methods has thrown a flood of light on geochemical reactions, such as the origin of rocks and minerals, the processes of weathering and decomposition, and other kinds of transformations going on within the earth.

Any reaction, chemical or physical, may be characterized by an equation representing the transition from one state to another. If the reactants are in equilibrium with the products, ΔG for the reaction is zero. A large negative value of ΔG means that the reaction as written tends to proceed nearly to completion; a large positive value means that the reaction tends to proceed in the opposite direction. As a simple example, we may consider the solubility of a solid in a liquid; if the reaction is represented by the equation $S \rightarrow L$, then at saturation ΔG is zero, whereas a large negative value of ΔG means undersaturation, that is, a strong tendency to dissolve, and a positive value of ΔG signifies supersaturation.

Equations have been derived to express the effect of temperature and pressure on equilibrium. For temperature changes the relevant equation is

$$\frac{d(-\Delta G/T)}{dT} = \frac{\Delta H}{T^2}$$

This equation signifies that, if ΔH is positive, an increase in temperature makes ΔG more negative; i.e., if heat is absorbed in a reaction, an increase in temperature causes the reaction to go more nearly to completion. If ΔH is negative, on the other hand, increasing the temperature tends to inhibit the reaction. The effect of pressure is characterized by the equation $d\,\Delta G/dP = \Delta V$, where ΔV is the aggregate change in volume that occurs when the reaction proceeds to completion in the direction indicated. Thus, if ΔV is negative, an increase of pressure makes ΔG more negative; that is, the reaction as written proceeds more nearly to completion. In other words, high pressure favors the existence of materials of small volume, that is, high density. These two equations give quantitative expression to Le Châtelier's principle, which can be stated thus: If a system is in equilibrium, a change in any of the factors determining the conditions of equilibrium will cause the equilibrium to shift in such a way as to nullify the effect of this change. An interesting conclusion is that in general increasing temperature produces the same kind of effect as decreasing pressure. The volume of a substance becomes greater at higher temperatures and lower pressure. Solubility of solids in liquids increases as a rule with increasing temperature but is usually diminished by high pressure.

For a chemical reaction to proceed spontaneously, it is necessary that the total free energy of the products be less than that of the reactants, that is, $dG < 0$. The value of dG is a measure of the driving force or *affinity* of the reaction. To predict whether a certain reaction can occur it is therefore necessary to determine its free-energy change. The factors influencing the free-energy change are (*a*) the nature of the reactants and of the products of reaction, (*b*) their state of aggregation, (*c*) the relative amounts present, and (*d*) the pressure and temperature. The experimental determination of the free-energy change of a chemical reaction is often extremely difficult. Nevertheless, if the discussion is confined to the standard states of the reactants and reaction products, and heat capacity data are available, the problem is often simplified, because the molal free energies of formation of many chemical compounds in their standard states are now known. Under these conditions the free-energy change is merely the sum of the free energies of the products of the reaction minus the sum of the free energies of the reacting substances. These free energies are obtained by multiplying the standard molal free energy of formation of each substance by the number of moles that enter into the reaction. Unfortunately, few data are yet available on the free energies of silicates, but this information is gradually being accumulated.

The major value of thermodynamics in geochemistry is that it provides a general approach to problems of stability, equilibrium, and chemical change. Even with qualitative data it enables predictions to be made regarding the probable course of all types of transformation. In any reaction for which the free energies of all possible phases are known under the specified conditions, thermodynamic equations permit the calculation of the relative amounts of reactants and products at equilibrium. If the amount of the products at equilibrium is found to be very small, then the reaction is not favored under the specified conditions. If the amount of the products is large, the suggested reaction is one that may be expected to go under the specified conditions. In geochemical processes, many of which proceed under conditions that cannot be reproduced in the laboratory, thermodynamics provides us with the means for predicting the conditions under which certain reactions may occur, even though we cannot reproduce them experimentally. It is important to realize, however, that thermodynamics cannot predict the *rate* at which a reaction will proceed and does not tell us anything of the mechanism of the reaction.

One of the first practical applications of thermodynamics to the solution of a geochemical problem was a study of the stability of jadeite, $NaAlSi_2O_6$ (Kracek, Neuvonen, and Burley, 1951). Jadeite occurs in metamorphic rocks but at that time had never been made in the laboratory; its comparatively high density suggested that it might only be stable under high pressures. Kracek and his coworkers examined the thermodynamics of the following reactions by which jadeite might be formed:

$$NaAlSi_3O_8 = NaAlSi_2O_6 + SiO_2$$

$$NaAlSiO_4 + NaAlSi_3O_8 = 2NaAlSi_2O_6$$

$$NaAlSiO_4 + SiO_2 = NaAlSi_2O_6$$

By measuring the heats of solution (in HF) of albite ($NaAlSi_3O_8$), nepheline ($NaAlSiO_4$), jadeite, and quartz, they were able to determine ΔH values for each of the above reactions. The entropies of these substances had been measured, so it was then possible to evaluate ΔG from the equation $\Delta G = \Delta H - T\,\Delta S$. The figures obtained indicated that at 25° and 1 atm pressure the first reaction would tend to proceed from right to left (i.e., jadeite would not be formed), whereas the other two reactions would tend to proceed from left to right, with the formation of jadeite. These results show that jadeite is more stable at ordinary temperature and pressure than mixtures of albite and nepheline, or of nepheline and quartz, and that its stability is not conditioned by high pressure. The difficulties of making jadeite in the laboratory are therefore to be ascribed to kinetic factors involving activation energies and rates of reaction.

Even if all the relevant thermodynamic data are not available, Miyashiro (1960) has pointed out that it is frequently possible to derive useful information from heats of reaction alone, if the reactions involve only solids, a situation common under geological conditions. If all the reactants and products are in the solid state the heat capacity changes are small and can be neglected. Under these circumstances the free energy $\Delta G_{T,P}$ of a reaction at temperature T°K and pressure P atm is given by the following equation

$$\Delta G_{T,P} = \Delta H^\circ_{298} - T\,\Delta S^\circ_{298} + P\,\Delta V$$

$$\Delta H^\circ_{298} = \text{heat of reaction at 298.16°K, i.e., 25°C, and 1 atm}$$

$$\Delta S^\circ_{298} = \text{entropy of reaction at 298.16°K and 1 atm.}$$

With all the phases in the solid state, ΔS_{298} and ΔV are usually quite small in comparison with ΔH°_{298}. Thus the heat of such a reaction at 25°C and 1 atm (ΔH°_{298}) is nearly equal to the free energy of the same reaction at any temperature and pressure ($\Delta G_{T,P}$). In other words, in reactions involving solid phases only, the free energy of reaction is usually nearly constant throughout a wide range of temperature and pressure. This is illustrated by the reaction

$$\underset{\text{(forsterite)}}{Mg_2SiO_4} + \underset{\text{(quartz)}}{SiO_2} = \underset{\text{(clinoenstatite)}}{2MgSiO_3} \qquad (1)$$

$$\text{heat of reaction } \Delta H^\circ_{298} = -2300 \text{ cal/mole}$$

$$\text{entropy of reaction } \Delta S^\circ_{298} = -0.35 \text{ cal/deg mole}$$

$$\text{volume change } \Delta V = -4.0 \text{ cm}^3\text{/mole}$$

$$\text{free energy of reaction } \Delta G_{T,P} = -2300 + 0.35T - 0.097P$$

(the ΔV term must be divided by 41.3 to convert it to cal/mole).

Clearly, the effects of temperature and pressure on the free energy of the reaction are very small. A change of temperature of 500° will change the free energy by 175 cal, and a change of pressure of 1000 atm will change it by less than 100 cal.

Similarly, for the reaction

$$\tfrac{1}{2}NaAlSiO_4 + SiO_2 = \tfrac{1}{2}NaAlSi_3O_8 \qquad (2)$$

(nepheline) (quartz) (albite)

$$\Delta G_{T,P} = -2500 - 0.05T + 0.016P$$

For the reaction

$$KAlSi_2O_6 + SiO_2 = KAlSi_3O_8$$

(leucite) (quartz) (orthoclase)

ΔH°_{298} is -4800 cal/mole, ΔV is -2.2 cc/mole, but ΔS°_{298} is not known; therefore $\Delta G_{T,P}$ cannot be evaluated. Nevertheless it is clear from the preceding information that $\Delta G_{T,P}$ cannot be greatly different from ΔH°_{298}. We can use the available data to make deductions of petrogenetic significance. If the SiO_2 content of a rock consisting of clinoenstatite, albite, and orthoclase were gradually decreased, the preceding data indicate that the first reaction to take place would be the conversion of clinoenstatite into forsterite, since of the possible reactions this involves the smallest increase in free energy. This would be followed by the conversion of albite into nepheline, and finally by the conversion of orthoclase into leucite. The order of reaction thus predicted agrees with petrological experience. In rocks slightly undersaturated in SiO_2 olivine, $(Mg,Fe)_2SiO_4$, appears instead of pyroxene, $(Mg,Fe)SiO_3$. At a greater degree of undersaturation nepheline appears in place of albite, and, with a still greater degree of undersaturation, leucite is formed instead of orthoclase. This explains why nepheline is found in association with olivine but not with $(Mg,Fe)SiO_3$ (hypersthene); the free energy relations favor the reaction of nepheline with hypersthene to give albite plus olivine.

However, the difference in the free energies of reaction (1) and reaction (2) above is not large, and high pressures, such as those at depths of 100–200 km, could be expected to reverse the order of the values of the free energies. Under these circumstances silica undersaturation might result in the conversion of albite into nepheline before the conversion of pyroxene into olivine. Considerations of this sort indicate that material of basaltic composition which crystallizes within the crust as a plagioclase-olivine rock might crystallize within the upper mantle as a nepheline-pyroxene rock. A unique advantage of thermodynamic data is that they facilitate the extrapolation of laboratory data to the more extreme conditions that can be expected in geological processes.

THE STATES OF MATTER

Geochemistry is to a large extent concerned with the transformation of matter from one state to another, as exemplified by the crystallization of magmas, the weathering of rocks, the deposition of salts from solution, and generally the formation of minerals over a wide range of temperatures, pressures, and chemical environments. These processes involve a change of state in all or part of the material. Three states of matter are recognized: solid, liquid, and gaseous. This division is a useful one, yet it should be realized that it is to some degree arbitrary. As we generally observe them, these states are sharply marked off from each other by distinctive properties, but under some conditions the boundaries lack definition, and the transitions solid $\rightleftharpoons$ liquid $\rightleftharpoons$ gas may be continuous rather than discontinuous.

In terms of the atomic theory, the state of matter ranges from complete atomic disorder in gases to complete order in crystals. However, complete order is an abstract concept and exists only in perfect crystals at absolute zero. At any temperature above absolute zero the kinetic energy of the atoms causes them to vibrate about their mean positions in the crystal lattice. If the kinetic energy of the atoms becomes sufficiently large, the crystal loses its rigidity, that is, it melts or decomposes. Usually, fusion takes place at a definite temperature, but theoretical considerations indicate that the melting point is a temperature range, possibly too small to measure but nevertheless finite in theory. Melting-point ranges have been observed experimentally. A special case of this phenomenon is illustrated by some complex organic substances that melt to give a liquid in which the molecules still maintain a one- or two-dimensional orientation, which is lost at still higher temperatures.

Liquids were formerly considered more akin to gases than to solids, and, indeed, beyond the critical point the distinction between liquid and gas ceases to have any validity. At temperatures and pressures well below the critical point, however, liquids resemble solids. They diffract X rays, and the diffraction effects show that a considerable degree of order exists in the arrangement of the atoms or molecules. A glass is simply a supercooled liquid held together by bonds extending throughout the structure, much as in the crystalline form of the same substance. The essential difference is that in the crystalline form the atoms are arranged in a symmetrical periodic network, whereas in glasses the degree of orientation and periodicity is much lower.

All matter strives to reach equilibrium with its environment. To accomplish this the atoms try to arrange themselves in such a way that the free energy of the system is a minimum, and in the solid state this arrangement is usually an ordered crystal structure or structures. In geological terms, those minerals are formed that are most in harmony with the physical environment and the bulk composition of the system. Glasses are metastable phases, and even though they may persist for an almost indefinite period they always tend to change into crystalline forms. Thus glasses are uncommon in rocks, and their occurrence signifies unusual conditions of composition and formation.

THE CRYSTALLINE STATE

The most obvious characteristic of crystals that have grown freely is their external form. Morphological crystallography, the study of the geometrical relationships of the faces of crystals, has shown that every crystal can be classified into one of 32 classes based upon symmetry. Haüy, at the beginning of the nineteenth century, conceived that the geometrical complex of crystal faces characteristic of a homogeneous substance must be determined by its internal structure, the molecular or atomic arrangement. In 1912 the truth of Haüy's conception was demonstrated experimentally when the discovery of X-ray diffraction by crystals showed that in the crystalline state there is an orderly, systematic arrangement of atoms. The atomic arrangement largely determines the chemical and physical properties of a crystalline compound and is thus a fundamental feature.

PRINCIPLES OF CRYSTAL STRUCTURE

Since 1912 the crystal structures of many substances have been determined. Since minerals provide a ready source of well-crystallized substances, many of the early workers in this field naturally used them for crystal structure investigations. As a result, and fortunately for the progress of geochemistry, the structures of many minerals were worked out comparatively soon. A valuable summary of the data on the atomic structure of minerals has been provided by Bragg and Claringbull (1965). Goldschmidt and his associates also made important contributions in this field; it was one of Goldschmidt's major services to geochemistry that he early realized the significance of crystal structure in controlling the distribution of elements in the earth's crust.

The basic unit in all crystal structures is the atom (the term *atom* also includes ion in this discussion), which may, however, be associated with other atoms in a group behaving as a single unit in the structure. We may consider atoms as being made up of electric charges distributed through a small sphere which has an effective radius of the order of 1 A (10^{-8} cm). The radius can be measured with considerable accuracy and depends not only upon the nature of the element but also on its state of ionization and the manner in which it is linked to adjacent atoms. For example, the radius of the sodium atom in metallic sodium is 1.86 A, but the radius of the sodium ion is sodium salts is 0.97 A.

The different kinds of interatomic linkage are classed into four bond types: the metallic bond responsible for the coherence of a metal; the ionic or polar bond, which is the linkage in salts such as sodium chloride; the homopolar or coordinate link present in crystals such as the diamond; and the residual or van der Waals' bond, which is responsible for the coherence of the inert gases when condensed to solids at low temperatures. These four types of bonds all impart characteristic properties to the substances in which they occur and provide a convenient basis for the classification of crystal structures. More than one type of bond may occur in a single compound; Evans terms such substances *heterodesmic*, and those in which only one bond type is present *homodesmic*. In heterodesmic structures the physical properties, such as hardness, mechanical strength, and melting point, are in general determined by the weakest bonds, which are the first to suffer disruption under increasing mechanical or thermal strain.

It should always be realized that although these four types of bonding have well-defined properties the classification is arbitrary insofar as the bonding in many compounds may be more or less intermediate. The silicon-oxygen bonds in silica and the silicates are neither purely ionic nor purely covalent but are intermediate in nature. The structure assumed by any solid is such that the whole system of atomic nuclei and electrons tends to arrange itself in a form with minimum potential energy. The energy of a configuration can theoretically be calculated by applying the principles of quantum mechanics, and no distinction between the different types of bonds appears in the rigid mathematical expressions.

Nearly all the common minerals can be looked on as ionic structures, and we can consider them as compounds of oxygen anions with practically all the other elements (except the halogens) acting as cations. The oxygen ion is so large in comparison to most cations that a mineral structure is mainly a packing of oxygen ions with the cations in the interstices. The radii of common ions are given in the appendix and illustrated in Figure 4.1. Hydrogen is not included, because it has unique properties. The hydrogen ion or proton is so small that it can hardly be considered as having any spatial extension; instead it acts rather like a dimensionless center of positive charge. The radius of the OH^- ion is essentially the same as that of the O^{2-} ion; the hydrogen is embedded in the oxygen atom, and the OH group is effectively a sphere.

Because the radius of an ion depends upon its atomic structure, it is related to the position of the element in the periodic table. The following rules are generally valid:

1. For elements in the same group of the periodic table, the ionic radii increase as the atomic numbers of the elements increase; e.g., Be^{2+} 0.35, Mg^{2+} 0.66, Ca^{2+} 0.99, Sr^{2+} 1.12, Ba^{2+} 1.34. This is, of course, to be expected, since for elements in the same group of the periodic table the number of electron orbits around the nucleus, and hence the effective radius, increases in going down the column.

2. For positive ions of the same electronic structure the radii decrease

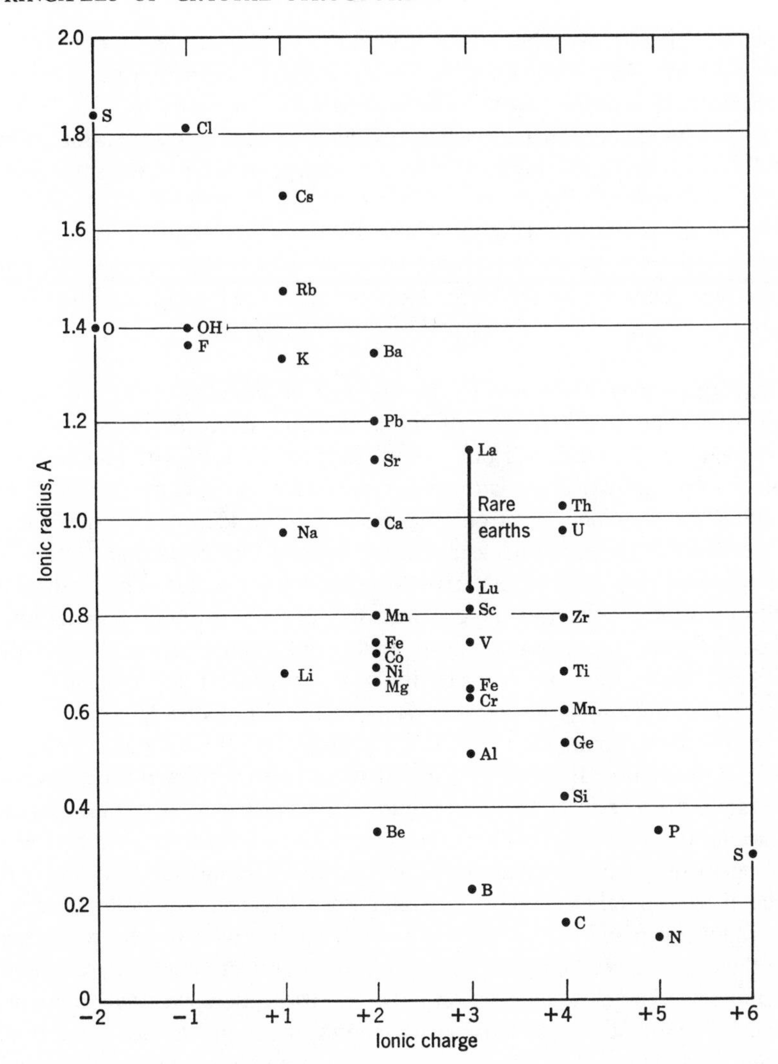

Figure 4.1 The relationship between ionic radius and ionic charge for some of the elements.

with increasing charge. For an example, we may take the elements in the second horizontal row in the periodic table, all of which have two electrons in the inner orbit and eight in the outer orbit:

Na^{+}	Mg^{2+}	Al^{3+}	Si^{4+}	P^{5+}	S^{6+}
0.97	0.66	0.51	0.42	0.35	0.30

Thus in going across a horizontal row in the periodic table the radii of the ions decrease. As electrons are lost the nucleus exerts a greater pull on those remaining, thus decreasing the effective radius of the ion.

3. For an element that can exist in several valence states, i.e., form ions of different charge, the higher the positive charge on the ion, the smaller the radius, for example, Mn^{2+} 0.80, Mn^{3+} 0.66, Mn^{4+} 0.60. The same reason given in the previous rule applies here also; the loss of an electron causes the remaining electrons to be more strongly attracted by the nucleus, thus effectively contracting the outer electron orbits and decreasing the ionic radius.

An apparent contradiction to the first rule is provided by the rare earth elements. The trivalent ions of these elements decrease in radius with increasing atomic number, from 1.14 for La^{3+} to 0.85 for Lu^{3+}. This remarkable feature, known as the lanthanide contraction, is the consequence of the building up of an inner electron shell, instead of the addition of a new shell; as a result the increasing nuclear charge produces an increased attraction on the outer electrons and an effective decrease in ionic radius. The lanthanide contraction also influences the geochemistry of the elements following lutetium; hafnium and tantalum have ionic radii almost identical with the elements above them in the periodic table—zirconium and niobium—and therefore show almost identical crystallochemical properties.

In an ionic structure each ion tends to surround itself with ions of opposite charge; the number that can be grouped around the central ion depends upon the radius ratio between the two. Figure 4.2 is a planar representation of the relationship. Assuming that ions act as rigid spheres of fixed radii, the stable arrangements of cations and anions for particular radius ratios can be calculated from purely geometric considerations (Table 4.1). Table 4.2 gives the radius ratio and predicted coordination number with respect to oxygen for the commoner cations, together with the coordination actually observed in minerals. The close correlation between observation and prediction confirms the assumption that ions act as spheres of definite radius; for cations larger than oxygen, however, the coordination number is less well defined, since the grouping is less regular.

Many cations occur exclusively in a particular coordination; others, for example, aluminum, which has a radius ratio lying near the theoretical boundary between two types of coordination, may occur in both. In such cases the coordination is to some extent controlled by the temperature and pressure at which crystallization took place. High temperatures and low pressures favor low coordination, and low temperatures and high pressures favor higher coordination. High coordination is evidently more economical of space. Aluminum is a good example; in typically high-temperature

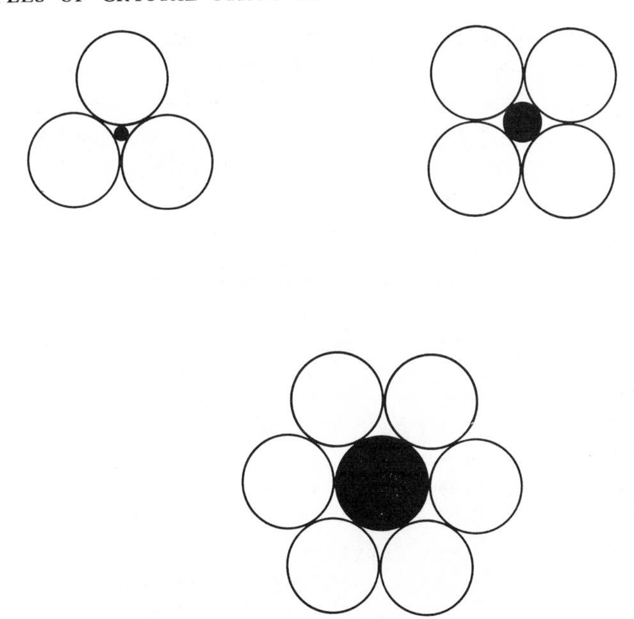

Figure 4.2 Planar representation of the relationship between radius ratio and coordination number.

minerals it tends to assume fourfold coordination and substitute for silicon, whereas in minerals formed at lower temperatures it tends to occur in sixfold coordination. The effects of high pressure on coordination have been dramatically illustrated by the synthesis of stishovite, a polymorph of SiO_2, in which Si is in sixfold coordination instead of the fourfold coordination otherwise universal in silicate minerals.

The structure of ionic compounds is determined primarily by the demands of geometrical and electrical stability. The relative sizes of the ions and the mode of packing must result in the ions being more or less rigidly

Table 4.1. Relationship between Radius Ratio and Coordination Number for Ions as Rigid Spheres

Radius Ratio (R_{cation}/R_{anion})	Arrangement of Anions Around Cation	Coordination Number of Cation
0.15–0.22	Corners of an equilateral triangle	3
0.22–0.41	Corners of a tetrahedron	4
0.41–0.73	Corners of an octahedron	6
0.73–1	Corners of a cube	8
1	Closest packing	12

Table 4.2. The Relationship between Ionic Size and Coordination Number with Oxygen for the Commoner Cations

Ion	Radius (R)	$R/R_{O^{2-}}$	Predicted Coordination Number	Observed Coordination Number
Cs^{+}	1.67	1.19	12	12
Rb^{+}	1.47	1.05	12	8–12
Ba^{2+}	1.34	0.96	8	8–12
K^{+}	1.33	0.95	8	8–12
Sr^{2+}	1.12	0.80	8	8
Ca^{2+}	0.99	0.71	6	6, 8
Na^{+}	0.97	0.69	6	6, 8
Mn^{2+}	0.80	0.57	6	6
Fe^{2+}	0.74	0.53	6	6
V^{3+}	0.74	0.53	6	6
Li^{+}	0.68	0.49	6	6
Ti^{4+}	0.68	0.49	6	6
Mg^{2+}	0.66	0.47	6	6
Fe^{3+}	0.64	0.46	6	6
Cr^{3+}	0.63	0.45	6	6
Al^{3+}	0.51	0.36	4	4, 6
Si^{4+}	0.42	0.30	4	4
P^{5+}	0.35	0.25	4	4
Be^{2+}	0.35	0.25	4	4
S^{6+}	0.30	0.21	4	4
B^{3+}	0.23	0.16	3	3, 4

held in the structure, just as in a house built of blocks, where each block must support its neighbors. More than one structure may fulfill this requirement, in which case the most stable will be that for which the potential energy of the ions is lowest. The requirement of electrical stability means that the sum of positive and negative charges on the ions must balance. This is not achieved by pairing off individual cations and anions; the positive charge on a cation must be considered as divided equally between the surrounding anions, the number of which is determined by relative size of the ions and not by their valency. Pauling stated this in the following rule: "In a stable structure the total strength of the valency bonds which reach an anion from all the neighboring cations is equal to the charge on the anion." This expresses the tendency of any structure to assume a configuration of minimum potential energy, whereby the charges on the ions are as far as possible neutralized by their immediate neighbors.

This rule may appear more or less self-evident, but it is highly significant in the rigorous conditions it imposes on the geometrical configuration of a structure, especially of complex substances such as the silicates. It may be called the cardinal principle of mineral chemistry. It often explains the non-existence of certain types of compounds, although their formulas would be quite possible according to the requirements of valency. For example, the feldspars have linked silicon-oxygen and aluminum-oxygen groups with Si and Al in fourfold coordination. An oxygen linked to Si and Al has a valency of only one fourth left unsatisfied. This cannot be balanced by Mg or Fe in sixfold coordination, for their contribution would be at least one third. It can, however, be balanced by large univalent and bivalent cations having coordination numbers of eight or more, since they can supply the necessary small fractions. Hence we find the feldspars are compounds of Ca, Na, and K and do not contain Mg or Fe.

The above principles are the basis for the crystal chemistry of minerals. They express the conditions for low potential energy and so for high stability. Only very stable compounds can occur as minerals; less stable compounds either do not form in nature or soon decompose. Artificial compounds have been made in which these general principles of crystal structure are not closely followed, but such substances are not found as minerals.

THE STRUCTURE OF SILICATES

In all silicate structures so far investigated (except those formed at extreme pressures) silicon lies between four oxygen atoms. This arrangement appears to be universal in these compounds, and the bonds between silicon and oxygen are so strong that the four oxygens are always found at the corners of a tetrahedron of nearly constant dimensions and regular shape, whatever the rest of the structure may be like. The different silicate types arise from the various ways in which these silicon-oxygen tetrahedra are related to each other; they may exist as separate and distinct units, or they may be linked by sharing corners (i.e., oxygens). Silicate classification is based on the types of linkages, which are as follows:

1. Independent tetrahedral groups: in this type the silicon-oxygen tetrahedra are present as separate entities. The resultant composition is SiO_4, and a typical mineral is forsterite, Mg_2SiO_4. This division of the silicates is known as the *nesosilicates*.

2. Finite linked tetrahedral groups: in this type the silicon-oxygen tetrahedra are linked by the sharing of one oxygen between each two tetrahedra. If two tetrahedra are linked in this way, the resulting composition is Si_2O_7; a typical mineral is akermanite, $Ca_2MgSi_2O_7$, and such substances are classed as *sorosilicates*. If more than two tetrahedra are so linked, closed units of a ring-like structure are formed, giving compositions Si_nO_{3n}. Rings containing up to six silicons are known. Typical examples are benitoite,

$BaTiSi_3O_9$, with three linked tetrahedra, and beryl, $Be_3Al_2Si_6O_{18}$, with six. This division of the silicates is known as the *cyclosilicates.*

3. Chain structures: tetrahedra joined together to produce chains of indefinite extent. There are two principal modifications of this structure yielding somewhat different compositions: (*a*) single chains, in which Si:O is 1:3, characterized by the pyroxenes, and (*b*) double chains, in which alternate tetrahedra in two parallel single chains are cross-linked and the Si:O ratio is 4:11, characterized by the amphiboles. These chains are indefinite in extent, are elongated in the *c*-direction of the crystal, and are bonded to each other by the metallic elements. This division of the silicates is known as the *inosilicates.*

4. Sheet structures: three oxygens of each tetrahedron are shared with adjacent tetrahedra to form extended flat sheets. This is the double-chain inosilicate structure extended indefinitely in two directions instead of just one. This linkage gives a ratio Si:O of 2:5 and is the fundamental unit in all mica and clay structures. The sheets form a hexagonal planar network responsible for the principal characteristics of minerals of this type—their pronounced pseudohexagonal habit and perfect basal cleavage parallel to the plane of the sheet. This division of the silicates is known as the *phyllosilicates.*

5. Three-dimensional networks: every SiO_4 tetrahedron shares all its corners with other tetrahedra, giving a three-dimensional network in which the Si:O ratio is 1:2. The various forms of silica—quartz, tridymite, cristobalite—have this arrangement. The quadrivalent silicon is balanced by two bivalent oxygen atoms. In silicates of this type the silicon is partly replaced by aluminum so that the composition is $(Si,Al)O_2$. The substitution of Al^{3+} for Si^{4+} requires additional positive ions in order to restore electrical neutrality. The feldspars and zeolites are examples of this division of the silicates, which is known as the *tektosilicates.*

All the silicate minerals can be placed in one of the types listed in Table 4.3 (a few have more than one type of linkage in the structure).

The other constituents of a silicate structure, such as additional oxygen atoms, hydroxyl groups, water molecules, and cations, are arranged with the silicate groups in such a way as to produce a mechanically stable and electrically neutral structure. Aluminum, after silicon the most abundant cation in the earth's crust, plays a unique role. As discussed earlier, it is stable both in fourfold and in sixfold coordination. It can replace silicon in the SiO_4 groups and also the common six-coordination cations—Mg^{2+}, Fe^{2+}, Fe^{3+}, etc.

The valence charge on the silicate unit, which determines the number and charge of the other ions that may enter the structure, can easily be calculated if it is remembered that each silicon has a positive charge of four and each

Table 4.3. The Structural Classification of the Silicates

Classification	Structural Arrangement	Silicon: Oxygen Ratio	Examples
Nesosilicates	Independent tetrahedra	1:4	Forsterite, Mg_2SiO_4
Sorosilicates	Two tetrahedra sharing one oxygen	2:7	Åkermanite, $Ca_2MgSi_2O_7$
Cyclosilicates	Closed rings of tetrahedra each sharing two oxygens	1:3	Benitoite, $BaTiSi_3O_9$ Beryl, $Al_2Be_3Si_6O_{18}$
Inosilicates	Continuous single chains of tetrahedra each sharing two oxygens	1:3	Pyroxenes, e.g., enstatite, $MgSiO_3$
	Continuous double chains of tetrahedra sharing alternately two and three oxygens	4:11	Amphiboles, e.g., anthophyllite, $Mg_7(Si_4O_{11})_2(OH)_2$
Phyllosilicates	Continuous sheets of tetrahedra each sharing three oxygens	2:5	Talc, $Mg_3Si_4O_{10}(OH)_2$ Phlogopite, $KMg_3(AlSi_3O_{10})(OH)_2$
Tektosilicates	Continuous framework of tetrahedra each sharing all four oxygens	1:2	Quartz, SiO_2 Nepheline, $NaAlSiO_4$

oxygen a negative charge of two. Thus, the charge on a single SiO_4 unit is $[4 + 4(-2)] = -4$; on an Si_2O_7 unit, -6; on an SiO_3 unit, -2; on an Si_4O_{11} unit, -6; on an Si_2O_5 unit, -2; and on SiO_2, 0.

THE LATTICE ENERGY OF CRYSTALS

It has been pointed out that the structure of a crystal is determined by the tendency of the constituent atoms to take up positions whereby their total potential energy is reduced to a minimum. This tendency may be expressed in terms of lattice energies. The lattice energy of an ionic crystal, generally represented by U, is defined as the energy absorbed when a mole of the crystal is dispersed into infinitely separated ions. The lattice energy depends upon the balancing of (*a*) the electrostatic forces between ions of opposite charge, which give a resultant attraction falling off with the square of the distance, and (*b*) the internuclear repulsive forces, which fall off very rapidly with distance. The attractive and repulsive forces result in an equilibrium position of minimum potential energy, which summed over all the ions is numerically equal to the lattice energy of the crystal. The greater the lattice energy the greater the energy required to break up the crystal into its constituent ions.

For binary compounds lattice energies may be directly calculated from the properties of the ions by means of the following equation, originally derived by Born and Landé:

$$U = \frac{NAz_c z_a}{r}\left(1 - \frac{1}{n}\right)$$

where U = lattice energy.
N = Avogadro number.
A = Madelung constant, characteristic of the type of crystal structure.
z_c, z_a = charge on cation and anion, respectively.
r = shortest anion-cation distance.
n = a factor allowing for the internuclear repulsion (n is usually about 10).

From the form of the equation it is clear that U approaches zero as r approaches infinity.

Although the Born-Landé equation applies only to binary compounds, it does enable qualitative statements regarding the lattice energy of more complex substances. For a particular structure type lattice energies are greater the higher the charge on the ions, the smaller the ions, and the closer the packing. For an example of the first effect, we may cite two substances with the same crystal structure, NaCl (U = 183 Cal) and MgO (U = 939 Cal).

The energy U of the Born-Landé equation is equal to the amount of work per mole which must be expended to disperse the crystalline substance into an assemblage of widely separated ions. As such, it cannot be equated with any directly measurable quantity and is not to be identified either with the heat of sublimation, which is the energy necessary to disperse the substance into a molecular gas, or with the heat of solution, which also includes the heat of hydration of the ions, or with the heat of formation, which is the heat evolved by the formation of the substance from its elements. Born and Haber devised a thermochemical cycle by means of which the lattic energy can be related to measurable thermal data. This cycle is as follows, using NaCl as an example (square brackets indicate crystalline substances and parentheses indicate substances in the gaseous state):

$$\begin{array}{ccc} [\mathrm{NaCl}] & \xrightarrow{U} & (\mathrm{Na^+}) + (\mathrm{Cl^-}) \\ \uparrow -Q & & \downarrow -I + E \\ [\mathrm{Na}] + (\tfrac{1}{2}\mathrm{Cl_2}) & \xleftarrow{-S - D} & (\mathrm{Na}) + (\mathrm{Cl}) \end{array}$$

The diagram represents the following cycle:

1. One mole of the crystalline substance is dispersed into ions in the gaseous state.

2. The ions are converted into neutral atoms.

3. The neutral atoms, now in the form of monatomic gases, are converted into the standard states for the elements at 25° and 1 atm pressure.

4. The elements are then allowed to combine chemically to reform the crystalline substance.

The symbols are defined by the following thermochemical equations:

$$[NaCl] = (Na^+) + (Cl^-) \qquad \Delta H = U, \text{ lattice energy}$$

$$(Na^+) + e = (Na) \qquad \Delta H = I, \text{ ionization energy}$$

$$(Cl^-) = (Cl) + e \qquad \Delta H = E, \text{ electron affinity}$$

$$(Na) = [Na] \qquad \Delta H = S, \text{ heat of sublimation}$$

$$(Cl) = (\tfrac{1}{2}Cl_2) \qquad \Delta H = D, \text{ heat of dissociation}$$

$$[Na] + (\tfrac{1}{2}Cl_2) = [NaCl] \qquad \Delta H = Q, \text{ heat of formation}$$

Since the final state of the system is the same as its original state the net change in heat content is zero; hence

$$U = Q + S + I + D - E$$

The cycle given by this equation is to be regarded as isothermal at 25°.

The Born-Haber cycle thus provides us with a means of determining lattice energies for complex compounds from other thermodynamic quantities. Unfortunately, for the common silicate minerals these thermodynamic quantities are imperfectly known. The whole question of considering geochemical processes in terms of lattice energies has been a subject of particular interest to Fersman and other Russian geochemists. In an attempt to get a simple method of determining lattice energies Fersman (1935) introduced the *EK* concept, which is an empirical constant for each element, representing the contribution of that element to the lattice energies of its compounds. Thus for NaCl

$$U_{NaCl} = K(EK_{Na} + EK_{Cl}) \quad (K \text{ is an independent constant})$$

From known lattice energies Fersman was able to assign *EK* values to most of the elements. However, the application of these values to the calculation of lattice energies of silicate minerals and to the interpretation of geochemical processes has so far given ambiguous results. The differences in calculated lattice energies of corresponding amphiboles and pyroxenes, for example, are small, smaller than the probable error. One serious criticism of Fersman's approach is that it allows only for the energy associated with a specific number of ions of different elements but fails to consider the energy associated with the crystal lattice as such. For example, the preceding

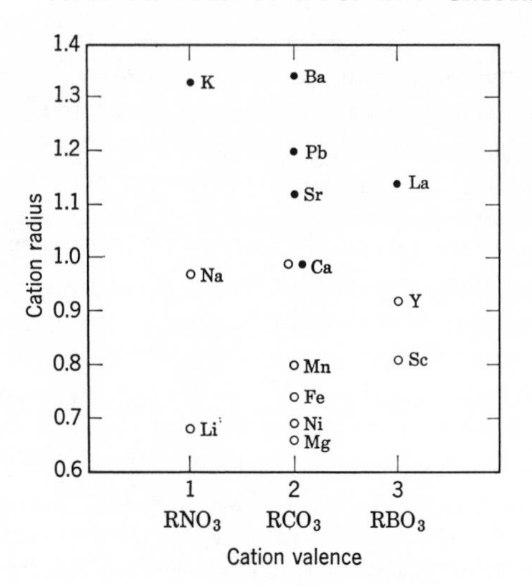

Figure 4.3 Effect of cation radius in determining crystal structure; ○ = trigonal; • = Orthorhombic. (After Fleischer, *J. Chem. Educ.* **31,** 450, 1954)

equation will clearly give the same lattice energies for all polymorphs of the same substance.

ISOMORPHISM

The term *isomorphism* is applied to the phenomenon of substances with analogous formulas having closely related crystal structures. The term was introduced by Mitscherlich in 1819, who prepared crystals of KH_2PO_4, KH_2AsO_4, $(NH_4)H_2PO_4$, and $(NH_4)H_2AsO_4$ and found that they showed the same forms and the interfacial angles between corresponding faces were very similar. By Mitscherlich's original definition, substances with analogous formulas and similar crystallography were said to be isomorphous. X-ray studies have shown that similar crystallography is a reflection of similar internal structure, hence the rewording of his original definition; sometimes the term *isostructural* or *isotypic* is used for the phenomenon.

Isomorphism is widespread among minerals and is one of the bases of their classification. Many isomorphous groups are recognized, for example, the spinel group, the garnet group, and the amphibole group. The basis of the phenomenon is that anions and cations of the same relative size (i.e., showing the same coordination) and in the same numbers tend to crystallize in the same structure type. This is well exemplified by some of the carbonate minerals (Figure 4.3). The anhydrous carbonates of the bivalent elements form two isomorphous groups, one orthorhombic and one trigonal. It can be seen that the nature of the structure is determined by the size of the bivalent cation; those minerals with cations larger then calcium crystallize in an orthorhombic structure, those with cations smaller than calcium crystallize in a trigonal structure. Calcium carbonate itself can crystallize in either structure, the phenomenon known as *polymorphism.*

Other substances with analogous formulas are isomorphous with these carbonates. Thus soda niter, $NaNO_3$, is isomorphous with calcite, whereas niter, KNO_3, is isomorphous with aragonite, reflecting the similar size of nitrate and carbonate groups and the larger size of potassium ions as compared to sodium ions. The borates of trivalent elements show similar relationships. Until the development of X-ray techniques for the determination of crystal structures, it was somewhat of an enigma that substances as different chemically as calcite and soda niter could show complete similarity in crystal form. Other isomorphous pairs at first sight do not even have analogous formulas. Thus the rare mineral berlinite ($AlPO_4$) is isomorphous with quartz; the true analogy is seen when the formula of quartz is written $SiSiO_4$. Both Al and P are similar in ionic size to Si and can exist in a crystal structure in four-coordination with oxygen; as a result $AlPO_4$ can crystallize with the same structure as quartz. Similarly, tantalite, $FeTa_2O_6$, is isomorphous with brookite, $TiO_2(TiTi_2O_6)$; the metallic ions are similar in size and all show sixfold coordination with oxygen.

The important factor in isomorphism is the similarity in size relations of the different ions rather than any chemical similarity. This explains many apparently unusual examples of isomorphism and its absence between many chemically similar compounds. Thus corresponding calcium and magnesium compounds are seldom isomorphous, although these elements are similar in chemical behavior; when it is noted that the radius of Ca^{2+} is 0.99 A and that of Mg^{2+} is 0.66 A it seems natural that the substitution of one for the other without producing a change in structure is improbable.

ATOMIC SUBSTITUTION

After the development of reliable methods for the analysis of minerals it was observed that many species are variable in composition. Substitution of one element by another is the rule rather than the exception. When this phenomenon was first observed it was described in terms of the concept of *solid solution* or *mixed crystals*, which implied the presence, in a single homogeneous crystal, of molecules of two or more substances. For example, common olivine may be described as a solid solution of Mg_2SiO_4(Fo) and Fe_2SiO_4(Fa), and the precise composition of any sample of olivine may be stated in terms of these *end-members*, such as $Fo_{85}Fa_{15}$, that is,

$(Mg_{0.85}Fe_{0.15})_2SiO_4$. This concept and terminology remain in general use, but the light thrown upon the structure of crystals by X-ray investigation has resulted in a revised interpretation. In an ionic structure there are no molecules, the structure being an infinitely extended three-dimensional network. Any ion in the structure may be replaced by another ion of similar radius without causing any serious distortion of the structure, just as a bricklayer, running short of red bricks, may incorporate yellow bricks of the same size here and there in his wall. Since minerals usually crystallize from solutions containing many ions other than those essential to the mineral, they often incorporate some foreign ions in the structure.

A solid solution (or mixed crystal) can be simply defined as a homogeneous crystalline solid of variable composition. It was early found that many isomorphous substances have the property of forming solid solutions. There has been a tendency to equate isomorphism and solid solution, in spite of marked inconsistencies. For example, many isomorphous substances show little or no solid solution, among them, calcite and smithsonite; and extensive solid solution may occur between components which are not isomorphous, for example, the presence of considerable amounts of iron in sphalerite, although FeS and ZnS have quite different crystal structures. On this account it must be emphasized that isomorphism is neither necessary to nor sufficient for solid-solution formation. Isomorphism and solid solution are distinct concepts and should not be confused.

In atomic substitution it is the size of the atoms or ions that is the governing factor, and it is not essential that the substituting ions have the same charge or valency, provided that electrical neutrality is maintained by concomitant substitution elsewhere in the structure. Thus, in passing from albite ($NaAlSi_3O_8$) to anorthite ($CaAl_2Si_2O_8$), Ca^{2+} substitutes for Na^+ and electrical neutrality is maintained by the coupled substitution of Al^{3+} for Si^{4+}; similarly, in diopside ($CaMgSi_2O_6$) Mg^{2+}—Si^{4+} may be replaced in part by Al^{3+}—Al^{3+}. Such coupled substitutions are especially common in silicate minerals and made the interpretation of their composition exceedingly difficult before this phenomenon was recognized and understood.

As a general rule, little or no atomic substitution takes place when the difference in charge on the ions is greater than one, even when size is appropriate (e.g., Zr^{4+} does not substitute for Mn^{2+}, nor Y^{3+} replace Na^+); this may be in part because of the difficulty in balancing the charge requirements by other substitutions.

The extent to which atomic substitution takes place is determined by the nature of the structure, the closeness of correspondence of the ionic radii, and the temperature of formation of the substance. The nature of the structure evidently has considerable influence on the degree of atomic substitution; some structures, such as those of spinel and apatite, are well known for extensive atomic substitution, whereas others, such as quartz, show very little. In part this phenomenon is due to the lack of foreign ions of suitable size and charge. Ionic size has, of course, a fundamental influence on the degree of substitution, since the substituting ion must be able to occupy the lattice position without causing distortion of the structure. From a study of many mixed crystals it has been found that, provided the radii of substituting and substituted ions do not differ by more than 15%, a wide range of substitution may be expected at room temperature. Higher temperatures permit a somewhat greater tolerance; in this respect solid solutions are analogous to solutions of salts in water, solubility increasing with temperature.

This property of increased atomic substitution at higher temperatures provides a means of determining the temperature of mineral deposition (*geological thermometry*). If for a specific mineral the degree of atomic substitution has been determined for different temperatures, the composition of the naturally occurring mineral may indicate the temperature of formation. Thus the amount of iron in solid solution is sphalerite as a function of temperature is known from laboratory investigations. Sphalerite is a common ore mineral. Provided the ore-forming solutions contained sufficient iron sulfide to saturate the sphalerite, the iron content of the mineral will indicate the temperature conditions during ore deposition.

The consequence of atomic substitution is that most minerals contain not only the elements characteristic of the particular species, but also other elements able to fit into the crystal lattice. For instance, dolomite is theoretically a simple carbonate of magnesium and calcium, but dolomites are found whose analyses show a considerable content of iron and manganese. Traditionally, these were described as solid solutions of the carbonates of all these elements, but it is more illuminating as well as more correct to consider them as products of the substitution of iron and manganese for magnesium. Nevertheless, we continue to use the traditional terms solid solution, mixed crystals, and solid solution series, since the terminology of atomic substitution has not yet provided expressions to take their place. The useful term *diadochy* has been introduced to describe the ability of different elements to occupy the same lattice position in a crystal; thus Mg, Fe, and Mn are diadochic in the structure of dolomite. The concept of diadochy, if used rigidly, always applies to a particular structure; two elements may be diadochic in one mineral and not in another.

Crystal structure investigations have also revealed two other types of solid solution besides that due to atomic substitution. One is known as *interstitial solid solution*, whereby atoms or ions do not replace atoms or ions in the structure but fit into interstices in the lattice. This type is very common in metals, which take up hydrogen, carbon, boron, and nitrogen, all small atoms, in interstitial solid solution. If a substance has an open structure, interstitial solid solution may take place even with atoms or ions of a considerable size. Thus cristobalite, the high-temperature form of SiO_2, has been found with a considerable content of sodium and aluminum; the Al^{3+} replaces Si^{4+}, and the Na^+ needed to maintain electrical neutrality occupies large openings in the cristobalite lattice. The other type of solid solution is that associated with *defect lattices*, in which some of the atoms are missing, leaving vacant lattice positions. It has been called *omission solid solution.* A good example is the mineral pyrrhotite, in which analyses always show more sulfur than corresponds to the formula FeS. This was for a long time described as solid solution of sulfur in FeS. Actually, the excess of sulfur shown by analyses is due to the absence of some iron atoms from their places in the lattice; there is a deficiency of Fe, not an excess of S. Just as in building a wall, where a brick may be omitted here and there without affecting the stability of the structure, so it is possible to omit some of the Fe atoms in FeS without the lattice collapsing. More and more defect structures are being recognized among minerals, thereby explaining otherwise puzzling deviations of chemical compositions from those predicted by the law of constant proportions.

POLYMORPHISM

An element or compound that can exist in more than one crystal form is said to be polymorphous. Each form has different physical properties and a distinct crystal structure; that is, the atoms or ions are arranged differently in different polymorphs of the same substance. Polymorphism is an expression of the fact that crystal structure is not exclusively determined by chemical composition, that there is often more than one structure into which the same atoms or ions in the same proportions may be built up. Different polymorphs of the same substance are formed under different conditions of pressure, temperature, and chemical environment; hence the presence of one polymorph in a rock will often tell something about the conditions under which that rock was formed. For example, marcasite can be formed only from acid solutions at temperatures below 450°, and the presence of marcasite in a deposit thus puts some limits on the conditions of origin.

Two types of polymorphism are recognized, according to whether the change from one polymorph to another is reversible and takes place at a definite temperature and pressure, or is irreversible and does not take place at a definite temperature and pressure. The first type is known as *enantiotropy* and is exemplified by the relationship between quartz and tridymite (quartz $\underset{1\ \text{atm}}{\overset{867°}{\rightleftharpoons}}$ tridymite). The second type is known as *monotropy;* an example is the marcasite-pyrite relationship, in which marcasite may invert to pyrite but pyrite never changes to marcasite. With monotropic polymorphs one form is always inherently unstable and the other inherently stable; the unstable form always tends to change into the stable form, but the stable form cannot be changed into the unstable form without first completely destroying its structure by melting, vaporization, or solution.

This distinction between enantiotropic and monotropic polymorphs is useful, but the recognition of monotropic polymorphs is usually based on experimental evidence, and investigation over wide ranges of temperature and pressure or determination of energy relationships of the different polymorphs sometimes indicates that supposedly monotropic polymorphs actually have an enantiotropic relationship under conditions far removed from those usually attainable. Thus studies of the energy relationships between calcite and aragonite indicate an enantiotropic transition between them at about −60°. The diamond-graphite relationship is particularly interesting in this respect, both from the geological significance of the occurrence of these two polymorphs and from the practical aspect of developing ways to made diamond synthetically. For a long time it was unknown whether diamond and graphite were enantiotropic or monotropic polymorphs; the latter conclusion was favored because under laboratory conditions the transition was always diamond → graphite, never the reverse. However, it has been established that the relationship is enantiotropic, and the actual conditions of the diamond ⇌ graphite equilibrium have been worked out (Figure 4.4). This figure shows that the practical problem of making diamond synthetically

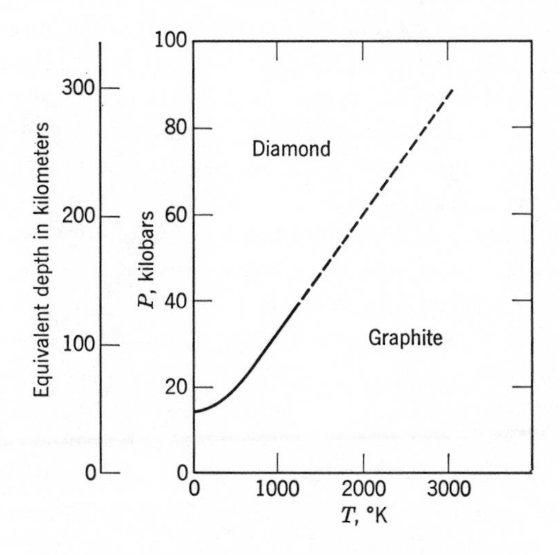

Figure 4.4 Graphite-diamond equilibrium curve, calculated to 1200°K, extrapolated beyond. (*Nature* **176,** 835, 1955)

lies in maintaining pressures within the stability field of diamond at temperatures for which the reaction velocity for its formation is appreciable; this has been achieved by the development of special equipment capable of withstanding great pressures at high temperatures. It also indicates that the natural occurrence of diamond in igneous rocks implies an origin at considerable depths in the earth, where the combination of temperature and pressure is within the diamond stability field. Diamond is actually unstable under the physical conditions in which it is found (and worn); that it does not change spontaneously into graphite is due solely to the infinitesimal rate of a reaction which the energy relations nevertheless favor.

The rate of change from one polymorph to another may be very slow or very rapid. Sometimes the change does not involve the breaking of bonds between neighboring atoms or ions, but simply their bending, for example, low-quartz $\rightleftharpoons$ high-quartz, low-leucite $\rightleftharpoons$ high-leucite. Such transformations are almost instantaneous at the transition temperature, and the high-temperature form cannot be preserved at lower temperatures (however, original crystallization as the high-temperature form can often be recognized from the nature of the crystals or from the twinning that so often results from inversions of this type). High-low polymorphs are also characterized by the fact that the high-temperature form is always more symmetrical than the corresponding low-temperature form. Transformations other than the high-low type require the breaking of bonds in the structure and the rearrangement of atomic or ionic linkages. They are often sluggish and may require the presence of a solvent in order to obtain an appreciable rate of change. These changes have been termed reconstructive transformations by Buerger and are exemplified by the quartz $\rightleftharpoons$ tridymite $\rightleftharpoons$ cristobalite inversions.

The high-temperature polymorph of a substance is generally more open-packed than a low-temperature form. The open character of the structure is dynamically maintained at high temperatures by thermal agitations. It may also be statically maintained by the incorporation of foreign ions in the interstices of the lattice. These foreign ions will buttress the structure and prevent its transformation to a different polymorph when the temperature is lowered. Their complete removal is usually necessary to permit inversion to the closepacked form stable at low temperatures. Thus impure high-temperature polymorphs may be formed and may survive indefinitely far below the normal stability range of pure compounds. This situation is likely to arise in nature. Buerger points out that the phenomenon is probably responsible for the formation and survival of cristobalite and tridymite under conditions in which the stable form of SiO_2 is quartz. As mentioned previously, cristobalite has been found with a considerable amount of sodium in interstitial solid solution, and the sodium atoms presumably stabilize the open structure of this mineral. The occurrence of a high-temperature polymorph at ordinary temperatures is not necessarily to be interpreted as indicating metastability; the polymorph may be simply a stable impure form.

Transformations between polymorphs show a close analogy to the changes between the liquid and solid states, being amenable to treatment by the same thermodynamic principles. Buerger discusses the thermodynamics of polymorphism at considerable length. Under a given set of conditions, each of several polymorphs of a substance is characterized by its free energy; all possible polymorphic forms tend to transform into the one with the minimum free energy, and that form is the stable one under those conditions. The free energy G is given by the equation $G = E - TS + PV$. For changes not involving a gas phase the PV term is small and can be omitted. When T is zero, $G = E$, that is, the free energy of a substance is equal to its internal energy. Hence at absolute zero the polymorph with the lowest internal energy will be the stable form. At temperatures other than zero the entropy term is significant, and the relative magnitudes of S for the different polymorphs may determine which form has the lowest free energy. At a transition point between two forms, the free energies of both are equal (Figure 4.5). Because the entropy of the high-temperature form is greater than that of the low-temperature form, it follows that the internal energy of the former must also be greater than that of the latter. The entropy involves the volume over which the atoms may be disordered; hence there is a tendency for the forms of higher entropy to have greater open space available for thermal

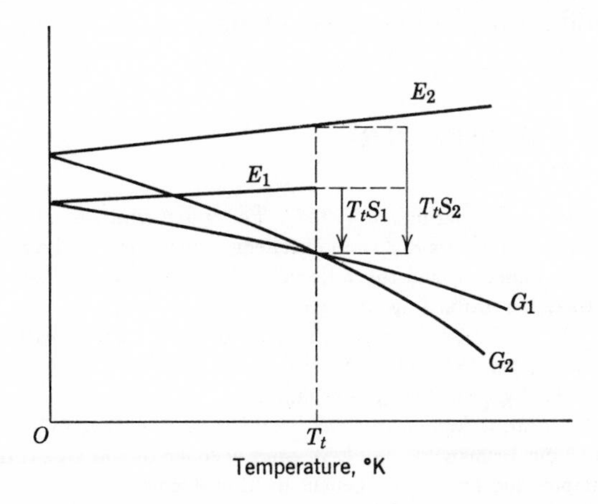

Figure 4.5 Energy relations between polymorphs 1 and 2 with transition temperature T_t. E = internal energy; G = free energy; S = entropy. (After Buerger, *Am. Mineralogist* **33,** 103, 1948)

motion. Although this does not necessarily involve openness of the whole structure, it often does; high-temperature forms therefore tend to be less dense than low-temperature ones. Temperature and pressure tend to impose opposite conditions: high temperature promotes open structures, high pressures compact structures.

An interesting transformation that may be considered as a variety of polymorphism is the order-disorder type. It has been most studied in alloys, as it has important effects on their physical properties, but it is probably more common in minerals than is generally realized. A simple example is an alloy of 50% Cu, 50% Zn. Two distinct phases of this alloy exist. In the disordered form the copper atoms and the zinc atoms are randomly distributed over the lattice positions, whereas in the ordered form each element occupies a specific set of positions. The structures of the two forms are related, but the ordered one has lower symmetry than the disordered one. There is no definite transition point between the two forms; perfect order is achieved only at absolute zero, and with increasing temperature the degree of order gradually decreases to complete disorder above a certain temperature characteristic of the structure and the composition of the crystal. It has been suggested that the relationship between microcline and sanidine is an order-disorder transformation, the one aluminum and three silicon atoms in $KAlSi_3O_8$ being in disorder in sanidine but ordered in microcline. Buerger points out that this accounts for the monoclinic-triclinic feature of the polymorphism, and the typical twinning of microcline is reminiscent of the twinning often observed in ordered forms.

SELECTED REFERENCES

Bragg, W. L., and G. F. Claringbull (1965). *The crystal structures of minerals.* 409 pp. Bell and Sons, London. Chapter 2 is an excellent presentation of the factors governing the structures of minerals, and the following chapters provide a succinct account of mineral structures determined up to 1963.

Buerger, M. J. (1951). Crystallographic aspects of phase transformations. *Phase transformations in solids*, pp. 183–211. John Wiley and Sons, New York. An account of the thermodynamic background to polymorphism.

Eitel, W. (1952). *Thermochemical methods in silicate investigation.* 132 pp. Rutgers University Press, New Brunswick, N. J. A brief account of the application of thermochemical principles and practice to geochemical problems.

Evans, R. C. (1964). *An introduction to crystal chemistry* (third edition). 410 pp. Cambridge University Press, Cambridge, England. A description of the general principles of crystal chemistry and their application.

Fersman, A. E. (1935). The EK system. *Compt. rend. acad. sci.* U.R.S.S. 2, 564–566. A brief account (in English) of the EK concept which was developed more fully in his great work on geochemistry.

Fyfe, W. S. (1964). *Geochemistry of solids.* 199 pp. McGraw-Hill Book Co., New York. A clear and concise account of the general principles of chemical bonding and related topics.

Goldschmidt, V. M. (1954). *Geochemistry.* Chapter 6.

Kracek, F. C., K. J. Neuvonen, and G. Burley (1951). Thermochemistry of mineral substances, I: A thermodynamic study of the stability of jadeite. *J. Wash. Acad. Sci.* **41,** 373–383. A pioneer application of thermodynamical principles and measurements to the study of mineral formation.

Miyashiro, A. (1960). Thermodynamics of reactions of rock-forming minerals with silica. *Japanese J. Geol. Geog.* **31,** 71–78. A discussion of the free energies of solid-state reactions and their petrogenetic significance.

Pauling, L. (1960). *The nature of the chemical bond* (third edition). 644 pp. Cornell University Press, N. Y. An exhaustive account of the types of bonding in elements and compounds and their significance for crystal structure and chemical properties.

Ramberg, H. (1955). Thermodynamics and kinetics of petrogenesis. *Geol. Soc. Amer., Spec. Paper* 62, 431–448. Discusses the major types of processes in the earth and their corresponding states of equilibria.

Turner, F. J., and J. Verhoogen (1960). *Igneous and metamorphic petrology.* 694 pp. McGraw-Hill Book Co., New York. Chapter 2 is an admirable account of the application of thermodynamic principles to chemical equilibrium in rocks.

Winkler, H. G. F. (1955). *Struktur und Eigenschaften der Kristalle.* 314 pp. Springer-Verlag, Berlin. A comprehensive account of the principles of crystal chemistry and crystal physics, with special reference to their applications to mineralogy.

CHAPTER FIVE

MAGMATISM AND IGNEOUS ROCKS

WHAT IS MAGMA?

Magma is the parent material of igneous rocks. Lava is magma poured out through volcanic vents and thereby accessible to our observation. However, much magma solidifies below the surface and can only be observed as the end product, an igneous rock, from which the original nature of the magma must be inferred. Magma has been defined as molten rock material, but this definition is not entirely satisfactory. It fails to bring out that magma contains volatile components which are lost as it solidifies but which nevertheless play a significant part in determining the course of crystallization. The definition also fails to emphasize that the solidification of a magma does not take place at a definite temperature, like, for example, the solidification of molten lead, but is generally drawn out in time and place by fractional crystallization. The end products of such crystallization include not only the igneous rock, but also a gas phase and possibly a watery solution. Hence the history of a magma may be long and complex, and the termination is often not easily defined. When does a magma cease to be a magma? When the first solid phase appears? When a quarter, or half, or three quarters of it has solidified? When all that remains is a watery solution? The diversity of answers to this question is in part responsible for the vigorous controversies regarding the origin of some plutonic rocks, such as granites and granodiorites. Magma is characterized (*a*) by composition, in that it is predominantly silicate; (*b*) by temperature, in that it is hot (although the range in temperature may be great, say from 500 to 1200°); and (*c*) by mobility, in that it will flow. Although a magma is fluid, only a minor part of it need be liquid. In this book the concept of magma is essentially that proposed by Turner and Verhoogen (1960, p. 50): ". . . the term magma is used to include all naturally occurring mobile rock matter that consists in noteworthy part of a liquid phase having the composition of silicate melt."

The question as to whether or not there is a "primary" magma is one that has been extensively discussed. Many years ago Bunsen decided that there were two primary magmas, granitic and basaltic, and that igneous rocks of other compositions were mixtures of these two types. Bunsen's contention is apparently supported by the abundances of different rock types. The igneous rocks of the globe belong chiefly to two types: granite and basalt. Daly estimates that the granites and granodiorites together comprise at least 95% of all intrusives and that basalts and pyroxene andesites make up about 98% of all extrusives (the latter figure seems excessive, since silica-rich extrusives—rhyolites and dacites—are fairly abundant). These relations are clearly of fundamental significance in a consideration of magmatism. The trend of geological thought is toward the idea that material of basaltic compositon may well be the single primary magma from which most igneous rocks have been derived. Major arguments in favor of this are (*a*) throughout geological time magma of this composition has broken through the crust and poured out on the surface in great floods; (*b*) the rocks of the great ocean basins are almost entirely basalts; (*c*) experimental work has shown that differentiation of basaltic magma can yield rocks of widely varying chemical and mineralogical composition, and such differentiation has been observed in natural occurrences. The assumption of a primary basaltic magma in no way implies that magmas of other compositions are nonexistent. This is clearly evident from the wide variety of lavas expelled by volcanoes. The formation of magmas of different compositions can be explained by fractional crystallization of a basaltic magma and the separation of residual magmas of different types, by assimilation of material of different composition, and by the escape of volatile substances, all of which either singly or together are adequate to produce significant changes in composition.

THE CHEMICAL COMPOSITION OF MAGMAS AND IGNEOUS ROCKS

Clarke and Washington calculated the average composition of igneous rocks, with the following results:

SiO_2	Al_2O_3	Fe_2O_3	FeO	MgO	CaO	Na_2O	K_2O	H_2O	TiO_2
59.14	15.34	3.08	3.80	3.49	5.08	3.84	3.13	1.15	1.05

All others less than 0.30 each

It must, of course, be borne in mind that these figures are the average of several thousand analyses and are of no greater or less significance than is implied in that statement; they do not represent the composition of a primary magma from which all igneous rocks may be derived, nor the composition of any particular magma; minor constituents are omitted, as are the volatiles known to be present in all magmas. Nevertheless, the figures do indicate that the following elements predominate: O, Si, Al, Fe, Mg, Ca, Na, and K; that is, magmas are multicomponent systems of these and other elements.

The different components do not vary randomly but are interdependent. The frequency relations of the commonest rock-forming oxides were worked out by Richardson and Sneesby on the basis of Washington's collection of 5000 superior analyses of fresh rocks (Figures 5.1 and 5.2). The dominant oxide is silica; most igneous rocks contain between 30 and 80% SiO_2. Its variation is not at all regular, as is clearly seen from Figure 5.1, there being two frequency maxima, at 52.5% SiO_2 and 73.0% SiO_2; i.e., rocks with about 52.5% SiO_2 and 73.0% SiO_2 are the commonest. This corresponds with field experience that the most abundant igneous rocks are granite and basalt. Alumina varies commonly between 10 and 20%. Low Al_2O_3 is characteristic of rocks with little feldspar or feldspathoid, i.e., ultrabasic types; high Al_2O_3 is characteristic of anorthosites and rocks containing much nepheline. Soda shows a very symmetrical variation, the common range being from 2 to 5%; Na_2O rarely exceeds 15%. The curve for K_2O is less regular, but most rocks have less than 6% K_2O, and only rarely does it exceed 10%. The curves for FeO and Fe_2O_3 are similar; both show a frequency peak at about 1.5% and fall off more or less regularly towards higher percentages. The sum of iron oxides in igneous rocks seldom exceeds 15%, except in magmatic iron ores. The frequency curve for magnesia percentage is very asymmetrical; most rocks have a low MgO content, and only ultrabasic types rich in pyroxene and/or olivine have more than 20%.

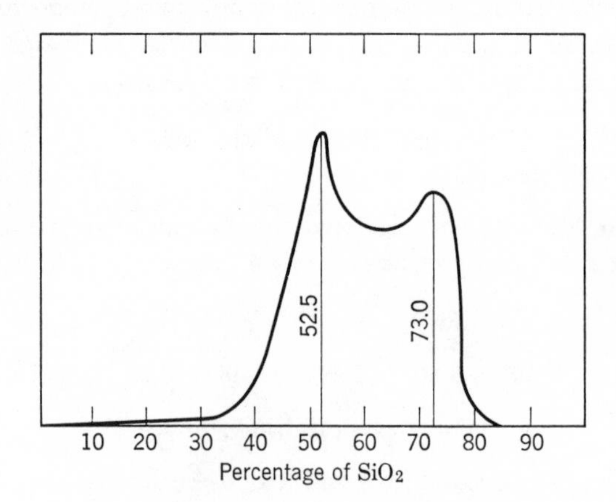

Figure 5.1 The frequency distribution of silica percentage in analyses of igneous rocks. (Richardson and Sneesby, *Mineralog. Mag.* **19,** 306, 1922)

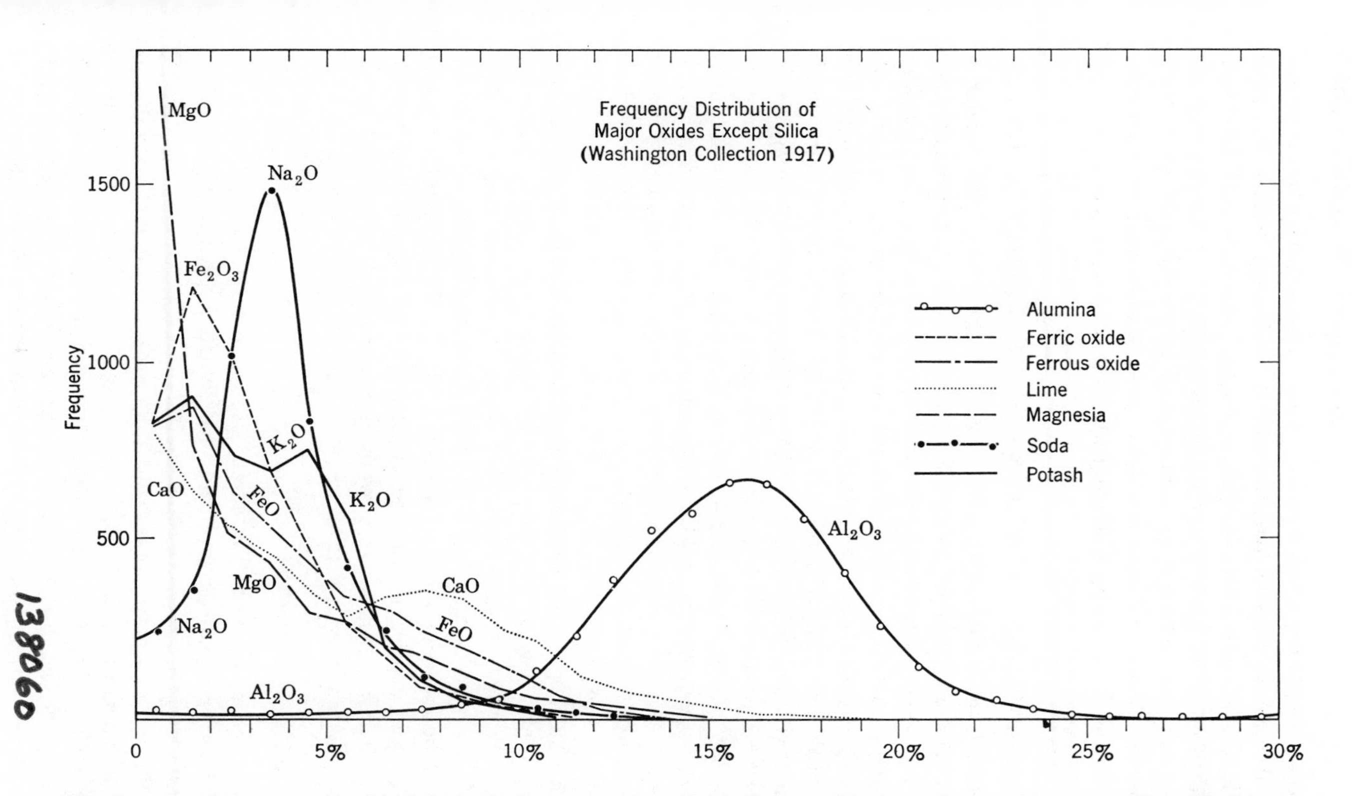

Figure 5.2 The frequency distribution of the percentages of the major oxides in analyses of igneous rocks. (Richardson and Sneesby, *Mineralog. Mag.* **19,** 309, 1922)

The curve for CaO resembles that for MgO; most rocks have less than 10% CaO, although in some pyroxenites CaO exceeds 20%. Water may reach 10% in a few volcanic glasses, but generally an igneous rock containing more than 2% H_2O has acquired the excess by alteration. Three minor constituents that should be determined in a good rock analysis are TiO_2, P_2O_5, and MnO, all of which are present in most igneous rocks. Many other elements may be found in small amounts; their occurrence is discussed later in this chapter.

THE MINERALOGICAL COMPOSITION OF IGNEOUS ROCKS

Although more than 1000 different minerals are known, the number of species present in more than 99% of the igneous rocks is very small. Apart from the seven principal minerals or mineral groups (the silica minerals, feldspars, feldspathoids, olivine, pyroxenes, amphiboles, and micas), only magnetite, ilmenite, and apatite are commonly found, and then usually in very small amounts. A statistical study of about 700 petrographically described igneous rocks (Clarke, 1924, p. 423) gave the following average mineralogical composition: quartz 12.0%, feldspars, 59.5%, pyroxene and hornblende 16.8%, biotite 3.8%, titanium minerals 1.5%, apatite 0.6%, other accessory minerals 5.8%. It is interesting to compare this average *mode* with the *norm* calculated from the average composition of igneous rocks. This is Q 10.02, or 18.35, ab 32.49, an 15.29, di 6.45, hy 8.64, mt 4.41, il 1.98, ap 0.67.

The mineralogical composition is a basic criterion in the classification of igneous rocks. Several hundred rock types have been named, but of these only a few are of common occurrence. Figure 5.3 is a diagrammatic representation of the common types, showing the mineralogical composition in terms of the important rock-forming minerals. The diagram serves to emphasize that rock types grade into each other, and a rock name is a convenient pigeonhole rather than a species of fixed composition.

The principal minerals and mineral groups are now discussed in detail.

THE SILICA MINERALS

Silica occurs in nature as seven distinct minerals: quartz (including chalcedony), tridymite, cristobalite, opal, lechatelierite, coesite, and stishovite. Of these, quartz is very common; tridymite and cristobalite are widely distributed in volcanic rocks and can hardly be called rare; opal is not uncommon; lechatelierite (silica glass) is very rare. Coesite and stishovite are high-pressure forms first made in the laboratory and later recognized in the shocked sandstone at Meteor Crater in Arizona, where they were evidently formed by the instantaneous high pressure of the meteorite impact.

Quartz, cristobalite, and tridymite are the forms of silica that may occur

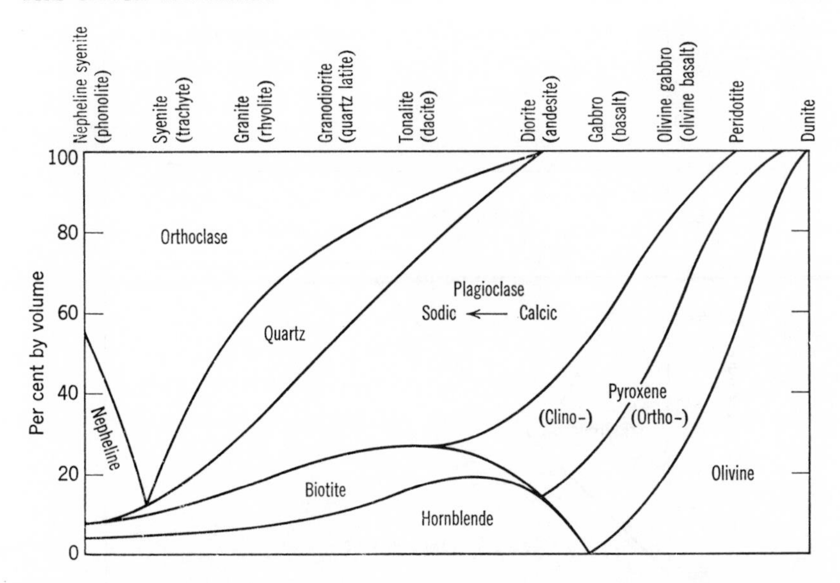

Figure 5.3 Approximate mineralogical composition of the commoner types of igneous rocks (effusive types in brackets).

in igneous rocks, and their relationships will be discussed in detail. These three forms illustrate the phenomenon of enantiotropism. Each has its own stability field; at atmospheric pressure quartz is the stable form up to 867°, tridymite between 867 and 1470°, and cristobalite from 1470° up to the melting point at 1713°; from 1713° to the boiling point liquid silica is the stable phase. The one-component system SiO_2 has been extensively studied under varying conditions of temperature and pressure; the equilibrium diagram is given in Figure 5.4.

The addition of small amounts of H_2O to this system produces remarkable results. Tuttle and England have investigated it at temperatures up to 1300° and pressures of water vapor up to 2000 kg/cm². The results are indicated with dotted lines on Figure 5.4. The liquidus is lowered by several hundred degrees. Thus, at pressures above 1400 kg/cm², quartz melts at about 1125°, a depression of some 600° in the melting point; the liquid so formed is a hydrous silica melt containing about 2.3% H_2O. The stability field of tridymite is greatly contracted, and, whereas in the dry system tridymite has no stable melting point, in the presence of water vapor it melts to a hydrous liquid at pressures above 400 kg/cm². Water vapor under pressure thus has a tremendous fluxing power for SiO_2 and clearly will greatly influence the transportation of silica in a fluid form. The large effects produced by a small weight per cent of water are a reflection of its low molecular

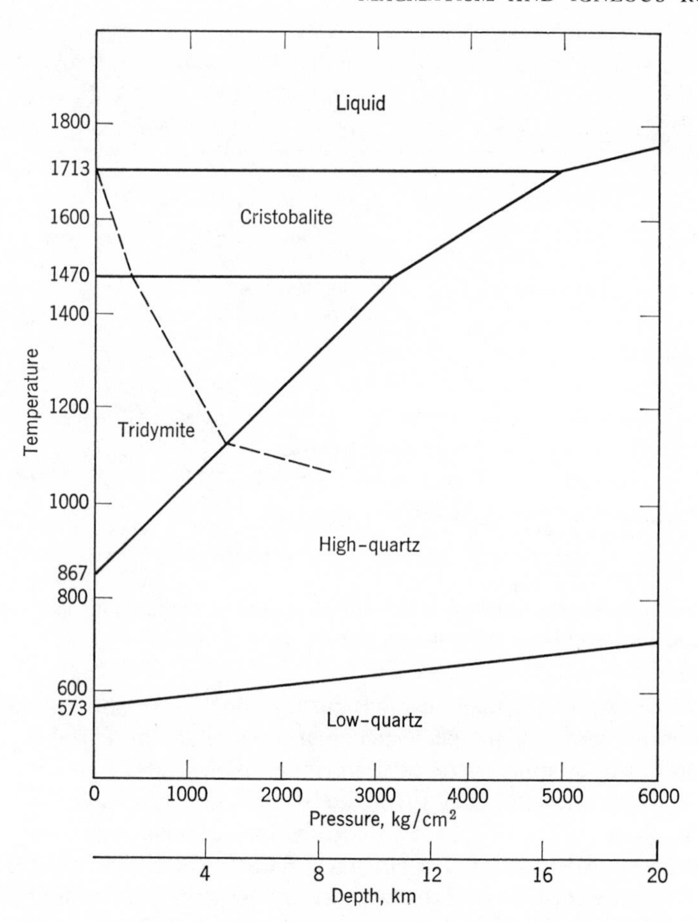

Figure 5.4 Stability relations of the different forms of SiO_2. Solid lines, dry system (*Ann. Rep. Geophys. Lab.* 1952–53, p. 61); dotted lines, liquidus under water-vapor pressure (*Bull. Geol. Soc. Amer.* **66,** 149, 1955).

weight, which results in its mole fraction in solution being comparatively large.

These three polymorphs of silica are all built of tetrahedral groups of four oxygen atoms surrounding a central silicon atom. The silicon-oxygen tetrahedra are linked together to form a three-dimensional network, but the pattern of linkage is different for each of the three forms; hence the difference in their crystal structures and their properties. Cristobalite and tridymite have comparatively open structures, whereas the atoms in quartz are more closely packed. This is reflected in the densities and refractive indices, which are much lower for cristobalite and tridymite than for quartz:

	Density	Mean Refractive Index
Quartz	2.65	1.55
Cristobalite	2.32	1.49
Tridymite	2.26	1.47
Lechatelierite	2.20	1.46

Each of these three polymorphs has a high- and low-temperature modification. In quartz, for example, the change from the one to the other takes place at 573° at atmospheric pressure. Similarly, high-tridymite changes into low-tridymite between 120 and 160° and high-cristobalite into low-cristobalite between 200 and 275°. The inversion of high- and low-temperature forms of the individual species is of quite another order to that between the species themselves. The three minerals have the SiO_4 tetrahedra linked together according to different schemes, and this linkage has to be completely broken down and rearranged for the transformation of one to another. On the other hand, the change from a high-temperature to a low-temperature form does not alter the way in which the tetrahedra are linked. They undergo a displacement and rotation which alters the symmetry of the structure without breaking any links. The high-temperature modification is always more symmetrical than the low-temperature modification.

The high-low transformation of each mineral takes place rapidly at the transition temperature and is reversible. The changes from one polymorphic form to another are extremely slow and sluggish, and the existence of tridymite and cristobalite as minerals shows that they can remain unchanged indefinitely at ordinary temperatures. Once formed, the type of linking in tridymite and cristobalite is too firm to be easily broken, and it is possible to study the high-low inversions of tridymite and cristobalite at temperatures at which they are really metastable forms. As pointed out in Chapter 4, the presence of foreign elements in the structure may have a stabilizing effect on tridymite and cristobalite. The few comprehensive analyses of these minerals show the presence of Na and Al, suggesting a substitution of NaAl for Si in the open structures; quartz on the other hand is generally very pure SiO_2. Two other phenomena of great significance should be mentioned here:

1. Even at temperatures below 867°, especially when crystallization takes place rapidly (for example, in the presence of mineralizers, such as hot gases), cristobalite and/or tridymite may crystallize, although quartz is the stable phase.

2. High-quartz and low-quartz are formed only within their stability fields, never at higher temperatures.

From these facts the following conclusions can be drawn: quartz in an igneous rock signifies that its crystallization from the magma took place below 867° (with due regard for the effect of pressure); the presence of cristobalite or tridymite, on the other hand, proves nothing as to the temperature of crystallization.

As already pointed out, at ordinary temperatures quartz is always present as low-quartz. By the crystal form, nature of the twinning, and other properties of less diagnostic importance, it is often possible to determine the original form. In this way it has been shown that in nearly all the quartz-bearing igneous rocks this mineral crystallized originally as high-quartz, that is, above 573°. In quartz veins and some pegmatites it crystallized originally as low-quartz. It may therefore be concluded that the crystallization of the magma corresponding to the commonest quartz-bearing rocks took place above 573° and the residual crystallization in part at least at lower temperatures.

THE FELDSPAR GROUP

The feldspars owe their importance to the fact that they are the most common of all minerals. They are closely related in form and physical properties, but they fall into two groups: the potassium and barium feldspars, which are monoclinic or very nearly monoclinic in symmetry, and the sodium and calcium feldspars (the plagioclases), which are definitely triclinic. An ammonium feldspar, buddingtonite, also monoclinic, was described in 1964 from a hot-spring deposit in California. A point of great interest is the solid solution between albite, $NaAlSi_3O_8$, and anorthite, $CaAl_2Si_2O_8$. The theory that feldspars of intermediate composition were mixed crystals of these two components was proposed by Tschermak in 1869. It is now known that NaSi often substitutes for CaAl, but Tschermak's theory is of historic importance as a first suggestion that so radical a substitution is possible.

The general formula for the feldspars can be written WZ_4O_8, in which W may be Na, K, Ca, and Ba, and Z is Si and Al, the Si:Al ratio varying from 3:1 to 1:1. Since all feldspars contain a certain minimum amount of Al, the general formula may be somewhat more specifically stated as $WAl(Al,Si)Si_2O_8$, the variable (Al,Si) being balanced by variation in the proportions of univalent and bivalent cations.

The structure of the feldspars is a continuous three-dimensional network of SiO_4 and AlO_4 tetrahedra, with the positively charged sodium, potassium, calcium, and barium situated in the interstices of the negatively charged network. The network of SiO_4 and AlO_4 tetrahedra is elastic to some degree and can adjust itself to the sizes of the cations; when the cations are relatively large (K,Ba) the symmetry is monoclinic or pseudo-monoclinic; with the smaller cations (Na,Ca) the structure is slightly distorted and the symmetry becomes triclinic.

The barium-containing feldspars are rare and of no importance as rock-forming minerals, and so we will omit them from further consideration. The feldspars may then be discussed as a three-component system, the components being $KAlSi_3O_8$ (Or), $NaAlSi_3O_8$ (Ab), and $CaAl_2Si_2O_8$ (An). Complexities are introduced by the solid solution relations existing among these three components and the occurrence of polymorphic forms.

The potash-feldspar minerals occur in several distinct forms having different but intergradational optical and physical properties. Sanidine, the monoclinic high-temperature polymorph, occurs in volcanic rocks. Common orthoclase, another monoclinic variety, and microcline (triclinic) are found in a wide variety of igneous and metamorphic rocks which have crystallized at intermediate to low temperatures. Adularia is the name given to a form (which may be either monoclinic or triclinic) with a distinctive crystal habit found in low-temperature hydrothermal veins.

Recent research has clarified the relationship between these forms. Microcline and sanidine are polymorphs with an order-disorder relationship, the Si and Al atoms being randomly distributed over their lattice positions in sanidine but ordered in microcline. The disordered form is the more stable polymorph above about 700°, and microcline has been transformed into sanidine by hydrothermal treatment at this temperature; the reverse transformation has not been achieved in the laboratory, evidently because of the high activation energy required for the ordering of the Si and Al atoms. Orthoclase and adularia are structurally intermediate between sanidine and microcline. Much orthoclase probably crystallized originally as sanidine. Adularia is evidently a metastable form which develops under conditions of rapid crystallization within the stability field of microcline; the rapid crystallization prevents the attainment of an ordered arrangement of Si and Al.

At high temperatures complete solid solution exists between $KAlSi_3O_8$ and $NaAlSi_3O_8$. The more potassic members of the series are monoclinic and are called soda-orthoclase; indeed, most of the orthoclase we identify in rocks is really soda-orthoclase with considerably more potassium than sodium. The more sodic members of the series are triclinic and are called anorthoclase. At lower temperatures solid solutions intermediate between orthoclase and albite are metastable and under conditions of slow cooling break down into an oriented intergrowth of subparallel lamellae, alternately sodic and potassic in composition. Such an intergrowth is called perthite or antiperthite. In the perthites the plagioclase occurs as uniformly oriented films, veins, and patches within the orthoclase (or microcline); in the antiperthites this relation is reversed. Many apparently homogeneous specimens of alkali feldspar on X-ray examination prove to be perthitic, the intergrowth being submicroscopic. Perthite, when heated for a long time at 1000°, becomes homogeneous once more. Not all perthites have been formed by exsolution; some are undoubtedly the product of partial metasomatic replacement of an originally homogeneous potash feldspar by sodium-bearing solutions.

X-ray examination of potash feldspar and of albite provides the following explanation of the perthite lamellar intergrowth. The framework of linked SiO_4 and AlO_4 tetrahedra, being similar for the monoclinic and triclinic forms, is continuous throughout the structure. At high temperatures the K and Na ions are randomly distributed in the framework, producing a homogeneous crystal. At lower temperatures ordering may occur with the formation of potassium-rich and sodium-rich lamellae, producing alterate sheets with monoclinic or pseudo-monoclinic and triclinic symmetry, respectively. The *a* repeat of potash feldspar and albite is markedly different (8.45 A and 8.14 A), whereas the *b* and *c* repeats are almost identical (12.90 A, 7.14 A and 12.86 A, 7.17 A). This accounts for the lamellae being approximately parallel to 100, since the *b* and *c* repeats coincide in the 100 plane, whereas the *a* repeat running through the lamellae shortens in the albite regions and lengthens in the potash feldspar regions.

The Or—Ab system has been carefully studied by Bowen and Tuttle (Figure 5.5). Great difficulty has been experienced in the laboratory investigation of this system on account of the viscosity of the melts and their extremely sluggish crystallization. Considerable advances have been made by the addition of water under pressure, which promotes crystallization and lowers its temperature without affecting the general equilibrium relations. A pressure of 2000 kg/cm^2 of water lowers the crystallization temperature by as much as 300°, thereby confirming the opinion long held among some petrologists as to the importance of the water content in reducing the temperature of magmatic crystallization. The system Or—Ab is not truly binary because of the incongruent melting of Or to give leucite; however, the field of leucite almost disappears at 2000 kg/cm^2 pressure.

Data have also been obtained on the form of the solvus curve *XY*. The shape and position of this curve are not significantly altered by the presence or absence of water, since the equilibrium is one between solids, in which water does not participate in any of the phases. This curve, dividing the two-feldspar field (i.e., the perthite field) from the one-feldspar field, has a maximum at 660° $\pm$ 10°, at a composition close to 55% Ab. Thus crystallization above 660° in the Or—Ab system gives a single feldspar at any composition. Any point on the solvus represents the minimum temperature of

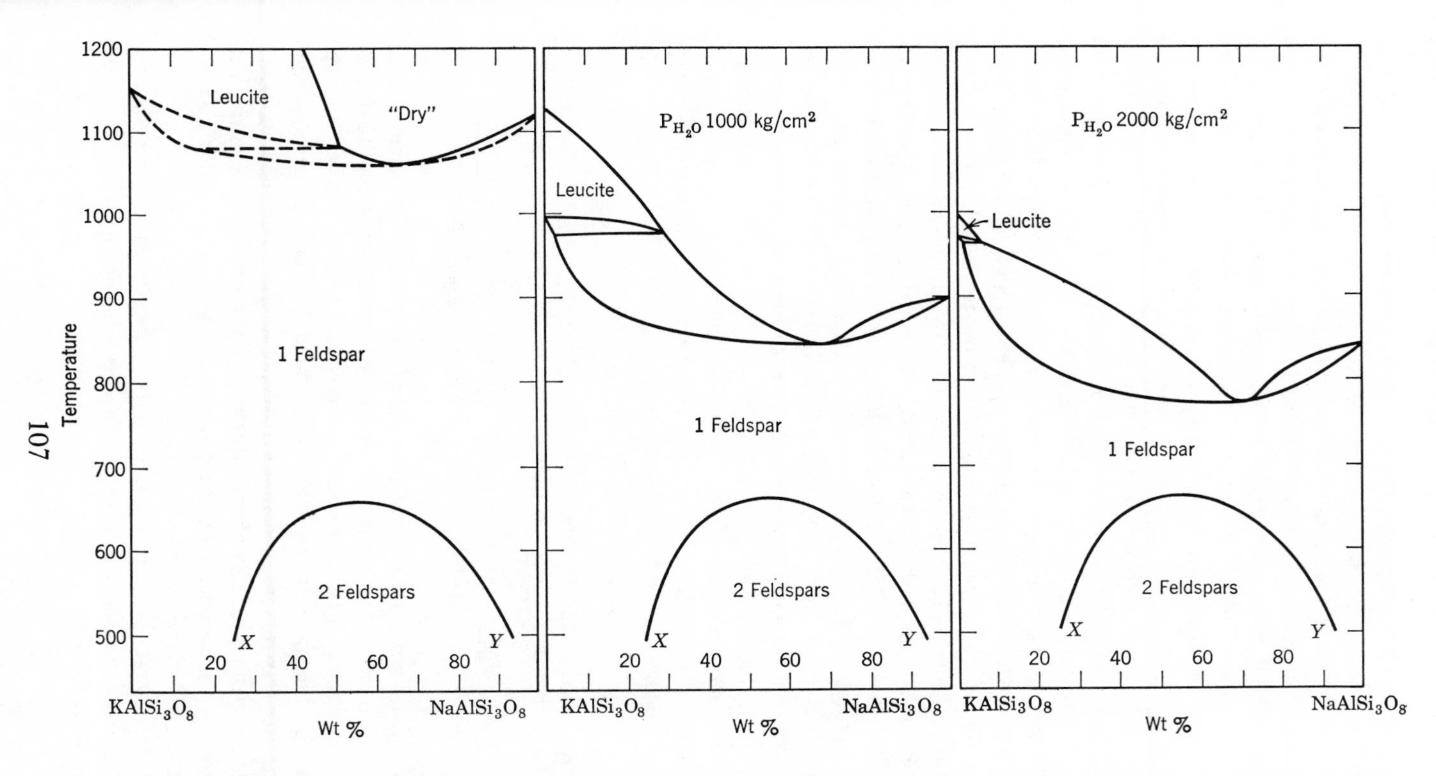

Figure 5.5 Equilibrium diagrams for the Or—Ab system in dry melts and at 1000 kg/cm² and 2000 kg/cm² pressure of H_2O. (Bowen and Tuttle, *J. Geol.* **58,** 497, 1950)

stable existence for homogeneous feldspar of that composition; if equilibrium is maintained below that temperature, the feldspar will begin to unmix.

The situation is complicated by the existance of low-temperature forms both of potash and of soda feldspar. Figure 5.5 represents equilibrium conditions for the high-temperature forms of these compounds. Bowen and Tuttle have found that in many instances the naturally occurring perthites are intergrowths of low-temperature feldspars. Different kinds of naturally occurring perthites may be recognized, according to the particular forms of the soda and potash feldspars which make up the intergrowth.

The system Ab—An, long quoted as a classic example of perfect solid solution, also shows unsuspected complications. True, the solidus-liquidus relations, one of the first fruits of research at the Geophysical Laboratory, remain unamended. The complete solid solution at high temperatures, however, is affected by the inversion of albite to a low-temperature form at about 700°; the relationship between the low and high forms is evidently an order-disorder one, similar to that between microcline and sanidine. The unit cell of anorthite has a doubled *c* axis with respect to that of albite, and two distinct structures are found, primitive and body-centered respectively. These complications were first suspected when discrepancies were observed between the optical properties of plagioclases of the same composition from volcanic and plutonic rocks, indicating distinct series of high-temperature and low-temperature plagioclases. The low-temperature series is not continuous but contains at least six structural divisions, the compositional ranges of which are approximately as follows:

Low-albite structure	Peristerite structure	Intermediate structure	Body-centered anorthite structure	Transitional anorthite structure	Primitive anorthite structure

An 0 — 1–5 — 21–25 — 70–75 — 80–85 — 90–95 — An 100

The low-albite structure can accept only a little calcium replacing sodium, and in the approximate range An_3 to An_{23} plagioclases are usually submicroscopic intergrowths, known as peristerite, of sodium-rich and calcium-rich regions. Plagioclases more calcic than about An_{23} crystallize in variants of the anorthite structure, and submicroscopic intergrowths of sodium-rich and calcium-rich regions may be present also. Thus many plagioclases which are optically homogeneous prove to consist of complex intergrowths when examined by X-ray techniques.

The mutual miscibility of Or and An is nearly zero at all temperatures; hence varieties intermediate between these two components do not occur.

Careful analyses of homogeneous plagioclases have shown that the potassium content is usually very low, indicating that the normal amount of the $KAlSi_3O_8$ component in solid solution is about 1% and that it seldom rises above 5%.

The composition of any feldspar is conveniently expressed by the use of the symbols Or, Ab, and An for the pure components: thus Ab_{32} An_{68} would be a plagioclase falling in the labradorite section of the series, and $Or_{26}Ab_{66}An_8$ is the composition of a possible anorthoclase.

THE FELDSPATHOIDS

The feldspathoids are a group of alkali-aluminum silicates which appear in place of the feldspars when an alkali-rich magma is deficient in silica. They are never associated with primary quartz. The following minerals are the more important members of this group:

Leucite	$KAlSi_2O_6$
Kaliophilite	$KAlSiO_4$
Kalsilite	$KAlSiO_4$
Nepheline	$NaAlSiO_4$
Sodalite	$Na_8Al_6Si_6O_{24}(Cl_2)$
Noselite	$Na_8Al_6Si_6O_{24}(SO_4)$
Cancrinite	$Na_8Al_6Si_6O_{24}(HCO_3)_2$

Analcime, $NaAlSi_2O_6 \cdot H_2O$, is sometimes included among the feldspathoids; it appears occasionally as a primary mineral of igneous rocks deficient in silica.

The feldspathoids are not a homogeneous series like the feldspars or the pyroxenes. They are grouped together more on petrographical than mineralogical similarities. Structurally, the feldspathoids belong to the tektosilicates, the SiO_4 and AlO_4 tetrahedra being linked as in the feldspars, whereas the metal ions (and the chloride, sulfate, and carbonate ions, when present) fit into cavities in this framework. The feldspathoids are readily attacked by acids. This characteristic is evidently due to the comparatively high Al:Si ratio; the aluminum is removed in solution and the lattice then collapses, often with the formation of gelatinous silica.

Leucite is the commonest potassium feldspathoid. It is abundant in the volcanic rocks of a few regions. It is not found in plutonic rocks, and laboratory investigations have indicated that leucite is not a stable phase in the K_2O—Al_2O_3—SiO_2 system at high pressures. Analyses of leucite show that a little sodium may replace potassium in the structure. Kaliophilite and kalsilite are mineralogical curiosities; kaliophilite has been recorded in

Vesuvian lavas, kalsilite in lavas from East Africa. Their optical properties are so similar to those of nepheline that misidentification could readily occur. On this account they may be of wider distribution than present records indicate.

Nepheline is the commonest of the feldspathoids and is found in both volcanic and plutonic rocks. It is isomorphous with high-tridymite (cf. the formulas $NaAlSiO_4$ and $SiSiO_4$). The analogy between $NaAlSiO_4$ and SiO_2 is further emphasized by the inversion at 1248° of nepheline to carnegieite, a high-temperature form which is isomorphous with high-cristobalite; carnegieite has not been found as a mineral. Analyses of nepheline generally show an excess of Si over the theoretical amount, the ratio Si:Al ranging up to 1.4 and the sodium being correspondingly deficient. The phenomenon is one of Si replacing Al and the consequent omission of sodium ions to preserve electrical neutrality. Nepheline always contains some potassium replacing sodium, the Na:K ratio often being close to 3:1; this reflects the atomic structure, in which one of the four positions that can be filled by an alkali ion is larger than the other three and preferentially accommodates potassium.

THE PYROXENE GROUP

The pyroxenes are a group of minerals closely related in crystallographic and other physical properties, as well as in chemical composition, although they crystallize in two different systems, orthorhombic and monoclinic. The group characteristics are the outward expression of common atomic structure. The tetrahedral SiO_4 groups are linked together into chains by the sharing of one oxygen atom between two adjacent groups; that is, in each group two oxygen atoms also belong half to the groups on each side, giving the Si:O ratio of 1:3. The silicon-oxygen chains lie parallel to the vertical crystallographic axis and are bound together laterally by the metal ions. The distinct prismatic habit of pyroxene crystals is a consequence of this internal structure, as is also the typical cleavage, which takes place between the silicon-oxygen chains.

The chemical composition of the pyroxenes can be expressed by the general formula $(W)_{1-p}(X, Y)_{1+p}Z_2O_6$, in which the symbols W, X, Y, and Z indicate elements having similar ionic radii and capable of replacing each other in the structure. In the pyroxenes these elements may be

W: Na, Ca
X: Mg, Fe^{2+}, Li, Mn
Y: Al, Fe^{3+}, Ti
Z: Si, Al (in minor amount)

On the basis of chemical composition and crystal system the following distinct species are recognized:

Orthorhombic Pyroxenes

Enstatite	$MgSiO_3$
Hypersthene	$(Mg,Fe)SiO_3$

Monoclinic Pyroxenes

Clinoenstatite	$MgSiO_3$
Clinohypersthene	$(Mg,Fe)SiO_3$
Diopside	$CaMgSi_2O_6$
Hedenbergite	$CaFe^{2+}Si_2O_6$
Augite	Intermediate between diopside and hedenbergite with some Al
Pigeonite	Intermediate between augite and clinoenstatite-clinohypersthene
Aegirine (acmite)	$NaFe^{3+}Si_2O_6$
Jadeite	$NaAlSi_2O_6$
Spodumene	$LiAlSi_2O_6$
Johannsenite	$CaMnSi_2O_6$

Spodumene, although a member of the monoclinic pyroxenes, is not closely related to the other species in the group, and no intermediate compounds are known. It is not an important rock-forming mineral, being confined to the complex granite pegmatites. Jadeite is a rare mineral of metamorphic rocks. Johannsenite is found only as a vein mineral. These species are not considered here.

The pyroxenes fall naturally into two divisions, those of orthorhombic symmetry and those of monoclinic symmetry.

The orthorhombic pyroxenes range in composition from pure $MgSiO_3$ to about 90% $FeSiO_3$. The compound $FeSiO_3$ is not a stable phase at high temperatures; from a melt of its composition silica and fayalite (Fe_2SiO_4) will crystallize. It has been reported in the monoclinic form from lithophysae of an obsidian, where it would have formed at moderate temperatures, but the identification is not certain. The common orthopyroxenes of igneous rocks are all magnesium-rich. The orthorhombic pyroxenes have been divided into a number of subspecies based on composition, similar to the division of the plagioclase series. The utility of such subdivision is dubious; subspecific names are a burden on the nomenclature and the memory, and in general a designation in the form hypersthene (En_x) is to be preferred, just as plagioclase (An_{25}) is preferable to a less precise identification as oligoclase.

When heated above 985° pure $MgSiO_3$ changes into protoenstatite, another orthorhombic phase with a different structure, which is stable up to

1557°, when it melts incongruently to give forsterite plus liquid. Protoenstatite has not been found as a mineral, but on rapid cooling it changes into clinoenstatite, a monoclinic phase. Hypersthene has a corresponding monoclinic phase known as clinohypersthene. Clinoenstatite and clinohypersthene are almost unknown in terrestrial rocks but have been recognized in some meteorites. Intermediate compounds between these substances and augite are, however, found in basic volcanic and hypabyssal rocks and are known as pigeonite. The few good analyses of pigeonite suggest a maximum of about 10 atom % Ca in $W + X + Y$. The diopside-hedenbergite series is well established throughout the whole range in composition. A limited amount of aluminum can enter into this series, and such aluminous pyroxenes are called augite. The introduction of Al into a diopside means that this Al must be divided between the Y and the Z lattice positions, otherwise the valence demands would not be satisfied. The aegirine-jadeite series may be continuous throughout the Al-Fe range. Aegirine is connected with the diopside-hedenbergite series and augite by members of intermediate composition, which are grouped under the name aegirine-augite.

The compositions of the naturally occurring pyroxenes are extremely variable because of the possibilities of atomic substitution. In the general formula p is zero or near zero in the diopside-hedenbergite and aegirine-jadeite series and 1 or close to 1 in the orthorhombic pyroxenes, their monoclinic dimorphs, and in the pigeonites. Magnesium and ferrous iron are completely interchangeable. Reliable analyses of the common pyroxenes, the diopside-hedenbergite-augite group, show a maximum of about 7% Fe_2O_3 (corresponding to 10% Fe^{3+} in Y); about 8% Al_2O_3 (corresponding to 10% or a little more Al replacing Si and a smaller Al replacement in Y); and about 1.5% TiO_2. Minor constituents recorded in pyroxene analyses include chromium (up to 1.2% Cr_2O_3 in some diopsides and augites), vanadium (4% V_2O_3 in aegirine from Libby, Montana), manganese, and potassium. The miscibility gap between augite and $(Mg,Fe)SiO_3$ is considerably less at high than at moderate temperatures; in volcanic rocks the pyroxenes may be low-calcium augite (sometimes termed subcalcic augite) and/or pigeonite, whereas in plutonic rocks the pyroxenes are normal augites and/or orthorhombic pyroxenes almost free of calcium (Figure 5.6).

THE AMPHIBOLE GROUP

The amphibole group comprises a number of species, which, although falling both in the orthorhombic and monoclinic systems, are closely related in crystallographic and other physical properties, as well as in chemical composition. They form isomorphous series, and extensive replacement of one ion by others of similar size can take place, giving rise to very complex

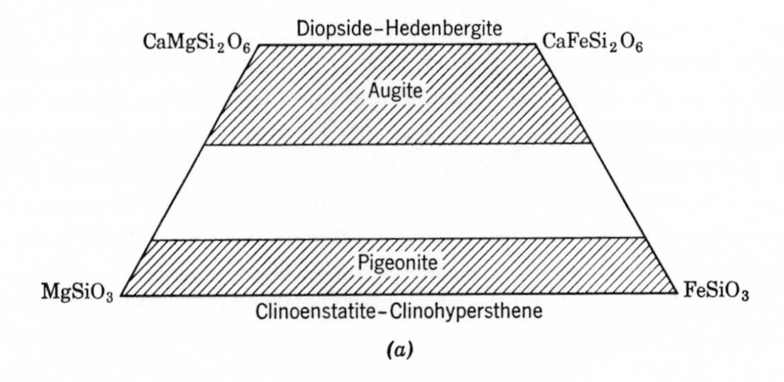

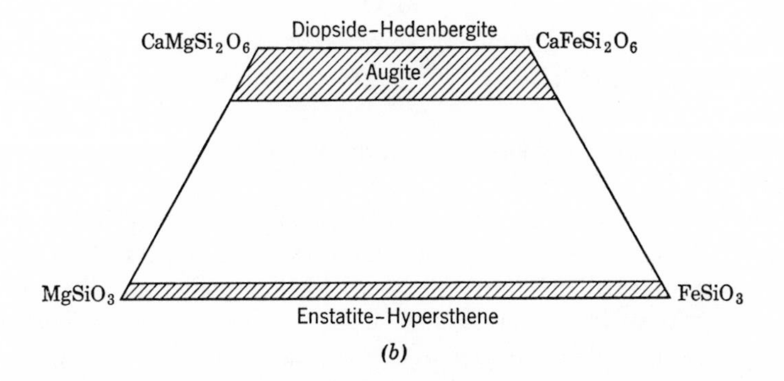

Figure 5.6 Solid solution relations and mineralogy in the pyroxene group: (*a*) at high temperatures; (*b*) at medium temperatures. Shaded areas are solid solution fields, blank areas, miscibility gaps.

chemical compositions. The species of the amphibole group form a series parallel to those of the allied pyroxene group; they were originally looked upon as complex metasilicates dimorphous with the corresponding pyroxenes. This is not so, however; the amphiboles contain essential (OH) groups in their structure, and the Si:O ratio is 4:11, not 1:3 as in the pyroxenes. The true nature of the amphiboles has been elucidated by X-ray studies, which have shown that the fundamental unit in their structure is a double chain of linked silicon-oxygen tetrahedra; in effect it is two single chains with alternate tetrahedra linked by the sharing of the oxygen, giving an Si:O ratio of 4:11, instead of 1:3 as in single chains. In the structure these double chains lie parallel to the vertical crystallographic axis and are bound together laterally by the metal ions. The binding force between the chains is not so strong as the Si-O bonds along the chain. This is reflected in the well-known fibrous or prismatic nature of the amphiboles parallel to the *c* axis and in the good prismatic cleavage. Another interesting feature of the amphibole structure is the presence of open spaces within the lattice into which an extra alkali atom for each two Si_4O_{11} groups will fit.

The difference in chemical composition between compounds of the amphibole type and corresponding compounds of the pyroxene type is not great, and it is understandable that it was overlooked for many years and that the amphiboles and pyroxenes of similar composition were considered to be polymorphs of the same substance. This is illustrated by the theoretical compositions of $MgSiO_3$ and $Mg_7(Si_4O_{11})_2(OH)_2$

	MgO	SiO_2	H_2O
$MgSiO_3$	40.0	60.0	—
$Mg_7(Si_4O_{11})_2(OH)_2$	36.2	61.5	2.3

A general formula for all members of the amphibole group can be written $(WXY)_{7\text{-}8}(Z_4O_{11})_2(O,OH,F)_2$, in which the symbols *W*, *X*, *Y*, *Z* indicate elements having similar ionic radii and capable of replacing each other in the structure: *W* stands for the large metallic cations Ca and Na (K is sometimes present in small amounts); *X* stands for the smaller metallic cations Mg and Fe^{2+} (sometimes Mn); *Y* for Ti, Al, and Fe^{3+}; and Z for Si and Al. In the general formula the degree of atomic substitution is as follows:

1. Al may replace Si in the Si_4O_{11} chains to the extent of $AlSi_3O_{11}$ (the amount of replacement is a function of the conditions of formation; high-temperature amphiboles can be more aluminous than low-temperature amphiboles).
2. Fe^{2+} and Mg are completely interchangeable.
3. The total (Ca, Na, K) may be zero or near zero or may vary from 2 to 3; however, total Ca never exceeds 2, and K is present only in minor amounts.
4. OH and F are completely interchangeable. The maximum is 2, but it may be less, presumably by O replacing OH or F.

With these possibilities it is clear that the composition of the amphiboles may be very complex. However, on the basis of chemical composition and crystal structure five distinct series are recognized.

	Orthorhombic
Anthophyllite series	$(Mg,Fe)_7(Si_4O_{11})_2(OH)_2$ (Mg predominant over Fe)
	Monoclinic
Cummingtonite series	$(Fe,Mg)_7(Si_4O_{11})_2(OH)_2$ (Fe predominant over Mg)
Tremolite series	$Ca_2(Mg,Fe)_5(Si_4O_{11})_2(OH)_2$
Hornblende series	$Ca_2Na_{0-1}(Mg,Fe,Al)_5[(Al,Si)_4O_{11}]_2(OH)_2$
Alkali amphibole series	$(Na > Ca)$, e.g.,
Glaucophane	$Na_2Mg_3Al_2(Si_4O_{11})_2(OH)_2$
Riebeckite	$Na_2Fe_3^{2+}Fe_2^{3+}(Si_4O_{11})_2(OH)_2$
Arfvedsonite	$Na_3Fe_4^{2+}Fe^{3+}(Si_4O_{11})_2(OH)_2$

The anthophyllite series corresponds to the orthorhombic enstatite-hypersthene series in the pyroxene group. The ratio Mg:Fe ranges from 7:0 up to about 1:1; Al is often present and can replace Si in (Si_4O_{11}) up to $(AlSi_3O_{11})$, with corresponding replacements in the (WXY) elements. Other substitutions are minor. Members of the anthophyllite series have been found only in metamorphic rocks.

The cummingtonite series corresponds to the clinohypersthene series in the pyroxene group. The ratio Fe:Mg ranges from 7:0 to about 1:2; that is, the cummingtonite series and the anthophyllite series overlap in the middle composition range. Rocks containing these two amphiboles side by side have been described. Members of the cummingtonite series are also confined to metamorphic rocks.

The tremolite series corresponds to the diopside-hedenbergite series in the pyroxene group. In tremolite magnesium is replaceable by ferrous iron and also in part by aluminum and ferric iron, silicon in part by aluminum; titanium and fluorine may be present; and an additional sodium ion for each two (Si_4O_{11}) groups may enter the structure. The product of all these substitutions is the hornblende series, analogous to augite in the pyroxene group but considerably more complex. Thus the mineral known as hornblende has a very wide range of composition and a correspondingly wide range in optical properties. Most hornblende is green, but there is a dark brown variety, long known as basaltic hornblende, which is sometimes considered as a separate series on account of its distinctive properties.

The alkali amphiboles can be considered as derived from the hornblende series by the partial or complete substitution of Na for Ca. The best known of these soda amphiboles are glaucophane and the related riebeckite and arfvedsonite, although other varieties have been described.

In view of the similarity in composition between corresponding amphiboles and pyroxenes the circumstances under which one or the other or both will crystallize from a magma have been the subject of considerable speculation. Amphiboles occur more often in plutonic rocks than in volcanic rocks, evidently because the incorporation of OH groups in the structure is favored by crystallization under pressure. However, the composition of the magma may also be significant. The Ca:Fe:Mg ratios of igneous hornblendes and pyroxenes show characteristic differences; many hornblendes fall in the composition gap between augite and the pigeonites and orthorhombic pyroxenes.

THE OLIVINE GROUP

The minerals of the olivine group are silicates of bivalent metals and crystallize in the orthorhombic system. There are a number of species:

Forsterite	Mg_2SiO_4
Fayalite	Fe_2SiO_4
Olivine	$(Mg,Fe)_2SiO_4$
Tephroite	Mn_2SiO_4
Monticellite	$CaMgSiO_4$
Glaucochroite	$CaMnSiO_4$

The only common rock-forming minerals are the magnesium-iron compounds, although monticellite is found in metamorphosed limestones and has been recorded from basic igneous rocks. The other species are known from ore deposits in metamorphosed limestones.

X-ray studies on olivine show that it is built up of independent SiO_4 tetrahedra and that the magnesium and ferrous ions lie between irregular groups of six oxygens and belong to two sets that are not structurally identical. In keeping with this structural type, which does not have extended chains or rings of SiO_4 tetrahedra, the minerals of the olivine group have no tendency to form fibrous or platy crystals and generally occur as equidimensional crystals. The close-packed structure of olivine is reflected in comparatively high density (forsterite, 3.22; cf. enstatite, 3.18; anthophyllite, 2.96; talc, 2.82) and high refractive indices.

The composition of olivine generally corresponds closely to $(Mg,Fe)_2SiO_4$, there being little replacement by other elements. A carefully collected and analyzed series of olivines from the Skaergaard intrusion (Wager and Deer, 1939, pp. 72–73) shows a little substitution by manganese, the amount increasing in the later-formed olivine (MnO 0.22–1.01%); the same series shows some calcium (2.18% CaO) in the last-formed olivine. Olivines from dunites generally contain some nickel, often about 0.3%. A noteworthy feature of olivine is practical absence of aluminum; evidently replacement of Mg and Si by Al does not occur in the olivine structure.

THE MICA GROUP

With their characteristic perfect basal cleavage the members of the mica group are readily recognizable. The composition of individual specimens may be very complex, but a general formula of the type $W(X,Y)_{2-3}Z_4O_{10}(OH,F)_2$ can be written for the group as a whole. In this formula W is generally potassium (Na in paragonite); X and Y represent Al, Li, Mg, Fe^{2+}, and Fe^{3+}; Z represents Si and Al, the Si:Al ratio being generally about 3:1. The different species form a typical isomorphous group, but the phase relations have not yet been fully worked out. Two members of the group frequently crystallize together in parallel position. Biotite crystallizes in this way with muscovite, muscovite with lepidolite, and so forth. In the following list the formulas have been simplified to an ideal type conforming to the structure established by X-ray investigation.

Muscovite	$KAl_2(AlSi_3O_{10})(OH)_2$
Paragonite	$NaAl_2(AlSi_3O_{10})(OH)_2$
Phlogopite	$KMg_3(AlSi_3O_{10})(OH)_2$
Biotite	$K(Mg,Fe)_3(AlSi_3O_{10})(OH)_2$
Lepidolite	$KLi_2Al(Si_4O_{10})(OH)_2$

The structural scheme of mica is as follows: The basic units, the SiO_4 tetrahedra, are each linked by three corners to neighboring tetrahedra, forming a sheet. Each SiO_4 tetrahedron thus has three shared and one free oxygen; hence the composition and valency are represented by $(Si_4O_{10})^{4-}$. Two of these sheets of linked tetrahedra are placed together with the vertices of the tetrahedra pointing inwards. These vertices are cross-linked by Al in muscovite or by Mg and Fe in phlogopite and biotite. Hydroxyl groups are incorporated and linked to Al, Mg, or Fe alone. A firmly bound double sheet is thus produced with the bases of the tetrahedra on each outer side. The structure is a succession of such double sheets, with the potassium ions placed between them.

The common mica of igneous rocks is biotite. Muscovite is present in some granites. Lepidolite has been recognized in a few granites, but its typical mode of occurrence is in granite pegmatites. Phlogopite is sometimes found in igneous rocks rich in magnesium and poor in iron, such as peridotites, but is more common in metamorphosed limestones and in some pegmatites. Paragonite is a rare mineral of schists.

The common occurrence of biotite in igneous rocks in contrast to the limitation of muscovite to pegmatites and some granites has been clarified by the work of Yoder and Eugster (Figure 5.7). They have found that the stability curve for phlogopite lies about 300° higher than that for muscovite

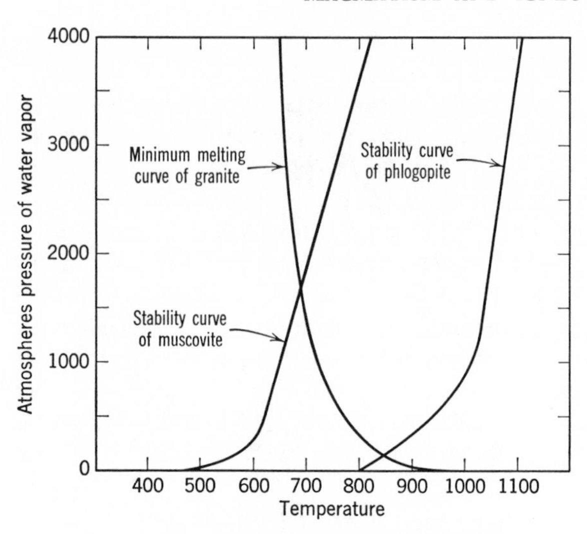

Figure 5.7 Relationship between the stability curves of muscovite and phlogopite and the minimum melting curve of granite. (After Yoder and Eugster, *Geochim. et Cosmochim. Acta* **6,** 195, 1954)

and is well above the minimum melting curve for granite. This means that phlogopite (and biotite) can form directly from a magma at normal crystallization temperatures. The stability curve for muscovite, on the other hand, lies below the minimum melting curve for granite at low pressure, intersecting it at about 700° and 1500 atm of water-vapor pressure; hence muscovite in a granite implies crystallization at high water-vapor pressure, that is, considerable depth, or later introduction into the already crystallized rock.

Biotites from igneous rocks are exceedingly variable in chemical composition. Magnesium and ferrous iron are completely diadochic, and all types are known, from iron-free (i.e., phlogopite) to varieties in which nearly all the magnesium is replaced by iron. Ferric iron may replace half or more of the six-coordinated aluminum. Part of the hydroxyl may be replaced by fluorine, although most analyses of igneous biotites show only small amounts of this element. Minor amounts of Mn, Ti, Li, Na, and Ca are recorded in analyses of biotite, and less commonly Ba, Cr, Ni, Rb, and Cs. There is a general trend from magnesium-rich biotites in ultrabasic rocks to iron-rich biotites in granites and nepheline syenites (Figure 5.8). Aluminum is greatest in biotites from granites and pegmatites and is lowest in biotites from ultrabasic rocks; silicon shows an inverse relationship to aluminum.

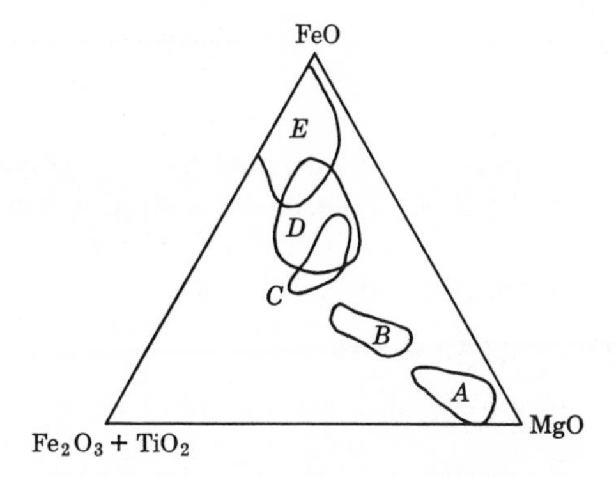

Figure 5.8 Variation of chemical composition of biotite with rock type; (*A*) in ultrabasic rocks; (*B*) in gabbros; (*C*) in diorites; (*D*) in granites; (*E*) in pegmatites. (After Heinrich, *Am. J. Sci.* **244,** 836, 1946)

THE NATURE OF A SILICATE MELT

Barth and Rosenquist (1949) and Wasserburg (1957) have discussed the thermodynamics of silicate melts. The entropy of fusion of silicates is not high, and so the atoms or ions in the molten silicate evidently have a degree of order not greatly different from that in the solid. It thus appears that a good deal of the structural arrangement is preserved on melting. Since the silicon-oxygen bonds are much stronger than other links in silicate structures, it is reasonable to assume that these bonds are present to some extent in the liquid also; that is, the anions in the melt are predominantly polymerized silicon-oxygen tetrahedra linked together by sharing oxygens into one-, two-, or three-dimensional networks, similar to those in crystalline silicates but more irregular. The high viscosity of silicate melts is evidently due to the presence of these complex silicon-oxygen groupings.

The degree of polymerization of the silicate anions is markedly affected by the ratio of Si (and Al) to O. The linking of the silicon-oxygen tetrahedra will be greatest at Si:O ratio of 1:2 and will decrease as the proportion of Si decreases. This is borne out by the increase in viscosity of silicate melts as the SiO_2 content increases. The presence of other ions is also significant. In alkali silicate melts the viscosity decreases in the sequence $K > Na > Li$; evidently sodium is more effective than potassium in disrupting the Si-O bonds and thereby breaking up the polymerized anions into smaller units, and lithium even more so. Probably bivalent cations are even more effective in this respect; iron-rich melts are of notably low viscosity even at high SiO_2 contents. Laboratory experiments have shown, however, that small amounts of H_2O have a particularly remarkable influence in decreasing the viscosity, evidently because of the strong tendency for the reaction $H_2O + O^{2-} = 2(OH)^-$ to take place. The oxygen links between silicon-oxygen tetrahedra are thereby destroyed and the polymerized anions broken down into simpler groups. Since the "molecular weight" of OH is low, a small weight percentage of H_2O in a silicate melt is very effective in eliminating links between silicon-oxygen tetrahedra.

CRYSTALLIZATION IN SILICATE MELTS

Through the study of crystallization in artificial silicate melts of known composition great advances have been made in the understanding of the geochemistry of igneous rocks. The principles of heterogeneous equilibrium govern crystallization from a liquid; they express the conditions under which only one of the possible crystallization products is to be expected and under what conditions more than one may appear simultaneously. The basic relation is, of course, the phase rule, which states that in any system the number of phases (P) plus the number of degrees of freedom (F) are equal to the number of components (C) plus two, or $P + F = C + 2$. The phase rule has been extended in the consideration of geochemical processes in the following way: In any system the maximum number of phases can be reached only when the number of degrees of freedom are at a minimum. This state can be realized by fixing both temperature and pressure. However, it is extremely unlikely that such conditions will occur during magmatic crystallization, since it proceeds as a rule over a great P-T range; that is, these factors remain variable and thus give two degrees of freedom. Under these circumstances the phase rule becomes $P = C$; that is, in a system of n components at arbitrary temperature and pressure no more than n phases (minerals) can be mutually stable. This extension of the phase rule is due to Goldschmidt and is sometimes known as the *mineralogical phase rule*.

Earlier in this chapter it was remarked that despite their wide range in chemical composition igneous rocks have a comparatively simple mineralogy in that probably 99% are made up essentially of seven minerals or mineral groups. This limitation to the number of phases normally formed by the crystallization of a magma is clearly understandable in terms of the phase rule. Actually, considering a magma as an eight-component system of O, Si, Al, Fe, Mg, Ca, Na, and K, it is improbable that as many as eight phases would crystallize from any composition in that system because the individual components are not completely independent, some being capable of replacing each other atom for atom in minerals.

The Geophysical Laboratory has been responsible for many of the laboratory investigations on equilibria in silicate melts. Since 1905 a steady stream of publications dealing with successively more complex systems has been issued by that institution. Since the results of this work are fully described and discussed in standard texts on petrology, no attempt is made to cover this field here. However, the following systems are selected as illustrating the major principles that have been developed.

THE PYROXENE-PLAGIOCLASE SYSTEM

This system is particularly significant in petrogenesis, since compositions within it are reasonably close to basalts and gabbros, which are often essentially pyroxene-plagioclase rocks. We may start with the three-component system diopside (Di)–anorthite (An)–albite (Ab). The phase relations as worked out by Bowen in 1915 are given in Figure 5.9. It is now known that the system is not strictly ternary, because some Al_2O_3 from the plagioclase enters the diopside; however, the phase relations as established by Bowen are still valid. The equilibrium diagram is divided into a plagioclase field and a diopside field by a boundary curve joining the eutectic points in the Di—An and Di—Ab systems. This boundary curve is known either as a

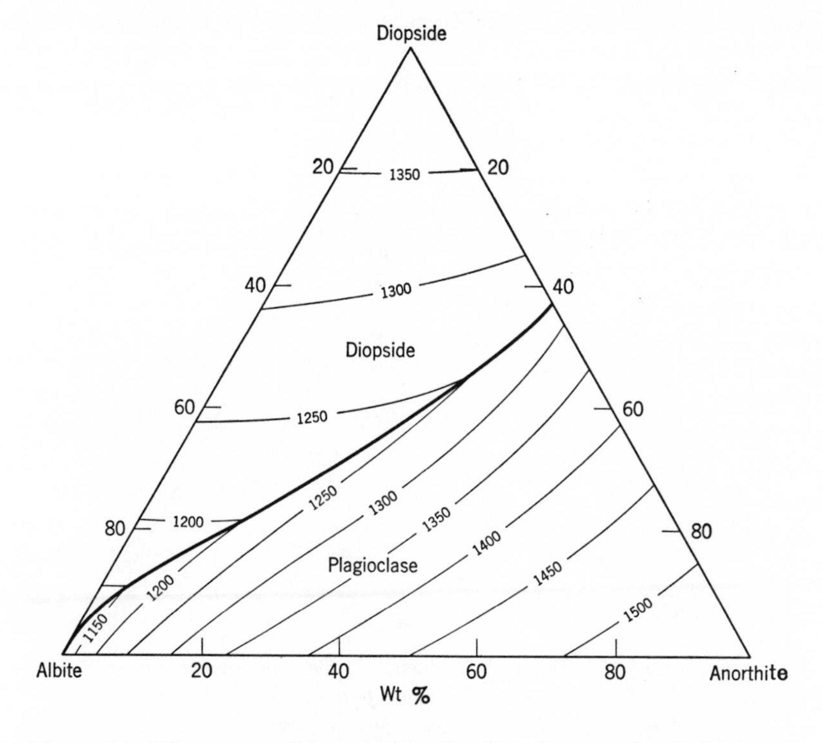

Figure 5.9 The system albite-anorthite-diopside. (Bowen, *Am. J. Sci.* **190,** 167, 1915)

reaction curve or a *cotectic line;* it marks the bottom of a valley on the liquidus surface, as can be seen from the temperature contours. From any melt with a composition that places it in the plagioclase field plagioclase crystallizes first, and the composition of the liquid moves in the directions of the reaction curve; when it reaches this curve, diopside begins to crystallize; thereafter diopside and plagioclase crystallize together, and the composition of the melt follows the reaction curve towards the albite–diopside eutectic. Solidification is complete at a point that varies according to the initial composition of the melt and the extent to which the composition of the feldspar changes during crystallization. The course of solidification is similar for a melt with an initial composition placing it in the diopside field, except that diopside is then the first phase to crystallize.

By the addition of $FeSiO_3$ (Fs) as a fourth component compositions resembling those of basalts and gabbros can be represented. This system has not been completely worked out in the laboratory, but considerable data are available for it, and Barth has established a tentative equilibrium diagram in the form of a tetrahedron in which each corner represents one component (Figure 5.10). In this diagram *A* represents Ab, *B* An, *C* Di, and *D* Fs; the

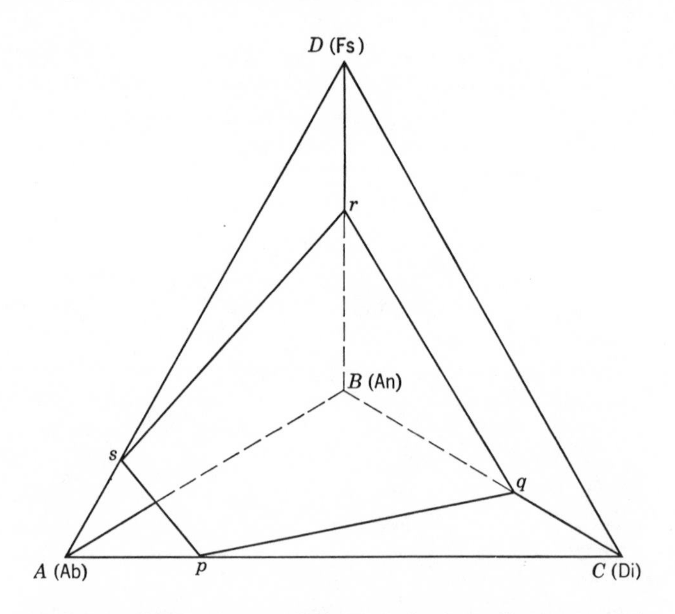

Figure 5.10 Tetrahedron illustrating the approximate composition of basaltic lavas. The plane *pqrs* represents the boundary surface separating those lavas that precipitate pyroxene first from those that precipitate plagioclase first. (Barth, *Am. J. Sci.* **231,** 331, 1936)

point *p* is the Di—Ab eutectic, *q* the Di—An eutectic, and *r* and *s* the (probable) eutectics between Fs and An and Ab. The system consists of two binary solid solution series, the plagioclase series *AB* and the pyroxene series *CD*; four binary eutectic systems *AC*, *BC*, *BD*, and *AD*, with binary eutectics at *p*, *q*, *r*, and *s*; and four ternary systems each divided into two fields by the reaction curves *pq*, *qr*, *rs*, and *sp*; the quaternary system itself is divided into two parts by the *reaction surface*.

From a melt with any composition between *AB* and the reaction surface the first phase to crystallize is plagioclase. The melt thereby becomes richer in components *C* and *D*, and eventually its composition lies in the reaction surface *pqrs*. When this surface is reached, pyroxene begins to crystallize together with the plagioclase. From then on the composition of the melt lies in the reaction surface and moves down the temperature gradient, and the crystals of both plagioclase and pyroxene react with the melt and change in composition continuously until solidification is complete. The same possibilities for fractional crystallization as in the simpler systems are present. If fractional crystallization takes place, the early-separated crystals will be rich in An and Di and the last-formed crystals richer in the Ab and Fs components.

Barth has studied the composition and sequence of crystallization of a number of basalts and has shown that the ideas developed from a consideration of the quaternary system discussed above are borne out by the petrographic data. He was able to derive approximate coordinates for the position of the reaction surface by plotting in the tetrahedron the compositions of those basalts that showed evidence of simultaneous crystallization of plagioclase and pyroxene. The points thus obtained defined the reaction surface; he found that they lay almost exactly in one plane and that the position of this plane agreed with the position that would be expected on the basis of the (incomplete) laboratory study of this system.

The experimental study of the crystallization of natural and synthetic melts of basaltic or near-basaltic composition has been greatly extended since the pioneer work of Bowen and Barth. The principles outlined above remain valid, but the detailed information is much more extensive, and possible complications and variations on the simple system have been elucidated. A comprehensive summary is provided by the monograph of Yoder and Tilley (1962).

THE $KAlSiO_4$—$NaAlSiO_4$—SiO_2 SYSTEM

Compositions within this system are close to some igneous rock—the granites, syenites, nepheline syenites, and the corresponding rhyolites, trachytes, and phonolites. Investigation of laboratory melts has shown that fractional crystallization of complex silicate melts containing potassium, sodium, and aluminum leads always to a residual liquid enriched in alkali-aluminum silicate. Bowen therefore refers to the $NaAlSiO_4$—$KAlSiO_4$—SiO_2 system as "petrogeny's residua system." In the equilibrium diagram (Figure 5.11) the $KAlSi_3O_8$—$NaAlSi_3O_8$ join divides the system into two portions. Compositions between this boundary and the SiO_2 apex approach granitic compositions; compositions below this boundary (i.e., towards the $NaAlSiO_4$—$KAlSiO_4$ join) approach the compositions of alkaline magmas which may crystallize as alkali feldspar plus nepheline or leucite. The most significant feature is the low-temperature trough within the 1100° isotherm. Fractional crystallization of any melt in this system will result in a residual liquid, the composition of which lies in this low-temperature region. Bowen pointed out that, if fractional crystallization has been of fundamental importance in the differentiation of magmas, then those igneous rocks which are the products of the crystallization of residual melts should have salic components with bulk compositions lying in this low-temperature trough. He was able to produce many such examples from among the rhyolites, trachytes, and phonolites, and plots of the salic portion of the average granite, rhyolite, syenite, etc., all fall in the low-temperature region, as shown by Figure 5.11.

Further work on the $NaAlSi_3O_8$—$KAlSi_3O_8$—SiO_2 part of this system under water-vapor pressure has established these relations even more clearly (Figure 5.12). The phase relations in this system resemble those in the albite-anorthite-diopside system, there being two composition fields, one in which quartz is the first mineral to crystallize, the other in which alkali feldspar crystallizes first. These two fields are separated by the reaction curve *AB*. Under a water-vapor pressure of 1000 kg/cm^2, crystallization ultimately leads to a low-temperature trough around the middle part of *AB*. When the normative composition of analyzed granites is calculated in terms of SiO_2, $NaAlSi_3O_8$, and $KAlSi_3O_8$ and these points are plotted on a composition diagram, the frequency clearly reflects the position of this low-temperature trough, i.e., the felsic portion of most granites corresponds in composition to the late liquid fraction produced by crystallization in the SiO_2—$NaAlSi_3O_8$—$KAlSi_3O_8$ system. This observation supports a process of fractional crystallization for the formation of most granites and is consistent with a magmatic origin for them. However, the evidence also shows that in fractional melting of rocks of appropriate composition the initial liquid would have a similar granitic composition.

THE CRYSTALLIZATION OF A MAGMA

From the study of crystallization in artificial silicate melts and the coordination of the results thereby obtained with the observations of igneous petrology, important conclusions have been established regarding magmatic

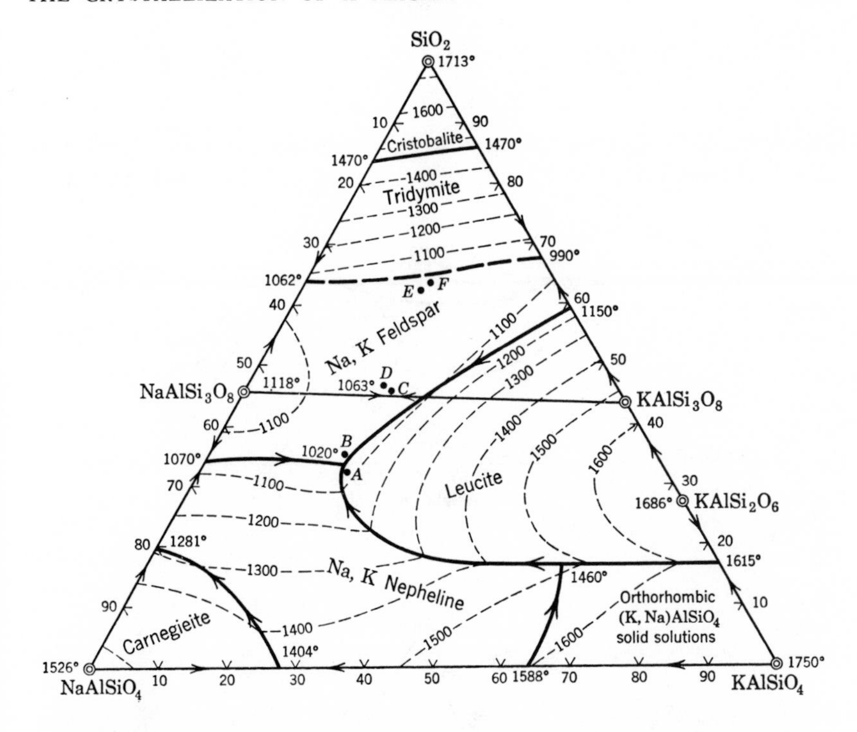

Figure 5.11 Equilibrium diagram for the system $NaAlSiO_4$—$KAlSiO_4$—SiO_2. The black dots represent plots of the normative compositions (except anorthite) of the averages of 15 tinguaites (*A*), 25 phonolites (*B*), 32 alkaline syenites (*C*), 19 alkaline trachytes (*D*), 546 granites (*E*), and 102 rhyolites (*F*). (Bowen, *Am. J. Sci.* **233,** 20, 1937, and Schairer, *J. Geol.* **58,** 514, 1950)

crystallization. Simple eutectic crystallization, once believed to be common and important in magmas, probably never occurs. Nearly all rock-forming minerals are solid solution series. Crystallization of systems containing such compounds takes place over a range of temperature, and the phases separating have a considerable range of composition; eutectic crystallization is not possible under such conditions. The course of crystallization is dependent upon the rate of solidification and the presence or removal of early-formed crystals, whereas in eutectic crystallization the rate of solidification has no influence, and the final condition is always the same, whether or not early-formed crystals are removed. Another significant feature in silicate systems is the frequent occurrence of incongruent melting, when one solid phase will be converted into a different phase by reaction with the liquid. Thus crystallization in a magma is characterized by reaction of two kinds: continuous reaction in a solid solution series, whereby early-formed crystals change

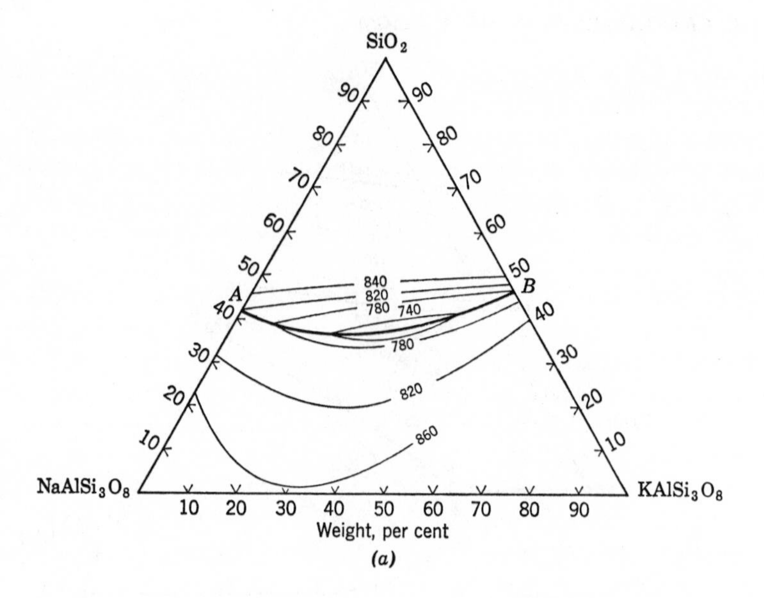

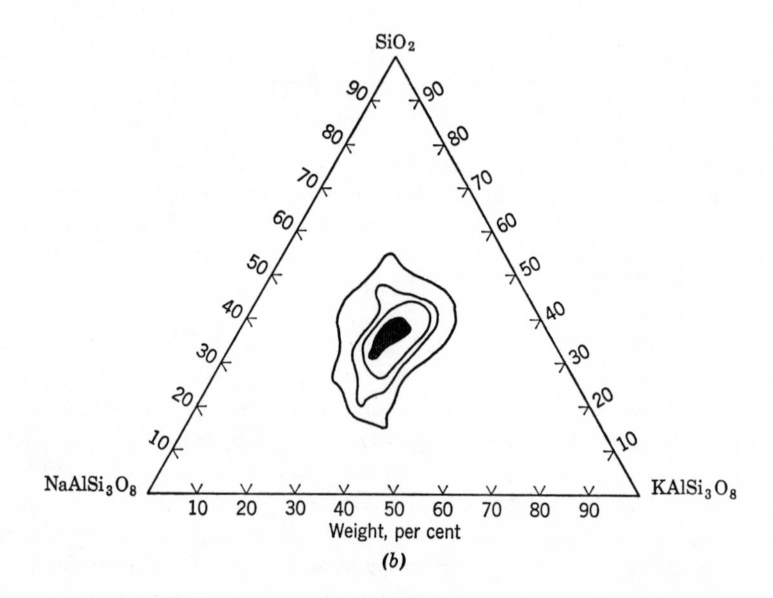

Figure 5.12 Composition of natural granites and its relation to the low-melting region of the system SiO_2—$NaAlSi_3O_8$—$KAlSi_3O_8$: (*a*) system under 1000 kg/cm^2 water vapor pressure (*AB* is the reaction curve separating the field of quartz from that of feldspar); (*b*) contoured frequency-distribution diagram of the normative composition of over 500 analyzed granites containing 80 per cent or more of these components (over 90 per cent of these analyses fall within the solid black area). Tuttle and Bowen, *Geol. Soc. Amer. Mem.* **74,** 1958)

uninterruptedly in composition by reaction with the melt, and discontinuous reaction, whereby an early-formed phase reacts with the melt to give a new phase with a different crystal structure and a different composition. This concept of *reaction* as the fundamental phenomenon of magmatic crystallization is due to Bowen and was developed by him into the *reaction principle.*

Bowen showed that the common minerals of igneous rocks can be arranged into two series, a discontinuous reaction series comprising the ferromagnesian minerals and an essentially continuous reaction series of the feldspars (Figure 5.13). In effect, each of the ferromagnesian minerals is itself a continuous reaction series, since all are solid solutions. The reaction series that Bowen set up, on the basis both of the laboratory study of silicate melts and the petrographic evidence, parallel the general sequence of magmatic crystallization as indicated by the petrology of the igneous rocks.

The petrological significance of the reaction principle may be illustrated by considering briefly the crystallization of a basaltic magma of such composition that olivine and bytownite are the first phases to form. As the temperature falls they react more or less completely with the melt and are

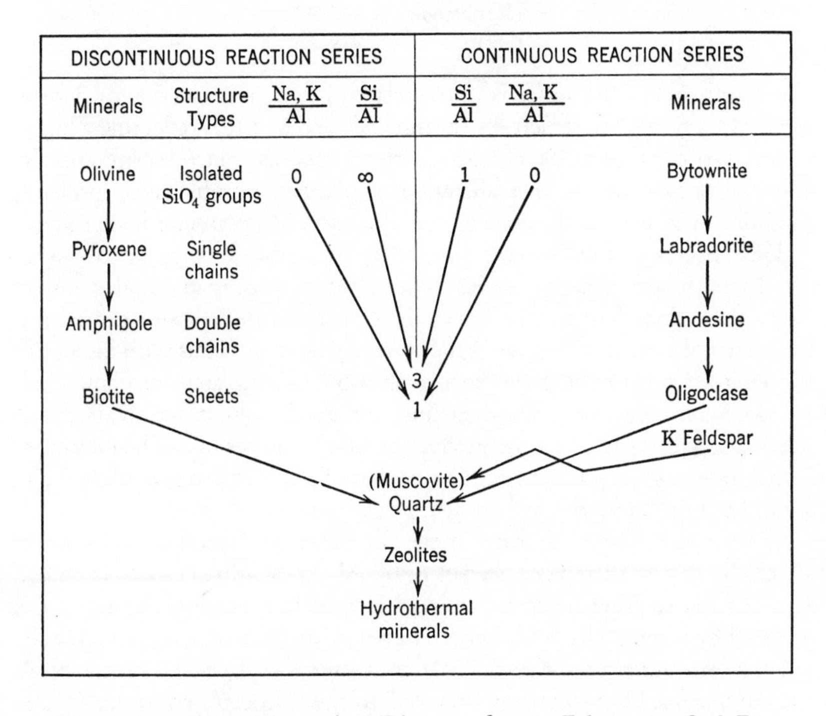

Figure 5.13 The reaction series. (Diagram after von Eckermann, *Geol. Fören. i Stockholm Förh.* **66**, 286, 1944)

converted into pyroxene and labradorite, and if no fractionation takes place the melt solidifies as a pyroxene-labradorite rock, a basalt or gabbro. If fractionation takes place and some of the early-formed olivine and bytownite is removed from the system, the reaction process will continue further, and the remaining melt will act upon pyroxene and labradorite to form hornblende and andesine. The greater the degree of fractionation, the more extensive the reaction process. With a high degree of fractionation, the whole reaction series is gone through, and the final liquid is a watery solution rich in silica.

At this stage it is of interest to examine some of the aspects of the continuous and discontinuous reaction series of Figure 5.13. First, the discontinuous reaction series is marked by increasing complexity of silicate linkage, the sequence being isolated tetrahedra–single chains–double chains–sheets. This sequence is also synonymous with greater size of the structural units, the unit cell volumes increasing as follows:

Forsterite	294 A^3
Diopside	434 A^3
Hornblende	925 A^3
Biotite	979 A^3

The latter part of the series is characterized by the introduction of fluorine and hydroxyl into the structures (hornblende and biotite), reflecting a higher concentration of volatiles as fractionation proceeds. Since biotite contains essential potassium, its crystallization in place of hornblende is probably conditioned in part by increasing concentration of potassium in the liquid.

This sequence of decreasing complexity of silicate linkage in passing up the discontinuous reaction series reflects an increasing thermal stability. Any of the linked structures lower in the series can be disintegrated into fragments of structures higher in the series by heating. Thus with increasing temperature a mica sheet can conceivably be disintegrated into amphibole double chains, pyroxene single chains, and finally single tetrahedra, each step being accompanied by the production of a liquid residue. The melting of hornblende to give pyroxene plus liquid and of pyroxene to give olivine plus liquid has been observed in laboratory experiments.

Another noteworthy feature in the discontinuous reaction series is the increasing degree of replacement of Si by Al. In the olivine there is no evidence of any such replacement; in most magmatic pyroxenes the amount of replaced Si is generally low, but it is greater in the magmatic amphiboles, and in biotite at least one-fourth of the Si is always replaced by Al (the Si:Al ratio in igneous biotites ranges from 6:2 to 5:3). In contrast to this feature the Al:Si ratio in the continuous reaction series of the plagioclases shows a steady decrease from 1:1 in anorthite to 1:3 in albite and potash feldspar.

The Na + K:Al ratio also shows a progressive change from 0 to 1 in both the continuous and discontinuous branches.

The structural significance of the Al—Si substitution is evidently connected with the distribution of bond energies. If aluminum proxies for silicon in an SiO_4 group, the lower charge on the aluminum ion results in lesser neutralization of the negative charges on the oxygen anions, and this increment of negative charge leads to stronger bonding between the silicate units. Substitution of aluminum for silicon in a structure thereby tends to increase its disintegration temperature and consequently raises its position in the reaction series. This is true not only in the continuous reaction series, but also in individual minerals of the discontinuous reaction series. Aluminum substituting silicon in amphiboles and pyroxenes evidently increases their thermal stability.

The reaction series provides us with a concise statement of the segregation of the major elements during magmatic crystallization. The first-formed minerals, olivine and calcic plagioclase, are low in silica, and the liquid is thereby enriched in this component; the olivine is rich in magnesium, the plagioclase in calcium, and so the concentration of these elements in the liquid is decreased. The crystallization of olivine also changes the Mg:Fe ratio in the liquid, since this ratio is always higher in the crystals than in the liquid from which they separate. In effect, olivine and all the other ferromagnesian minerals are individual continuous series in which the early-formed crystals are magnesium-rich, the late-formed, iron-rich. As crystallization proceeds, pyroxene becomes the stable phase instead of olivine, and calcium may now be removed from the melt both as plagioclase and as augite. The liquid becomes relatively enriched in Na, K, and Si. As the concentration of calcium in the liquid falls the sodium content of the plagioclase progressively increases; with the appearance of hornblende as a stable phase in the discontinuous reaction series some sodium may be incorporated in this mineral also. Potassium remains in the liquid until a late stage, since it can be removed in appreciable amounts only in biotite and in potash feldspar.

According to the reaction principle, therefore, the fractional crystallization of a basaltic magma under suitable conditions can lead to the successive formation of more siliceous rocks until ultimately a granitic composition is reached. This sequence has been confirmed in many areas of igneous rocks, for example, the Oslo region, the Caledonian plutonic rocks of western Scotland, and the batholith of southern California. Nevertheless, as Bowen himself pointed out, fractional crystallization is a very flexible process, and two magmas of approximately the same initial composition may produce very different rock types. One of the best documented examples of the fractional crystallization of a basaltic magma is the Skaergaard intrusion (Wager and Deer, 1939). In this intrusion differentiation has resulted in the late crystallization of iron-rich rock consisting essentially of plagioclase (An_{30}), olivine (Fa_{97}), and clinopyroxene close to hedenbergite. Here the trend of differentiation has been towards an enrichment in iron instead of an enrichment in silica. Osborn (1962) has shown that we can distinguish two distinct reaction series. The one leading to a silica-rich residual liquid, as described by Bowen, is characteristic of the igneous rocks of orogenic belts. The other reaction series leads to a residual liquid rich in FeO, and is typified by layered intrusions such as Skaergaard, the Bushveld intrusion, and the Stillwater complex. Each of these differentiation series is equally the product of fractional crystallization and reaction, but the very different results reflect a difference in the partial pressure of oxygen during crystallization. The original reaction series of Bowen describes what happens during the fractional crystallization of a basaltic magma under a relatively high oxygen pressure. Under these circumstances much of the iron is removed as magnetite, the silica content of the liquid is thereby enhanced, and the residual liquid is silica-rich. On the other hand, under relatively low oxygen pressure little or no ferric iron is present to form magnetite and the FeO accumulates in the liquid until it is removed at a late stage by the crystallization of iron-rich olivine and pyroxene.

Recent work on the crystallization of natural and synthetic melts under high pressures has greatly extended our understanding of the scope and variety of magmatic differentiation. For example, the incongruent melting of $MgSiO_3$ to give Mg_2SiO_4 and a silica-rich liquid is eliminated at quite moderate pressures, indicating that although a reaction relation between olivine, orthopyroxene, and liquid may be an important factor in near-surface magmatic differentiation, it will be greatly modified in the deeper levels of the crust. Green and Ringwood (Table 5.1) have demonstrated experimentally that the field of primary crystallization of orthopyroxene from basaltic magma is substantially increased at high pressure. With increasing pressure the amount of olivine decreases and that of orthopyroxene increases.

Table 5.1. Crystallization of Basaltic Melt under Differing Conditions of Temperature and Pressure

P (kb)	T	Phases Present	Remarks
10	1250	ol + opx + liq	ol ≫ opx
$12\frac{1}{2}$	1250	ol + opx + ?cpx + liq	opx > ol
15	1300	ol + opx + cpx + liq	opx ≫ cpx > ol
20	1300	opx + liq	

ol = olivine,
opx = orthopyroxene,
cpx = clinopyroxene.
Green and Ringwood, *Nature* **201,** 1276, 1964.

The composition of the orthopyroxene changes with increasing pressure, becoming progressively richer in aluminum. Green and Ringwood have also studied the crystallization of a basaltic glass at 1100° and different pressures. Up to 15 kb the glass crystallizes to pyroxene and plagioclase, but the plagioclase decreases in amount and becomes more sodic with increasing pressure, evidently through the incorporation in the pyroxene of calcium and aluminum from the anorthite component. Above 15 kb garnet (pyrope-almandine) appears and increases in amount with increasing pressure, being formed by the breakdown o aluminous pyroxene. Plagioclase disappears above 20 kb, and quartz appears, evidently produced by the decomposition of the albite component into jadeite and SiO_2, the jadeite going into solid solution in the pyroxene. This sequence is petrologically equivalent to the sequence gabbro—garnet granulite (garnet, pyroxene, sodic plagioclase)—eclogite. The different behavior under high pressure of the anorthite and albite components of plagioclase is particularly noteworthy. In this connection potash feldspar provides a further contrast; there is no corresponding compound to jadeite ($KAlSi_2O_6$—leucite—is a feldspathoid and unstable above about 2 kb), and potassium cannot be accommodated in the pyroxene structure. Under high pressures in the lower crust and upper mantle potassium is probably contained in hornblende or phlogopite.

Thus the solidification of a magma, although governed by the simple principles of phase equilibria and fractional crystallization, is capable of a remarkable versatility under the variety of conditions in the crust and upper mantle. The distribution of the major elements is controlled by the sequence of mineral separation as the temperature drops and as the composition of the melt changes by the removal of some elements in the solid phases. The principles governing magmatic crystallization are, however, essentially descriptive, not explanatory. They describe how the major elements distribute themselves among the different minerals but not why they act in this way, nor do they tell us anything regarding the fate of the minor elements. The explanation must ultimately depend upon thermodynamic considerations, the energy changes involved in assembling ions of differing size and charge in particular crystal lattices.

THE THERMODYNAMICS OF MAGMATIC CRYSTALLIZATION

The thermodynamics of magmatic crystallization is controlled by the nature of the ions present, their concentration, the temperature and pressure, and the type of crystal lattices formed. In effect, the question posed is, "How do changes in the composition and physical conditions of the magma affect the solubility of the many different possible compounds?" According to thermodynamics a solid A will crystallize from a liquid containing A if the chemical potential of A is less in the solid than in the liquid. However, although we can state the problem, we cannot solve it, since the necessary thermodynamic data are not available.

Crystallization is essentially an ordering phenomenon. In a crystal the degree of order is great and the ions are arranged in a regular lattice, whereas in a melt the arrangement of the ions, although not entirely random, is of a low degree of order. From the viewpoint of thermodynamics magmatic crystallization is characterized by the energy and entropy changes involved in removing ions from the melt and packing them in an orderly fashion in a crystal lattice. The fate of an element during magmatic crystallization is linked with its concentration in the magma and the nature of the structural lattices that may form. The silicon and aluminum content of the magma and the temperature are the factors controlling the sequence of crystal lattices. These crystal lattices act as a sorting mechanism for the cations. A cation can enter a crystal lattice only if it is of suitable size and can attain its appropriate coordination number. Since in general a number of different ions can satisfy this requirement, that ion enters the lattice in largest amount, relative to its concentration in the liquid, which holds its position in the lattice with greatest tenacity. From studies of crystal structures and independently of energy considerations, Goldschmidt formulated the following empirical rules as a general guide to the course of an element during liquid → crystal formation in a multicomponent system:

1. If two ions have the same radius and the same charge, they will enter a given crystal lattice with equal facility.
2. If two ions have similar radii and the same charge, the smaller ion will enter a given crystal lattice more readily.
3. If two ions have similar radii and different charge, the ion with the higher charge will enter a given crystal lattice more readily.

These rules have wide application in the geochemistry of igneous rocks; we have already seen examples in considering the reaction principle. Calcium enters the feldspar lattice more readily than does sodium on account of its higher charge and so is concentrated in the early-formed plagioclase. The magnesium ion is somewhat smaller than the ferrous ion, and magnesium is always concentrated in the early-formed ferromagnesian minerals, whether olivine, pyroxene, amphibole, or biotite. However, Goldschmidt's rules have had their greatest utility in predicting the order of removal from a magma not only of the major elements but of the minor elements also.

MINOR ELEMENTS IN MAGMATIC CRYSTALLIZATION

When a minor element has the same charge and an ionic radius similar to a major element, we speak of it as being *camouflaged* in the crystal lattice containing the major element. Thus Ga^{3+} (0.62 A) is camouflaged in aluminum minerals, and Hf^{4+} (0.78 A) is camouflaged in zirconium minerals. When a minor element has a similar ionic radius but a higher charge than that of a major element (or the same charge but a lesser radius), it is said to be *captured* by the crystal lattice containing the major element. Thus Ba^{2+} (1.34 A) is captured by potassium minerals (K^+ 1.33 A). Finally, when a minor element has a similar ionic radius but a lower charge than that of a major element (or the same charge but a greater radius), it is said to be *admitted* into the crystal lattice containing the major element. Thus Li^+ (0.68 A) is admitted into magnesium minerals. In capture and admission of ions of different charge the charge balance is maintained by concomitant substitution elsewhere in the crystal lattice.

Goldschmidt's rules are a useful guide to the distribution of the elements during magmatic crystallization, but they are not universally valid and have been criticized on that account. The major source of this lack of universal validity seems to lie in the fact that the bonding in most minerals in not exclusively ionic, whereas the rules are predicated on a purely ionic basis. This has been carefully considered by Ringwood (1955), who shows that the electronegativity of an element, a measure of its tendency to form covalent bonds, has an important influence on the extent to which it will proxy for another element of similar size. Elements of the B subgroups of the periodic table have considerably higher electronegativities than corresponding elements of the A subgroups, and this must be considered when comparing diadochy between elements belonging to different subgroups (Table 5.2). Ringwood expresses the electronegativity factor in the form of a rule: Whenever diadochy in a crystal is possible between two elements possessing appreciably different electronegativities the element with the lower electronegativity will be preferentially incorporated because it forms a stronger

Table 5.2. Electronegativities of the Commoner Cations

Cation	Electronegativity	Cation	Electronegativity
Rb^+	0.8	Mn^{2+}	1.4
K^+	0.8	Cr^{3+}	1.6
Ba^{2+}	0.85	Pb^{2+}	1.6
Na^+	0.9	Fe^{2+}	1.65
Sr^{2+}	1.0	Ni^{2+}	1.7
Ca^{2+}	1.0	Co^{2+}	1.7
Li^+	1.0	Zn^{2+}	1.7
Rare earths	1.05–1.2	Fe^{3+}	1.8
Mg^{2+}	1.2	Cu^+	1.8
Sc^{3+}	1.3	Cu^{2+}	2.0
V^{3+}	1.35		

and more ionic bond than the other. In practice this rule is found to apply to substitutions involving elements differing in electronegativity by more than 0.1. The significance of bond type on the distribution of the elements has been extensively reviewed by Ahrens (1964).

The fate of the individual elements during magmatic crystallization is now discussed briefly, and is illustrated in Tables 5.3 and 5.4.

Rubidium. The only major element Rb can replace should be K, and this is found to be true. Rubidium forms no minerals of its own, being always incorporated in potassium minerals; in igneous rocks it is present in biotite and potash feldspar. Since Rb^+ is considerably larger than K^+, rubidium is admitted into potassium minerals, and accordingly the Rb:K ratio increases with increasing differentiation; this ratio is highest in pegmatite feldspars and micas.

Barium. Barium is too large (1.34 A) to replace Ca or Na. The only major element of comparable ionic size is potassium, and barium appears, therefore, in biotite and potash feldspar. On account of its higher charge barium should be captured by potassium compounds. The available data indicate that this is generally true, barium being relatively enriched in early-formed potassium minerals.

Lead. Traces of lead are found in the silicate material of many igneous rocks, particularly in granites. It is evidently present at the Pb^{2+} ion (1.20 A), diadochic with K^+. From the ionic charge alone it might be expected

Table 5.3. Trace Elements (ppm) *in Minerals from Hypersthene Olivine Gabbro, Skaergaard Intrusion*

	Plagioclase	Hypersthene	Olivine
Ga	55	3	3
Cr	<1	350	<1
V	10	250	<5
Mo	<1	2	10
Li	4	2	3
Ni	<2	150	350
Co	<2	50	125
Cu	40	35	20
Sc	<10	30	<10
Zr	<10	50	<10
Mn	<10	2100	1700
Sr	2000	<10	10
Ba	110	5	5

Mitchell, in Bear, *Chemistry of the soil.* Courtesy of Reinhold Publishing Corp.

Table 5.4. Distribution of Minor and Trace Elements (ppm) *in the Minerals of the Rubideaux Mt. Granite, California*

	Plagio-clase and quartz 64%	Potash-feldspar 32%	Hypers-thene ½%	Horn-blende 1%	Biotite 2%	Magnetite ½%
Ga	15	10	5	20	25	100
Cr	*	*	4	30	5	250
V	*	*	*	175	55	1100
Mo	*	*	*	3	*	100
Li	15	10	7	15	260	—
Ni	*	*	10	5	<5	20
Co	*	*	12	10	15	40
Sc	*	*	30	200	15	*
Zr	*	*	*	250	45	*
Y	*	*	*	2000	*	*
La	*	*	*	300	*	*
Sr	75	130	*	10	*	—
Ba	125	2500	*	60	2000	—
Rb	*	600	*	10	1750	—

Sen, Nockolds, and Allen, *Geochim. Cosmochim. Acta* **16,** 77, 1959. An asterisk indicates that the element is below its limit of sensitivity.

that the lead in a magma would tend to be captured by potassium minerals, but the much greater electronegativity of lead evidently results in a weakening of this tendency, and lead is admitted rather than captured by potassium minerals.

Strontium. The size of the strontium ion indicates that it can proxy for either calcium or potassium, being admitted to calcium minerals (higher radius) or captured by potassium minerals (higher charge). The data indicate that strontium in igneous rocks is present mostly in the plagioclase and potash feldspar and that its concentration increases as crystallization proceeds; this implies that admittance in place of calcium is the dominant process of removal of strontium from the magma. However, this process does not operate with all calcium minerals; strontium is present in only insignificant amounts in augite, suggesting that the pyroxene structure does not readily accommodate the strontium ion.

The Rare Earths (including Y). The charge and the comparatively large radius (Lu 0.85 A–La 1.14 A) of the rare earths, coupled with their general low concentration, suggest that they would show relatively little tendency to replace the major elements during magmatic crystallization, and this is borne out by the concentration of rare earths as individual minerals in pegmatites. However, some replacement of Ca^{2+} by rare earths can take place in apatite and hornblende. Titanite often contains rare-earth elements. In granites and pegmatites formed at sufficiently low temperatures for the epidote structure to be stable we find the mineral allanite, in which some of the calcium of epidote is replaced by rare earths.

Manganese. Manganese is present in magmas as the Mn^{2+} ion and in this form may be expected to proxy for Fe^{2+} or for Ca^{2+}. However, manganese is much more electronegative than calcium and probably on this account is seldom found replacing this element (except in the apatite of pegmatites). The manganese in igneous rocks replaces ferrous iron, and a relative increase in the Mn:Fe ratio has been noted in later differentiates, indicating that the larger size of the manganese ion causes it to be admitted to ferromagnesian minerals. Table 5.3 shows clearly the absence of manganese from the calcium-bearing plagioclase and its essentially uniform distribution in the ferromagnesian minerals hypersthene and olivine.

Zirconium. The combination of high charge and comparatively high radius (0.79 A) sets zirconium apart from any of the major elements of igneous rocks. On this account zirconium does not enter into the common rock-forming minerals to any degree but appears in a specific phase, usually zircon. Zircon is most abundant in later differentiates, evidently because the original zirconium concentration of a magma is generally less than saturation for zircon.

Hafnium. Hafnium, having the same charge as Zr and about the same radius, always occurs camouflaged in zirconium minerals, and the Zr : Hf ratio remains almost constant throughout any process of fractional crystallization.

Scandium. Scandium has a radius (0.81 A) close to that of ferrous iron, and, in view of its higher charge, scandium would be expected to be captured by ferromagnesian minerals. This is true for magmatic pyroxenes, which generally show a relative concentration of scandium (Table 5.3); it may also be present in hornblende and biotite. Scandium is not concentrated in the even earlier-formed olivines, evidently because of the difficulty in balancing the excess positive charge thus introduced by other suitable replacements.

Cobalt. The bivalent cobalt ion is practically the same size (0.72 A) as the ferrous ion (0.74 A), and cobalt should thus be camouflaged in ferrous compounds. It is found, however, that the Co:Fe ratio is greatest in early-formed minerals and decreases steadily with increasing fractionation. The effective radius of cobalt is, therefore, somewhat less than the radius given above and is apparently almost identical with that of Mg, since Nockolds and Allen found the Co:Mg ratio to be almost constant throughout a rock series. The major part of the cobalt in a magma is removed in the early-formed magnesian minerals, especially olivine.

Nickel. The nickel ion has essentially the same radius (0.69 A) and the same charge as magnesium, and therefore should be camouflaged in magnesium minerals. However, the ratio Ni : Mg is highest in early-formed crystals (especially olivine) and shows a steady decline in the later-formed rocks and minerals. Nickel behaves as if its effective radius were somewhat less than that of magnesium (resulting in capture in Mg minerals); that this is actually so can be seen by comparing the sizes of the unit cells of Ni_2SiO_4 (281 A^3) and Mg_2SiO_4 (292 A^3).

Lithium. On the basis of chemical properties it might be expected that lithium would follow the other alkali elements in magmatic crystallization. However, ionic size, not chemical properties, is decisive in crystallization, and the lithium ion is very much smaller than any of the other alkali ions (Li^+ 0.68 A, Na^+ 0.97 A, K^+ 1.33 A). Hence lithium follows magnesium, since the ionic sizes are nearly identical; because the lithium ion has a lower charge than the magnesium ion it should be admitted into magnesium minerals, and this is found to be true. The Li:Mg ratio shows a steady increase in later-formed rocks and minerals. Strock, who made careful measurements of this ratio for different igneous rocks, suggested that it could be used as an index to the stage of differentiation reached by a given rock. Lithium is found in the pyroxenes, amphiboles, and particularly the micas. However, a considerable amount remains in the liquid until a very late stage of differentiation, since pegmatites often show a particular concentration of lithium, which in the practical absence of magnesium forms individual minerals, such as lepidolite, spodumene, amblygonite, and petalite.

Vanadium. Vanadium is probably present in magmas as the V^{3+} ion. It is largely removed in magnetite, in which it proxies for Fe^{3+}; its ionic radius is greater than ferric iron, but its electronegativity is much less, the latter factor evidently being responsible for the enrichment of vanadium in early-formed magnetite. Vanadium also occurs in pyroxenes, amphiboles, and biotite and has been noted in marked concentration in aegirine, a mineral with a high Fe^{3+} content.

Chromium. Chromium is also present in magmas as the Cr^{3+} ion. The radius of this ion is very close to that of Fe^{3+}, but chromium shows a high degree of preferential concentration relative to ferric iron, being largely removed from a magma in the early stage of crystallization as chromite. This can be ascribed to the smaller electronegativity of Cr^{3+} relative to that of Fe^{3+}. Chromium is also enriched in pyroxenes, especially those from ultrabasic rocks.

Titanium. In igneous rocks titanium is present mainly as ilmenite. It can replace Al in six-coordination and appears therefore in pyroxene, hornblende, and biotite; it is probably captured by such minerals on account of its higher charge (Ti^{4+}—Al^{3+}). (The latter argument is not valid, however, if titanium is present in magmas as the Ti^{3+} ion, as has been suggested.) Titanium does not appear in muscovite, probably because in highly siliceous magmas it is removed as titanite ($CaTiSiO_5$).

Gallium. Gallium, with the same ionic charge and radius close to that of aluminum, is camouflaged in aluminum-bearing minerals. The somewhat larger size of the gallium ion (Ga^{3+} 0.62 A, Al^{3+} 0.51 A) suggests that gallium would tend to be more abundant in later-formed aluminum minerals. Measurements of the Ga:Al ratio in igneous rocks and their minerals show that this ratio is nearly constant, indicating effective camouflage of gallium in aluminum minerals and little tendency to concentration in later differentiates.

Germanium. The germanium ion has the same charge and a somewhat higher radius than silicon (Ge^{4+} 0.53 A, Si^{4+} 0.42 A). Germanium replaces silicon, and measurements of the Ge:Si ratio in silicates generally show little variation, indicating that it is effectively camouflaged in such minerals; there is, however, evidence of some degree of concentration of germanium in late differentiates.

The remaining lithophile elements are those which do not substitute the major elements because of a great difference in ionic radius and ionic charge. On account of low concentration in the original magma they remain is solution and are enriched in the residual liquid of magmatic crystallization. These elements are B^{3+} (0.23 A), Be^{2+} (0.35 A), W^{6+}(0.62 A), Nb^{5+} (0.69 A), Ta^{4+} (0.68 A), Sn^{4+} (0.71 A), Th^{4+} (1.02 A), U^{4+} (0.97 A), and Cs^{+} (1.67 A) and also the rare earths (0.85–1.14 A), Li^{+} (0.68 A), and Rb^{+} (1.47 A). They are concentrated in pegmatites, the only economic source of many of them.

During the final crystallization of the residual liquid these elements may enter the structure of a common mineral (Rb in microcline) or a rarer mineral (Mn in apatite); they may enter or be trapped by a much less appropriate mineral structure (Sn in muscovite), followed in some instances by transfer to an exsolved mineral; they may concentrate in the residual fluid till a specific mineral is formed (Cs in pollucite, Be in beryl); or they may be bound to the surface of some mineral or minerals by adsorption. This last process evidently is effective in the crystallization of granitic rocks and results in the attachment of the ions of minor and trace elements to the crystal surfaces of common silicate minerals. The binding by adsorption is relatively weak, as demonstrated by experiments in which considerable amounts of minor and trace elements have been removed from granites by leaching with dilute acids.

It might be expected that the chalcophile elements in a magma would associate with sulfur and lesser amounts of arsenic, antimony, bismuth, selenium, and tellurium to form sulfides and related compounds. However, the only sulfides commonly present in normal igneous rocks are pyrite and pyrrhotite. The sulfur content of magmas is relatively low, and nearly all combines with the abundant iron. Many of the chalcophile elements become predominantly incorporated in silicates rather than in sulfides. This can be partly accounted for by the low thermal stability of most sulfides, so that sulfides, except for pyrite and pyrrhotite, characteristically segregate at a late stage and are ultimately deposited in veins formed at comparatively low temperatures. The relative concentration of sulfur and chalcophile elements probably increases in the late liquid fraction of a crystallizing magma as a result of the removal of the lithophile elements in the silicate minerals.

The work of Wager and Mitchell (1951) on the Skaergaard intrusion has given a useful account of the distribution of minor elements in different rocks which are clearly related on the basis both of field exposures and of their petrology. The sequence of differentiation is from gabbro to granophyre, and the amounts of the minor elements in various rocks of this sequence are given in Table 5.5. The amounts assigned to the initial magma are based on the analysis of material from the chilled margin of the intrusion. The correlation between the results presented in Table 5.5 and predictions on the

Table 5.5. The Content of Minor Elements in Differentiates of the Skaergaard Intrusion Compared to That in the Initial Magma

(figures in ppm)*

	Initial Magma	Gabbro Picrite and Eucrite (earliest)	Olivine Gabbro	Olivine-Free Gabbro	Ferrogabbros		Hedenbergite Granophyre	Granophyre (latest)
Rb	—	—	—	—	—	—	30	200
Ba	40	25	25	45	50	150	450	1700
Sr	300	200	700	450	700	450	500	300
La	—	—	—	—	—	—	25	150
Y	—	—	—	—	—	125	175	200
Zr	40	40	35	25	20	100	500	700
Sc	15	7	20	15	10	—	—	—
Cu	130	70	80	175	400	200	500	20
Co	50	80	55	40	40	5	10	4
Ni	200	600	135	40	—	—	5	5
Li	3	3	2	3	3	15	25	12
V	150	170	225	400	15	—	—	12
Cr	300	700	175	—	—	—	—	3
Ga	15	12	23	15	20	20	40	30
Proportion of rocks as percentage		65	14	10	$7\frac{1}{2}$	$2\frac{1}{2}$	$\frac{1}{2}$	$\frac{1}{2}$

* From Wager, *Observatory* **67,** 103, 1947.

basis of Goldschmidt's rules is striking and provides independent support for the belief that the rocks of the Skaergaard intrusion are the result of differentiation by fractional crystallization.

RESIDUAL SOLUTIONS AND PEGMATITES

The data in the preceding sections indicate that the residual melt from fractional crystallization of a magma will in general be a siliceous liquid rich in the alkalies and alumina, containing water and other volatiles, and with a concentration of those minor elements which are not incorporated in the lattices of the common minerals of igneous rocks. Such a residual liquid will probably be unusually fluid for a silicate melt on account of the concentration of volatiles. The pressure of volatiles will also provide the driving force to inject it along surfaces of weakness in the surrounding rocks, which may be the already solid part of the same igneous intrusion or the country rock. In this way pegmatites and hydrothermal veins may be formed. The literature on pegmatites is very extensive and has been critically summarized by Jahns (1956).

Pegmatites are found in association with most plutonic rocks but more commonly in connection with granites, as is to be expected if granites are the end product of fractional crystallization of a magma. Granite pegmatites consist essentially of quartz and alkali feldspar, generally with muscovite and biotite also, and are thus similar in composition to granite; the essential difference is in texture, the minerals of pegmatites being of exceedingly variable grain size. Pegmatites are typically coarse-grained, and some of the largest crystals known have been found in them. This is evidently due to low viscosity and concentration of volatiles, which promote the formation of large crystals.

Most pegmatites are chemically and mineralogically simple, but the complex ones are spectacular in their content of rare elements and unusual minerals; hence they have been intensively studied. Complex pegmatites are economically important and have been worked for lithium, beryllium, scandium, rubidium, yttrium, zirconium, molybdenum, tin, cesium, the rare earths, niobium, tantalum, tungsten, thorium, uranium, and industrial minerals, such as feldspar, muscovite, phlogopite, tourmaline, and quartz. They are often internally zoned, and a specific rare mineral may be confined to a particular zone; zoned pegmatites commonly have a core which consists almost entirely of massive quartz. The geochemistry of pegmatites is a fascinating subject. The major factors influencing the concentration of certain elements in them have already been outlined. Many interesting features, however, such as the localization of one or a few minor elements in a particular pegmatite or group of related pegmatites, are as yet little understood. Thus the granite pegmatites in southern Norway are noted for their content of rare earths, and tin, lithium, and the rare alkalies are absent. In the pegmatites of the Black Hills (South Dakota) lithium is particularly abundant, tin is common, and the rare earths are absent. Those of central Texas contain rare-earth minerals but carry no other unusual elements. The pegmatites on the island of Honshu in Japan are characterized by the presence of the rare earths and beryllium. These regional differences may, of course, reflect the original composition of the magmas from which they were derived, but the reasons for such compositional differences are unexplained. Another curious fact is that only relatively few pegmatites carry the unusual elements; over 90% consist essentially of quartz and feldspar and are quite barren of rare minerals. Many geologists who have studied the complex pegmatites believe that the unusual minerals were deposited during one or more phases of hydrothermal replacement following the original formation of a simple quartz–feldspar–mica pegmatite. The idea helps to explain the often erratic occurrence of a single rare-mineral pegmatite among many simple ones. Another explanation of the simple character of most pegmatites is that many of them, especially those present in strongly metamorphosed terrains, may have been formed by the differential fusion (anatexis) of pre-existing rocks. If pegmatites are the last fraction of a magma to crystallize, then it is reasonable to expect that the first fraction of rock to liquefy (provided suitable volatiles are present) will also have pegmatitic composition. The temperature of formation of pegmatites (probably 500–700°) has certainly been reached during regional metamorphism in many areas.

Pegmatites often contain minor amounts of sulfides and arsenides. Gold has been recorded from some. They show many of the characteristics of hydrothermal veins, and examples of pegmatites passing into sulfide-bearing quartz veins have been described. More frequently, however, hydrothermal veins are younger than cogenetic pegmatites. The relationship of hydrothermal solutions to the final stage in the crystallization of a magma is a subject of much difference of opinion. Some geologists believe that hydrothermal solutions are formed as a separate phase later than the pegmatite stage, and laboratory experiments have been cited as indicating that a residual magma of granitic composition would separate on cooling into two immiscible liquids, one of which would be a hydrothermal solution. On the other hand, it has been found that during crystallization in systems containing an excess of alkali over that required to combine with alumina and silica to give feldspar, the liquid phase varies continuously in composition from a silicate melt containing a small amount of water to a liquid consisting largely of water (i.e., there is a continuous passage from magmatic melt to hydrothermal solution). However, as Verhoogen (1949) states in a discussion of the thermodynamics of a magmatic gas phase, ". . . the number of factors which are relevant in any magmatic process is much larger than is usually recognized, and there are no simple means of deciding in a general way what a magma would do under the most general circumstances. . . . Each case must be examined separately and the idea must constantly be kept in mind that what is true for one magma under certain circumstances is not necessarily true for another magma under the same circumstances, or for the same magma under different circumstances."

THE VOLATILE COMPONENTS OF A MAGMA

All magmas contain volatile material in an amount and composition not well known, since no means of directly sampling it have yet been devised. As a result, the significance of this material has been the subject of extreme controversy. It has been maintained that magmas are essentially "dry." This opinion has been largely discarded, however, since water has been found to be universally present both in volcanic exhalations and in the gases evolved from fresh igneous rocks by heating in vacuo. Nevertheless, even this evidence has been subjected to doubt. It has been suggested that primary magma is rich in hydrogen, which by surface or near-surface oxidation has produced the water or a considerable part of it. This is quite improbable, however, because a primary magma rich in hydrogen would crystallize to give rocks containing metallic iron and no ferric iron, and such rocks are not found even among those which have crystallized within the crust (the occurrence of iron in basalt at Disco Island in Greenland is the result of a natural smelting process produced by the assimilation of coal by the magma). From thermodynamic data Krauskopf (1959) has calculated the compositions of possible magmatic gas systems and shows that observed gas compositions are explicable on the basis of a primary magma containing H_2O, CO_2, H_2S, and minor amounts of HCl, N_2, and HF. His estimate for the average composition of magmatic gas (by volume) is: H_2O, 1000; CO_2, 50; H_2S, 30; HCl, 10; N_2, 10; HF, 0.1.

The amount of volatile matter in magmas is also subject to considerable differences of opinion. Here we are once more dependent on indirect evidence, such as observations on volcanoes, petrographic examination and chemical analyses of igneous rocks, and experimental determination of the solubility of water and other volatiles in silicate melts. This evidence is in part contradictory. Some volcanologists have stated that in certain eruptions, such as that of Vesuvius in 1906, the mass of gas evolved outweighed the solid matter ejected; in other eruptions the proportion of gas in the material ejected has been estimated as low as 0.3%. Different volcanoes, and the same volcano at different times, evidently present very dissimilar appearances. The data on the solubility of water in artificial silicate melts have indicated that the limiting solubility of water in a granitic melt is about 6%;

this suggests that magmas probably contain less than this amount. All igneous rocks contain some volatile matter, mostly water, which can be extracted by heating in vacuo, and Clarke's average of the analyses of fresh igneous rocks shows 1.15% H_2O; this figure may perhaps be taken as a lower limit for the average water content of a magma. Current estimates run from about 0.5 to 8% of water in magmas.

As mentioned above, the composition of the volatile components cannot be directly determined. Once again we have to fall back on analyses of the gases evolved on heating igneous rocks, on observations of materials deposited in fumaroles, and on analyses of gases collected around volcanoes. The last procedure, although most direct, is technically difficult and personally uncomfortable or dangerous. The most satisfactory collections yet made are those of the gases evolved from the Kilauea volcano (Table 5.6). Averages of such variable figures cannot have more than qualitative significance; they show that the major constituents of the volatile matter at this place are H_2O, CO_2, and sulfur compounds and that H_2O is dominant. Observations on other volcanoes confirm this general picture. Water is always the major constituent, generally making up over 80% of the whole by volume; CO_2, H_2S, S, SO_2, HCl, and NH_4Cl are often abundant; and HF, N_2, H_2, CH_4, H_3BO_3, and CO have been recorded in lesser amounts. In addition, gaseous emanations from magmas collect, transport, and deposit many other elements. Such minerals as magnetite, hematite, molybdenite, pyrite, realgar, galena, sphalerite, covellite, sal ammoniac, ferric chloride, and many others have been found in the throats of fumaroles, where they have been deposited directly from gases. The Valley of Ten Thousand Smokes provides us with

Table 5.6. Analytical Data on Kilauea Gases

	Minimum and Maximum Mole Percentages as Collected	Minimum and Maximum Mole Percentages, Recalculated to Exclude O_2, N_2, and Ar
H_2O	4.7 — 99.8	97.2 — 99.1
H_2	0.0 — 0.43	0.00001 — 0.43
CO_2	0.00023 — 3.8	0.80 — 2.30
CO	0.0 — 0.086	0.0 — 0.087
CH_4	0.0 — 0.047	0.0 — 0.023
H_2S	0.0 — 0.12	0.0004 — 0.67
SO_2	0.0 — 0.16	0.001 — 0.43
CS_2	0.0 — 0.026	0.0 — 0.026
O_2	0.0 — 21	
N_2	0.0 — 78	
Ar	0.0 — 0.064	

Heald, Naughton, and Barnes, *J. Geophys. Res.* **68,** 546, 1963.

one of the best documented examples of the role of volcanic emanations. The work of Allen and Zies showed that the principal gas in all the fumaroles was water-vapor, the percentage by volume ranging from 98.8 to 99.99. Other gases recognized were CO_2, O_2, CO, CH_4, H_2S, H_2, N_2, HCl, and HF. Although these were present in very small percentages, the total amount evolved was enormous. Thus the average concentration of HCl was 0.117% and of HF, 0.032%, but the total quantity of HCl evolved in 1919 was estimated as 1,250,000 tons and of HF, 200,000 tons. One difficulty in discussing the geochemical significance of these figures is that of determining how much of the emanations is primary, that is, was contained in the original magma, and how much secondary, that is, assimilated by the magma from surrounding rocks during its rise to the surface. Allen and Zies believe that much of the water-vapor given off at the Valley of Ten Thousand Smokes was ground water heated by the igneous material. The same difficulty arises in assessing the amount of material in mineral springs that is magmatic in origin. Day and Allen suggest that perhaps 90% of the water in the Yellowstone geysers and hot springs is activated ground water. Precise information on these points would greatly advance our knowledge of the geochemical significance of magmatic emanations, but it is evident that igneous activity during geological time has brought vast amounts of water and other volatile matter to the earth's surface.

MAGMATISM AND ORE DEPOSITION

Much evidence exists for the belief that many ore deposits are genetically related to magmas. Some of this evidence lies in the geological association of ore bodies with igneous rocks, and often with particular types of igneous rocks, for example, tin deposits with granites; sometimes gradations can be traced from pegmatites to ore-bearing veins to barren quartz veins, and examples of direct segregation of ore minerals from magmas are well known. Nevertheless, much difference of opinion exists as to the processes whereby ore materials are extracted from a magma, transported, and deposited. Did the ore-forming material leave the magma as a gas phase or a liquid phase? Was the initial ore-forming fluid acid or alkaline? What were the conditions of temperature and pressure? All these questions and many others remain to be answered.

No definite criteria exist for telling whether a deposit was formed from a gaseous or liquid solution. One can visualize the following possibilities:

(*a*) If the temperature was above the critical temperature, transportation and deposition took place in a gas phase, no matter how high the pressure.

(*b*) If the temperature was below the critical temperature, the depositing solutions were most probably liquid, since pressures at quite moderate depths are greater as a rule than the critical pressure of moderately dilute solutions.

On proceeding outward from an intrusive body the temperature decreases, and material leaving the magma in a gas phase may be expected to condense to a liquid when it cools below the critical temperature. This suggests that the country rock immediately surrounding a cooling intrusion may be saturated with gaseous emanations, beyond which is an envelope of liquid solutions.

It must be emphasized that a gas phase associated with a magma at depth has a density of the same order of magnitude as that of liquid water at the earth's surface. This dense fluid may be capable of dissolving and transporting many substances not themselves volatile. Laboratory experiments have demonstrated this, and the evidence of fumaroles shows that such transportation and deposition takes place readily under geological conditions. Proponents of gaseous transfer generally favor the idea that solution and transportation of metallic compounds is aided by the presence of halogen compounds. Because the chlorides of the common metals have high vapor pressures and because the direct observation of volcanic sublimates shows their presence, it is considered probable that chlorides are the most important form in which metals are removed from a magma. The physicochemical principles governing the conditions under which metallic halides may exist in the gas phase together with such precipitating agents as H_2O and H_2S are expressed by the law of mass action. For a specific example

$$ZnCl_2 + H_2S \rightleftharpoons ZnS + 2HCl$$

the law states that $(p_{ZnS})\,(p_{HCl})^2/(p_{ZnCl_2})\,(p_{H_2S})$ is a constant for any given temperature (p symbolizes partial pressure). This means that the ratio of $ZnCl_2$ to ZnS increases as the square of the concentration of HCl and inversely as the concentration of H_2S; therefore, increase in HCl concentration has a much greater influence on the reaction than increase in H_2S concentration. Similarly, for the reaction

$$SiCl_4 + 2H_2O \rightleftharpoons SiO_2 + 4HCl$$

the effect of HCl in driving the reaction to the left increases as the fourth power of its concentration, whereas the effect of H_2O in promoting the reverse reaction varies as the square.

Considerations of this kind suggest that highly compressed water-rich magmatic gases containing halogen acids would be efficient transporting agents of many elements. Cooling and neutralization of the acid by reaction with country rock would then bring about precipitation of various sulfides and oxides. This may take place either before or after the condensation of the gases to an aqueous liquid solution. Incidentally, the change from gas to liquid, which probably occurs as soon as the ore fluid cools below its critical temperature, will have little effect on its chemical or physical properties. If the gases or liquids were originally acid, they must eventually become weakly alkaline by the combination of alkalies extracted from country rock with weak acids, such as H_2S, H_2CO_3, and H_3BO_3, derived from the magma. New types of reaction become possible, involving previously deposited ore minerals and whatever else is present in the channels followed by the solutions.

In this connection the drilling of a well for geothermal power near the Salton Sea in California in 1962 provided some very suggestive information. At a depth of some 5000 feet this well tapped a hot brine which had an unusually high potassium content, and perhaps the highest lithium and heavy-metal content known for natural waters. During a production test, the brine deposited material astonishingly high in silver, copper, and other elements normally concentrated in ore deposits, such as arsenic, bismuth, lead, and antimony. This brine may be the first sample of an "active" ore solution of the type that probably formed many of the world's economic metal concentrations in the geological past. The brine possibly represents the water-rich fluid residue from the magma that supplied the rhyolitic volcanoes of the area.

We are once again faced with the crux of so many geological problems; in general we can observe only the final product and not the steps by which it was formed. No certain criteria have yet been found to provide an unequivocal answer to the questions posed at the beginning of this discussion. The answers undoubtedly vary from one ore body to another and probably from place to place within the same ore body. Ore material may be carried by gaseous or liquid solutions, may be deposited by changes in tempertaure, pressure, or chemical environment or combinations of these factors, and may be subjected to further changes after deposition. Seldom can any one of the numerous determining factors be singled out as of unique importance; the question is generally to what degree each has been significant.

SELECTED REFERENCES

Ahrens, L. H. (1964). The significance of the chemical bond for controlling the geochemical distribution of the elements. *Phys. Chem. Earth* **5**, 1–54. An extensive study of the effect of varying degrees of covalency on the distribution of the elements in rocks and minerals.

Barth, T. F. W. (1962). *Theoretical petrology*. 416 pp. John Wiley and Sons, New York. A concise account of the interpretation of igneous rocks in terms of the physicochemical factors involved in their formation.

Barth, T. F. W., and T. Rosenquist (1949). Thermodynamic relations of immiscibility and crystallization of molten silicates. *Am. J. Sci.* **247,** 316–323. A discussion of the physicochemical nature of a silicate melt.

Bowen, N. L. (1928). *The evolution of the igneous rocks.* 334 pp. Princeton University Press, Princeton, N. J. (reprinted 1956 by Dover Publications Inc., New York). A classic work correlating the results of laboratory investigations on the crystallization of silicate melts with the data of igneous petrology.

Clarke, F. W. (1924). *The data of geochemistry.* Chapters 8, 9, 10, and 11.

Deer, W. A., R. A. Howie, and J. Zussman (1962–1963). Rock-forming minerals. 1788 pp. (5 vols.). John Wiley and Sons, New York. A comprehensive treatise, with excellent coverage on chemistry, crystal structure, and paragenesis.

Goldschmidt, V. M. (1954). *Geochemistry.* Chapter 3.

Jahns, R. H. (1956). The study of pegmatites. *Econ. Geol.*, 50th Ann. Vol., 1025–1130. A comprehensive account of the geology and geochemistry of pegmatites; 698 references.

Krauskopf, K. B. (1959). The use of equilibrium calculations in finding the composition of a magmatic gas phase. *Researches in Geochemistry*, pp. 260–278. John Wiley and Sons, New York. The application of thermodynamic data to the problem of the composition of magmatic gases.

Morey, G. W. (1964). *Phase-equilibrium relations of the common rock-forming oxides except water.* U. S. Geol. Surv. Prof. Paper 440-L, 159 pp. [*Data of Geochemistry* (sixth edition), chapter L]. A comprehensive account of the phase relations of the common rock-forming oxides and the binary, ternary, quaternary, and quinary systems formed by them.

Nockolds, S. R., and R. Allen (1953–6). The geochemistry of some igneous rock series. *Geochim. Cosmochim. Acta* **4,** 105–142; **5,** 245–285; **9,** 34–77. A series of papers describing the distribution of minor and trace elements in a considerable number of igneous rock series.

Osborn, E. F. (1962). Reaction series for subalkaline igneous rocks based on different oxygen pressure conditions. *Am. Mineral.* **47,** 211–226. The paper establishes the important effect of oxygen pressure in modifying the classic reaction series for the evolution of igneous rocks.

Rankama, K., and Th. G. Sahama (1950). *Geochemistry*, pp. 94–189.

Ringwood, A. E. (1955). The principles governing trace element distribution during magmatic crystallization. *Geochim. Cosmochim. Acta* **7,** 189–202, 242–254. A discussion of trace element distribution in igneous rocks, emphasizing the importance both of ionic size and of electronegativity.

Roedder, E. (1965). Report on S.E.G. symposium on the chemistry of the ore-forming fluids. *Econ. Geol.* **60,** 1380–1403. A summary of current thinking and experimentation on the geochemistry of ore deposits.

Shaw, D. M. (1964). *Interprétation géochemique des éléments en traces dans les roches cristallines.* 237 pp. Masson et Cie., Paris. An interesting account of the determination of abundances of trace elements, their distribution in igneous (and metamorphic) rocks, and the geochemical significance of these data.

Smith, F. G. (1963). *Physical geochemistry.* 624 pp. Addison-Wesley Publishing Co., Mass. This book includes an extensive review of phase diagrams and their application to magmatic and hydrothermal processes.

Taylor, S. R. (1965). The application of trace element data to problems in petrology. *Phys. Chem. Earth* **6,** 133–213. An extensive review article on the interpretation of trace element abundances in rocks and minerals; excellent bibliography.

Turner, F. J., and J. Verhoogen (1960). *Igneous and metamorphic petrology.* 694 pp. McGraw-Hill Book Co., New York. A comprehensive account of igneous and metamorphic petrology, with careful evaluation of both laboratory and field data.

Verhoogen, J., (1949). Thermodynamics of a magmatic gas phase. *Univ. Calif. Pub. Bull. Dept. Geol. Sci.* **28,** 91–136. A precise discussion of the factors governing equilibrium in magmatic processes.

Wager, L. R., and W. A. Deer (1939). The petrology of the Skaergaard intrusion, Kangerdlugssuag, East Greenland. *Medd. Grønland* **105,** no. **4,** 352 pp. A comprehensive account of an excellently exposed intrusion showing a remarkable sequence of differentiated rocks.

Wager, L. R., and R. L. Mitchell (1951). The distribution of trace elements during strong fractionation of basic magma—a further study of the Skaergaard intrusion, East Greenland. *Geochim. Cosmochim. Acta* **1,** 129–208. An extensive study of the content of minor and trace elements in a strongly differentiated intrusive body.

Wasserburg, G. J. (1957). The effects of H_2O in silicate melts. *J. Geol.* **65,** 15–23. A thermodynamic analysis of the role of water in silicate melts.

Yoder, H. S., and C. E. Tilley (1962). Origin of basalt magmas: an experimental study of natural and synthetic rock systems. *J. Petrology* **3,** 342–532. A thoroughly documented account correlating the results of laboratory studies on silicate melts with petrological experience.

CHAPTER SIX

SEDIMENTATION AND SEDIMENTARY ROCKS

SEDIMENTATION AS A GEOCHEMICAL PROCESS

Sedimentation is in effect the interaction of the atmosphere and hydrosphere on the crust of the earth. Various aspects of sedimentation are described in terms of weathering, erosion, deposition, and diagenesis, but no one of these processes works in isolation. The original constituents of the crust, the minerals of igneous rocks, are to a large extent unstable with respect to the atmosphere and hydrosphere. They have been formed at high temperatures and sometimes at high pressures also, and cannot be expected to remain stable under the very different conditions at the earth's surface. Of the common minerals of igneous rocks, only quartz is highly resistant to weathering processes. All the other minerals tend to alter; by the action of oxygen, carbonic acid, and water, they are more or less attacked, and new minerals are formed which are more stable under the new conditions. The altered rock rapidly crumbles under the mechanical effects of erosion, and its constituents are transported by wind, water, or ice and redeposited as sediments or remain in solution.

The core problem in the geochemistry of sedimentation is the chemical breakdown of some minerals and the formation of others. Of these, silicates are the most important, since they constitute more than 90% of the earth's crust (including quartz as a silicate). The processes by which silicate minerals are broken down chemically during weathering have long been a subject of speculation. It was early found that the univalent and bivalent cations readily go into solution, but the fate of aluminum and silicon has been less well understood. Older hypotheses assumed hydrolysis of aluminosilicates with the formation of colloidal silicic acid and aluminum hydroxide, which later reacted to give clay minerals. Recent investigations have shown, however, that during the initial attack a silicate mineral goes into ionic solution, and even the silica and the alumina are at least for a short time in true solution. At the surface of a crystal unsatisfied valencies exist which are the loci of reaction with water molecules. Hydration and hydrolysis follow, whereby strong bases, such as potassium, calcium, and magnesium, are removed and oxygen anions in the crystal lattice may be partly replaced by hydroxyl ions. Aluminum and silicon attract OH ions strongly; aluminum probably groups six OH around it to assume its preferred sixfold coordination, whereas silicon remains four-coordinated. When first set free these elements are in ionic solution, but the ions tend to aggregate and to form clusters of collodial size. When first formed these colloidal aggregates are probably amorphous, but, on ageing, orientation into definite crystal lattices, such as those of the clay minerals, takes place. Some silicate minerals may not undergo complete breakdown of the lattice during weathering; for example, biotite and muscovite may perhaps pass directly into clay minerals by ionic substitution, whereby fragments of the sheet structures may be directly incorporated in the new minerals. The ultimate fate of different elements thus depends largely upon the relative stability of their ions in water. The most stable are the alkali metal ions, followed by those of the alkaline earths, and they are for the most part carried away in solution. Silicon, aluminum, and iron, on the other hand, are generally soon redeposited as insoluble compounds; new minerals are formed from them at an early stage of weathering.

SOIL GEOCHEMISTRY

Weathering and sedimentation are the geochemical processes of greatest importance to man, since they provide us with our basic economic resource, the soil. Man's culture and civilization can be closely correlated with the pattern of soil fertility. This can be directly traced back to the geochemical processes which have been responsible for the formation of the soil from the parent rock materials.

The unique characteristic of soil is the organization of its constituents and properties into layers that are related to the present-day surface and change vertically with depth. The individual layers are referred to as soil horizons and may range in thickness from a few inches to several feet; collectively, they are known as the soil profile. Most soil profiles comprise three principal horizons, identified from the surface downwards as A, B, and C. The A horizon develops primarily as a result of partial loss of original material by leaching and mechanical removal (eluviation), resulting from the downward percolation of rain water. This eluviated material accumulates in the B horizon, which is thus a zone of accumulation. The B horizon is characteristically enriched in clay and often has a red-brown or yellow-brown color from an accumulation of iron oxides. Minor and trace elements are frequently enriched in the B horizon. The C horizon is the parent material for the overlying A and B horizons; it may be rock in situ, transported alluvial or glacial material, or even the soil of a preceding cycle.

Because soil results from the weathering of rocks, its composition must depend on the composition of the rock from which it was formed. This statement is obvious, but it may be misleading. Although soils do differ in composition, they are on the whole rather uniform, and the differences are primarily the result of environmental factors. The same parent rock may give rise to very different soils under different conditions. The environmental factors include climate, biological activity, topography, and time, and the most important of these is climate. This can be seen by comparing the productivity of soils formed from the same rock type in the normal temperate zone and in the humid tropics. In general, the soil in the humid tropics will be much less fertile, as a result of the intense leaching brought about by high rainfall, high temperature, and the almost complete removal of organic matter by microorganisms.

The components of a soil include not only the minerals present, but also organic matter, water, and air. These components are generally not in equilibrium with the environmental conditions, and are hence in a continual state of change. The transformation from parent rock to soil is generally accompanied by a marked decrease in Ca, Mg, Na, and K, relatively smaller losses of Al and Fe, and increase in Si. The most active part of the soil is the colloidal fraction, which consists mostly of clay minerals, together with organic matter. Many, although not all, of the important agronomic characteristics of soils depend upon the quantity and kind of colloids present and their state of aggregation. True soil cannot be formed without the presence of some organic matter. Mere chemical and physical weathering of rocks does not necessarily result in soil formation, as in shown by the absence of soils in true deserts, either tropical or arctic. Most soil processes are directly or indirectly biological in nature. Organisms are effective agents for extracting and dissolving many elements. Because of the enormous multiplication rate of microorganisms, their total effect can be considerable and is probably significant in the migration of minor and trace elements in soils.

The data on major and minor elements in the parent rock can be used as an indication of the background composition of the soil. However, background composition in soils is also subject to considerable variation according to soil type and soil horizon, particularly in well-differentiated profiles characterized by marked enrichment in iron oxides or organic matter. The range of values for some elements in normal soils is given in Figure 6.1.

The importance of trace elements in the nutrition of plants, and through plants of animals, has been increasingly realized in recent years. Good crops and healthy humans and domestic animals are closely connected with the

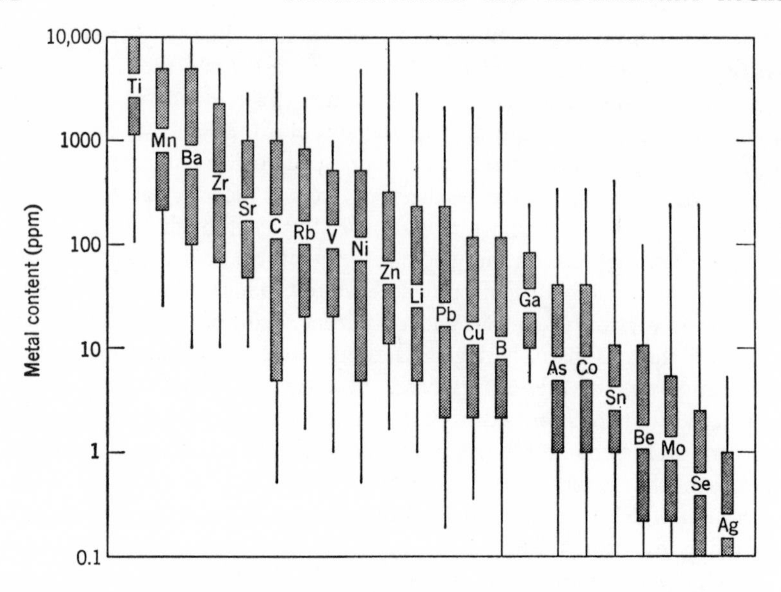

Figure 6.1 Range of the content of some minor elements in normal soils. Thin lines indicate more unusal values. (From Mitchell, in Bear, *Chemistry of the soil*. Courtesy of Reinhold Publishing Corp.)

presence in the soil of a number of minor and trace elements. The productivity of many regions has been greatly enhanced by the application of certain elements in minute amounts. This subject is essentially part of soil geochemistry, because plants derive these trace elements from the soil, not directly from the rocks. It is interesting to note that some trace elements, essential for animal health, seem not to be required by plants. Grass grows on soil deficient in cobalt as well as on soil with a normal cobalt content, but animals fed on such grass develop a deficiency disease that can be corrected by the feeding of trace amounts of cobalt, either directly or through the addition of cobalt-containing fertilizers to the soil. The possible correlation of the geographical distribution of disease and soil geochemistry is a field of extreme significance which as yet has been inadequately studied. In part this is because of the extreme complexity of the problem and the difficulty of isolating the numerous factors involved.

THE CHEMICAL COMPOSITION OF SEDIMENTARY ROCKS

The chemical composition of sedimentary rocks is exceedingly variable, more so than that of igneous rocks, since sedimentation generally leads to a further diversification. Considering compositions in terms of oxides, we find that SiO_2 may exceed 99% in some sandstones; Al_2O_3 may reach nearly 70% in bauxite; Fe_2O_3 up to 75% in limonite; FeO as high as 60% in siderite; MgO to 20% in dolomite; and CaO to 56% in pure limestones. In view of such variations, the determination of the average chemical composition of sedimentary rocks is not simple. The method used by Clarke and Washington for igneous rocks is hardly applicable because of a lack of analyses of sedimentary rocks and of inadequate sampling. There is little urge to analyze these rocks unless they have economic significance, and then they are generally of unusual composition.

Clarke estimated the average composition of the common sedimentary rocks—shale, sandstone, and limestone—by analyzing mixtures of many individual samples (Table 6.1). Then, by using an estimate for the relative amounts of these common sedimentary rocks, an average for sedimentary rocks as a whole can be calculated. Such an average is given in Table 6.1 using the figures shale 82%, sandstone 12%, and limestone 6%, given by Leith and Mead. There is considerable doubt as to the acceptability of these figures, since they seem to underestimate the amount of limestone. Poldervaart calculated an average composition of all sediments, based on what is known of their distribution and average composition (Table 6.1). When his figures are compared with those based on Leith and Mead's estimates the differences are clearly due largely to the greater amount of calcium carbonate allowed for by Poldervaart.

Table 6.1. Chemical Composition of Rocks

	Average Igneous Rock	Average Shale	Average Sandstone	Average Limestone	Average Sediment*	Average Sediment†
SiO_2	59.14	58.10	78.33	5.19	57.95	44.5
TiO_2	1.05	0.65	0.25	0.06	0.57	0.6
Al_2O_3	15.34	15.40	4.77	0.81	13.39	10.9
Fe_2O_3	3.08	4.02	1.07	0.54	3.47	4.0
FeO	3.80	2.45	0.30		2.08	0.9
MgO	3.49	2.44	1.16	7.89	2.65	2.6
CaO	5.08	3.11	5.50	42.57	5.89	19.7
Na_2O	3.84	1.30	0.45	0.05	1.13	1.1
K_2O	3.13	3.24	1.31	0.33	2.86	1.9
H_2O	1.15	5.00	1.63	0.77	3.23	
P_2O_5	0.30	0.17	0.08	0.04	0.13	0.1
CO_2	0.10	2.63	5.03	41.54	5.38	13.4
SO_3		0.64	0.07	0.05	0.54	
BaO	0.06	0.05	0.05			
C		0.80			0.66	
MnO						0.3
	99.56	100.00	100.00	99.84	99.93	100.0

* Shale 82, sandstone 12, limestone 6; after Leith and Mead.
† Poldervaart, *Geol. Soc. Amer. Spec. Paper* 62, 132, 1955.

It has been considered that the average composition of sedimentary rocks should correspond fairly closely to that of igneous rocks, since all sedimentary rocks have ultimately been derived from igneous rocks by processes of weathering. The only permanent change should be the loss of those elements, principally sodium, which tend to accumulate in solution in the ocean, and the addition of components from the atmosphere and hydrosphere, such as oxygen, carbon dioxide, and water. However, if Poldervaart's figures are reliable, the average sediment does not correspond to the average igneous rock. This apparent discrepancy in the geochemical balance sheet, especially for calcium, is a significant problem. It may be related, in part at least, to a marked increase of pelagic foraminifers in Tertiary and Recent times, leading to a greater precipitation of calcium carbonate in young sediments and sedimentary rocks.

Significant features of the chemical composition of sedimentary rocks are the dominance of potassium over sodium, alumina in excess of the 1 : 1 ratio to alkalies and calcium, high silica in sandy and cherty rocks, high lime and

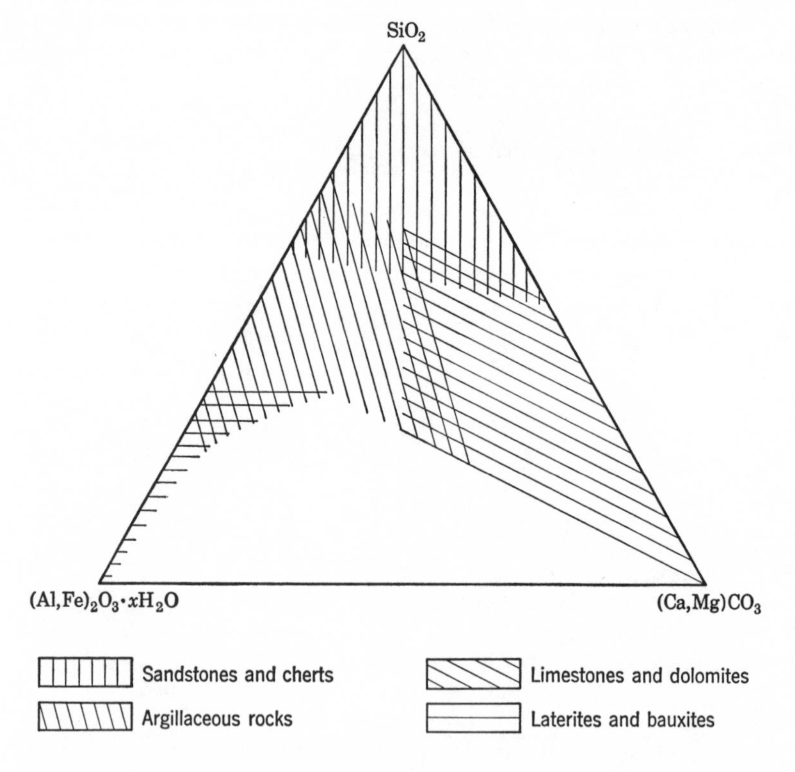

Figure 6.2 The chemical composition of common sediments. Sediments with compositions falling in the blank area are rare or nonexistent.

magnesia in the carbonates, and the presence of iron mainly in the ferric state. These are generalizations, and many exceptions may be found. The relationship between chemical composition and rock type in the sediments can be expressed in a gross fashion by a triangular composition diagram in which the apices are SiO_2, $(Al,Fe)_2O_3{\cdot}xH_2O$, and $(Ca,Mg)CO_3$ (Figure 6.2). Such a diagram neglects the alkalies, but they are generally low except in some argillaceous rocks. It is difficult, if not impossible, to define sedimentary rock types from chemical analyses alone by any such system as the norm for igneous rocks. The main groups overlap, as Figure 6.2 shows. Some tentative limits can be stated; argillaceous rocks with more than 50% SiO_2 generally contain free silica, whereas those with more than 40% Al_2O_3 contain free alumina. To call a rock a limestone or dolomite should at least imply that carbonate is the dominant component, and similarly in sandstone or chert free silica should exceed any other component. Imbrie and Poldervaart (1959) have developed a procedure for the calculation of mineralogical compositions from chemical analyses of sedimentary rocks.

THE MINERALOGICAL COMPOSITION OF SEDIMENTARY ROCKS

The mineralogy of sedimentary rocks is characterized by two distinct types of material: resistant minerals from the mechanical breakdown of the parent rocks, and minerals newly formed from the products of chemical decomposition. The latter minerals are generally hydrated compounds, as is to be expected in substances formed in a water-rich environment. Goldich (1938) has pointed out that the order of stability of minerals of igneous rocks towards weathering is the reverse of their order in the reaction series of Bowen; thus

Stability decreases ↓		
	Quartz	
	Muscovite	
	Potash feldspar	
	Biotite	
Alkalic plagioclase		
Alkali-calcic plagioclase		Hornblende
Calcic-alkalic plagioclase		
		Augite
Calcic plagioclase		
		Olivine

The arrangement does not of course imply a reverse reaction series; the minerals do not invert one into the other on weathering but are decomposed into their components. This identity of arrangement between Bowen's reaction series and Goldich's stability series indicates that the last-formed minerals of igneous rocks are more stable at ordinary temperatures than are the minerals formed at an early stage of crystallization. In other words, the differential between conditions at the time of formation and those existing at the surface reflects the order of stability of the common silicates of igneous rocks. Fieldes and Swindale (Figure 6.3) have presented a scheme which shows how primary minerals change to successive secondary minerals through weathering processes. The primary minerals are listed in order of increasing resistance to weathering, and the secondary minerals are arranged to indicate their relation to the minerals from which they originate.

The total number of minerals recorded from sedimentary rocks is very large, because almost any mineral of igneous or metamorphic origin may have at least a transitory existence in a sediment. Nevertheless, the common and abundant minerals of sedimentary rocks are few: quartz, feldspar, calcite, dolomite, and clay minerals. Some other minerals, such as glauconite, limonite (geothite and hematite), bauxite (gibbsite and boehmite), and

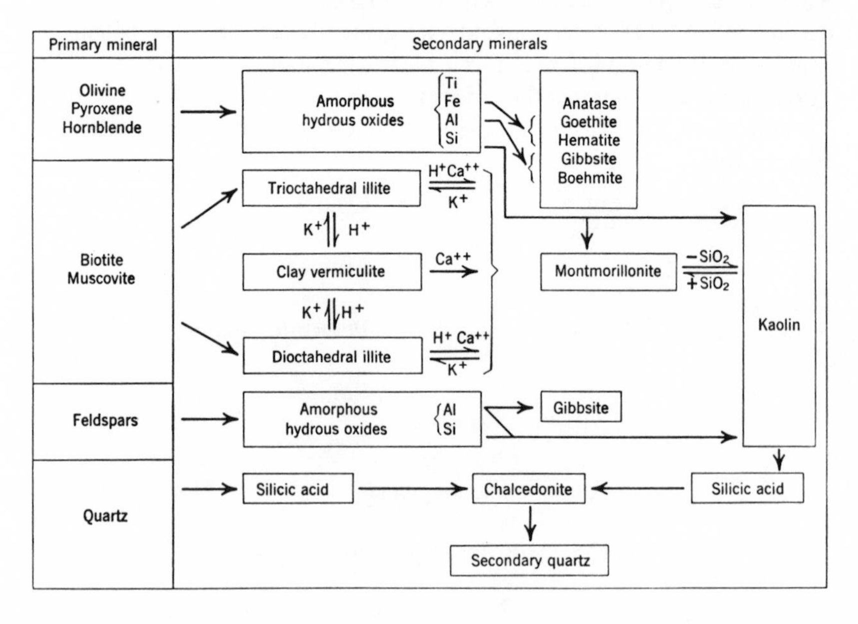

Figure 6.3 Weathering of primary rock-forming minerals. (After Fieldes and Swindale, *New Zealand J. Sci. Tech.* **36B**, 140, 1954)

collophane (sedimentary apatite), are abundant in sediments of restricted extent, and amorphous colloidal material, such as allophane, may be present.

Quartz and the feldspars are the abundant detrital minerals, that is, those set free by the mechanical breakdown of preexisting rocks. Quartz is very resistant to chemical attack under surface conditions. The feldspars are less resistant, for, although they may persist indefinitely in sedimentary rocks, they are chemically decomposed by prolonged weathering. On the other hand, the alkali feldspars may undoubtedly be formed in sediments under surface or near-surface conditions, since authigenic albite, orthoclase, and microcline have been recorded. This variability in behavior presumably reflects the environment; solution probably takes place in acid environments, whereas in alkaline environments authigenic feldspar can form. Anorthite is readily decomposed by weathering, but it is interesting to note that authigenic calcium zeolites (heulandite, chabazite, laumontite, and others) have been found in some sediments and sedimentary rocks.

Calcite is precipitated from solution either by physicochemical changes or by the vital processes of organisms. Aragonite is sometimes the form in which calcium carbonate is deposited, but in general it does not persist, for it inverts more or less readily into calcite, the more stable form. The origin of dolomite has been a subject of much discussion, and general agreement has not been reached. However, geological evidence indicates that many dolomites have been formed from limestones by the metasomatic action of magnesium-bearing waters. In many instances sea water acting on calcium carbonate during diagenesis has evidently been responsible. For the reaction

$$2CaCO_3 + Mg^{2+} = CaMg(CO_3)_2 + Ca^{2+}$$

the law of mass action predicts that equilibrium will be determined almost entirely by the relative concentration of calcium and magnesium ions in solution. Studies on the thermodynamics of the reaction have shown that with the conditions of temperature and concentration prevailing in the sea the free-energy change of the above reaction is negative, that is, dolomitization will proceed spontaneously.

Kaolinite, montmorillonite, illite, and chlorite, together with a number of less common species, make up the clay minerals of sediments and sedimentary rocks. Clay mineralogy has been the subject of intensive research in recent years, and a detailed account has been given by Grim (1953). The clay minerals are stable secondary products formed by the decomposition of other aluminosilicates. It is significant that all these new-formed minerals have layer-lattice structures, which seem to have greater stability than other types under surface conditions. Besides their characteristic crystal structure, the clay minerals have other features in common. They are all hydrous aluminosilicates. In most sediments their average grain size is very small, generally less than 0.005 mm in diameter and ranging down to colloidal dimensions; as grain size diminishes so does perfection of crystallinity, and probably no sharp break exists between imperfectly crystallized clay minerals and amorphous material which may perhaps be regarded as aluminosilicate gel. Very often more than one clay mineral is present in a particular sediment. Not only mechanical mixtures, but also "mixed" crystals, in which molecular layers of more than one clay-mineral species are interleaved in a single crystal, have been recognized. Thus the complications are many, and positive identification of the phases in the clay fraction of sediments can be the most exacting problem with which a mineralogist is faced.

The similarities and differences between the clay minerals can best be understood by consideration of their structures (Figure 6.4). They are all phyllosilicates, and their structures can be considered as produced by the stacking of two different units in the direction of the *c* axis. These units are (*a*) linked (SiO_4) groups, each with three oxygens shared with adjacent groups, giving (Si_4O_{10}) sheets; and (*b*) aluminum-hydroxyl units, consisting of aluminum ions between two sheets of close-packed hydroxyls or oxygens; each aluminum is surrounded by six oxygens or hydroxyls, that is, is in sixfold coordination. These two types of units are linked together in the clay mineral by oxygens common to both. The individual clay minerals differ in the relative number of the two types of units in their structures and in the possibility of replacement of silicon or aluminum by other elements.

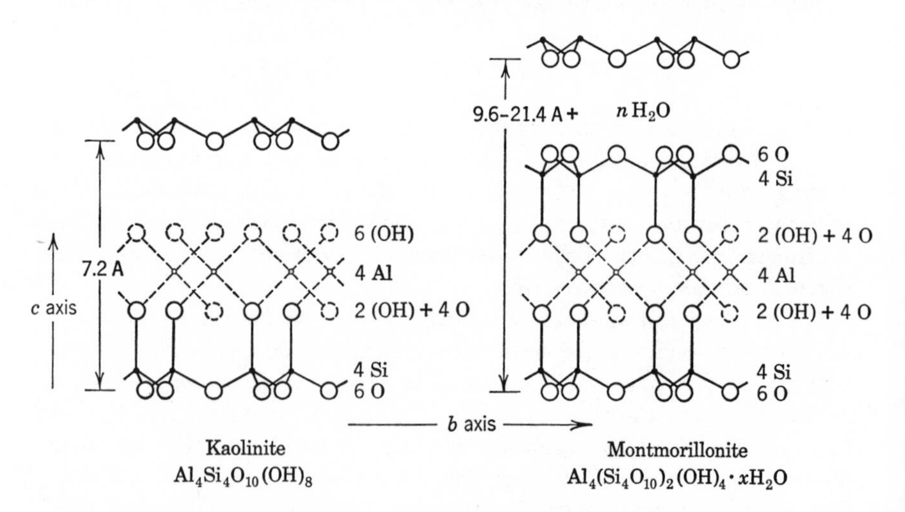

Figure 6.4 Schematic presentation of the structures of the principal clay minerals. (After Grim, *J. Geol.* **50,** 225–275, 1942)

The structure of kaolinite consists of one (Si_4O_{10}) sheet linked with one aluminum-hydroxyl sheet, that is, a two-layer structure. Replacement of silicon and aluminum by other elements does not occur, and so analyses of pure kaolinite always correspond closely with the ideal formula. Kaolinite is one of four polymorphs, the others being dickite, nacrite, and halloysite. Dickite and nacrite are generally of hydrothermal origin and rarely occur in sediments. Halloysite occurs in hydrothermal deposits, but it is also found occasionally in sedimentary rocks, where it has generally been precipitated from acid ground water carrying alumina and silica in solution.

The montmorillonite structure has layers consisting of one aluminum-hydroxyl unit sandwiched between two (Si_4O_{10}) sheets; these layers are stacked one above the other in the *c* direction, with water molecules between them. A characteristic feature of montmorillonite is the variable water content, which is reflected in the *c* repeat, varying from 9.6 A in dehydrated material to 21.4 A when the mineral is water-saturated. Montmorillonite is therefore said to have an expanding lattice; the characteristic swelling properties of bentonites in water are due to their montmorillonite content. Other polar liquids, such as ethylene glycol and glycerol, have a similar effect. Considerable atomic substitution is also possible in the montmorillonite structure; the aluminum can be partly or wholly replaced by ferric iron (nontronite), by magnesium (saponite), by zinc (sauconite), and by smaller amounts of lithium, trivalent chromium, manganese, and nickel; the silicon can be partly replaced by aluminum, giving the variety beidellite.

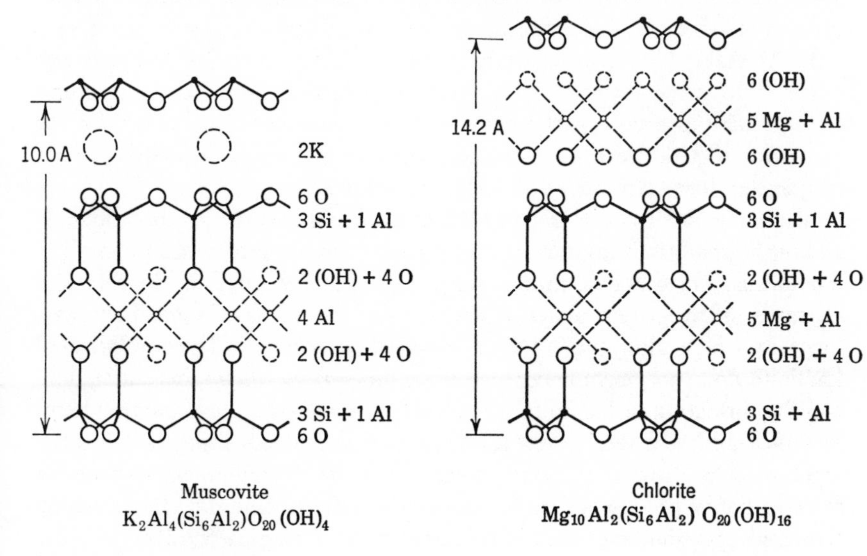

Figure 6.4 (continued)

The chemical composition of montmorillonite (in the group sense) is therefore exceedingly variable.

Montmorillonite always differs from the theoretical formula in that the three-layer unit has a net negative charge resulting from the substitutions noted above. Many analyses have shown this to be about two-thirds of a unit per unit cell. This charge deficiency is balanced by exchangeable cations adsorbed between the unit layers and around their edges; these adsorbed cations are often calcium and sodium, elements which are therefore found in analyses of montmorillonite, although the ideal structure has no lattice positions for them.

A common and important constituent in many clays and shales is the material known as illite or hydromica (Glimmerton—mica clay—in German). The occurrence of a micaceous clay material with less potassium and more water than the theoretical composition of muscovite has been recognized for many years. The nature of illite has been elucidated by Yoder and Eugster (1955), who have shown that the name is useful as a general term, not as a specific clay-mineral name. Some illite is essentially clay-size muscovite; some is a mixed-layer muscovite-montmorillonite; and some is a mechanical admixture of muscovite and montmorillonite, often with some clay-size quartz. It is evident from the work of Yoder and Eugster that muscovite is an authigenic clay mineral in many sediments.

The occurrence of chlorite as an important constituent of argillaceous material is a comparatively recent discovery. Small amounts of chlorite admixed with other clay minerals are particularly difficult to detect. Structurally, chlorite may be derived from montmorillonite by the insertion of a (Mg,Al)(OH) layer between each montmorillonite layer, just as muscovite is derived by the insertion of a layer of potassium ions. This structural relation may also be a paragenetic relation, since there is evidence for the formation of chlorite from montmorillonite in sea water, which is rich in magnesium ions to provide the (Mg,Al)(OH) layer.

Much remains to be learned of the mutual relations of the clay minerals and of the conditions favoring the formation of one in preference to another. The primary factors determining the nature of a clay are, first, the chemical character of the parent material and, second, the physicochemical environment in which alteration of this material takes place. The structure of kaolinite does not accommodate cations other than silicon and aluminum, and the formation of kaolinite is evidently favored by an acid environment, in which all bases tend to be removed in solution. Kaolinite also has the highest Al:Si ratio of the clay minerals, and its formation is promoted if the weathering processes tend to remove silica in solution, thereby enriching the residue in alumina. This is reflected in the geological evidence for the common formation of kaolinite by the alteration of alkali feldspar in an acid environment, the Na^+ and K^+ ions released tending to stabilize the silica in solution. Montmorillonite, on the other hand, has an Si:Al ratio about 2:1, and other cations, such as magnesium and iron, are probably essential to its formation; it seems to form most readily in a neutral or slightly alkaline environment from ferromagnesian minerals, calcic feldspars, and especially volcanic ash. Montmorillonite has a close structural relationship to illite (muscovite) and chlorite and evidently changes readily to these minerals, especially in the marine environment. The comparatively high concentration of potassium and magnesium in sea water promotes this change. Illite is the commonest clay material in marine sediments and sedimentary rocks. The fixation of potassium in illite and magnesium in chlorite is probably an important mechanism in regulating sea-water composition. In general, diagenesis promotes the formation of illite and chlorite and the disappearance of kaolinite and montmorillonite, so that shales and argillites consist largely of the former minerals.

PHYSICOCHEMICAL FACTORS IN SEDIMENTATION

The geochemistry of sedimentary processes is essentially the geochemistry of reactions taking place in the presence of water. Now water is by no means a typical liquid, and in this connection its unique properties deserve to be emphasized. As a solvent water is unequalled; no other liquid can compare with it in the number of substances it can dissolve nor in the amounts it can hold in solution. The structure of the water molecule is the key to these remarkable properties. The hydrogens are deeply imbedded in the oxygen atom so that the molecule is approximately spherical. Its radius is only slightly greater than that of the oxygen ion. The bond angle between the lines joining the hydrogen nuclei to the center of the oxygen atom is 105°. This structure results in a very uneven distribution of charge in the molecule. An excess of positive charge appears at or between the protons, and the opposite side of the molecule is negatively charged. Thus the water molecule is a dipole. The mutual attraction of these dipoles makes the cohesive forces between water molecules very much larger than for normal liquids, which owe their cohesion to weak van der Waals' forces. Hence water, for a liquid of low molecular weight, has extremely high melting and boiling points. The liquid range of water, 100° at 1 atm, is unexpectedly long because of the persistence of bonding between the molecules. The energy necessary to break the remaining bonds at the boiling point is reflected in the abnormally high heat of vaporization, 9720 cal/mole. The cohesive forces between water molecules are also reflected in the surface tension, which is 72.75 dynes/cm at 20°, compared with 26.77 for CCl_4 and 28.88 for C_6H_6. Another important effect of the dipole nature of the molecules is to give water its abnormally high dielectric constant, namely, 80.

The high dielectric constant is responsible for its activity as a solvent for ionic compounds, because the force of attraction between ions varies inversely as the dielectric constant of the medium, and solution of ionic compounds is essentially a dispersion of the ions by the molecules of the solvent.

IONIC POTENTIAL

Ions in solution attract water molecules to them; cations attract the negative ends of nearby water dipoles, anions the positive ends. The number of water molecules that can thereby be packed around a given ion depends on the size of the ion; the bigger it is, the greater the number of water molecules that can cluster around it. However, the degree of hydration depends not only on the size of the ion, but also on the intensity of the charge on its surface. For example, the lithium ion, with a radius of 0.68 A, exerts a much stronger attraction for water dipoles than the cesium ion, with a radius of 1.67 A, although the charge on each is the same. As a result, the lithium ion is hydrated, despite its small radius, whereas the cesium ion is not. The hydration of an ion is thus proportional to its charge (Z) and inversely proportional to its radius (r). The factor Z/r is known as the ionic potential and is of great significance not only for the hydration of an ion, but for many of its properties in the presence of water. The importance of ionic potential in sedimentary processes was first pointed out by Goldschmidt as providing a measure of the behavior of an ion towards water. The ionic potentials of a number of ions are given in Table 6.2. In effect, the ionic potential is a measure of electronegativity, since the smaller the radius of a positive ion and the higher its charge, the more acidic is its oxide; and, conversely, the larger the radius and the smaller the charge, the more basic the oxide. From

Table 6.2. Ionic Potentials

Ion	Z/r	Ion	Z/r
Cs^{+}	0.60	Th^{4+}	3.9
Rb^{+}	0.68	Ce^{4+}	4.3
K^{+}	0.75	Fe^{3+}	4.7
Na^{+}	1.0	Zr^{4+}	5.1
Li^{+}	1.5	Be^{2+}	5.7
Ba^{2+}	1.5	Al^{3+}	5.9
Sr^{2+}	1.8	Ti^{4+}	5.9
Ca^{2+}	2.0	Mn^{4+}	6.7
Mn^{2+}	2.5	Nb^{5+}	7.5
La^{3+}	2.6	Si^{4+}	9.5
Fe^{2+}	2.7	Mo^{6+}	9.7
Co^{2+}	2.8	B^{3+}	13
Mg^{2+}	3.0	P^{5+}	14
Y^{3+}	3.3	S^{6+}	20
Lu^{3+}	3.5	C^{4+}	25
Sc^{3+}	3.7	N^{5+}	38

the electrostatic viewpoint, the ionic potential is a measure of the intensity of positive charge on the surface of the ion. This concentration of positive charge on the surface of a cation repels the protons in the coordinated water molecules. If the repulsion is sufficiently great, some of these protons may be detached, thereby neutralizing the charge on the central cation and resulting in the precipitation of an insoluble hydroxide. With very high repulsive forces, that is, high ionic potentials, all the protons are expelled from the attracted water molecules and an oxyacid anion is formed.

The ionic potential of an element largely determines its place of deposition during the formation of sedimentary rocks and is significant in all mineral-forming processes in an aqueous medium. It provides an explanation for the similar behavior of dissimilar elements as, for example, the tendency of the hydrated ions of bivalent beryllium, trivalent aluminum, and quadrivalent titanium to precipitate together during sedimentation. Elements with low ionic potential, such as sodium, calcium, and magnesium, remain in solution during the processes of weathering and transportation; elements with intermediate ionic potential are precipitated by hydrolysis, their ions being associated with hydroxyl groups from aqueous solutions; elements with still higher ionic potentials form anions containing oxygen which are usually again soluble. When the elements are plotted on a diagram with ionic radius as ordinates and ionic charge as abscissae, the field can thus be divided into three parts: soluble cations, elements of hydrolysates, and elements of soluble complex anions (Figure 6.5).

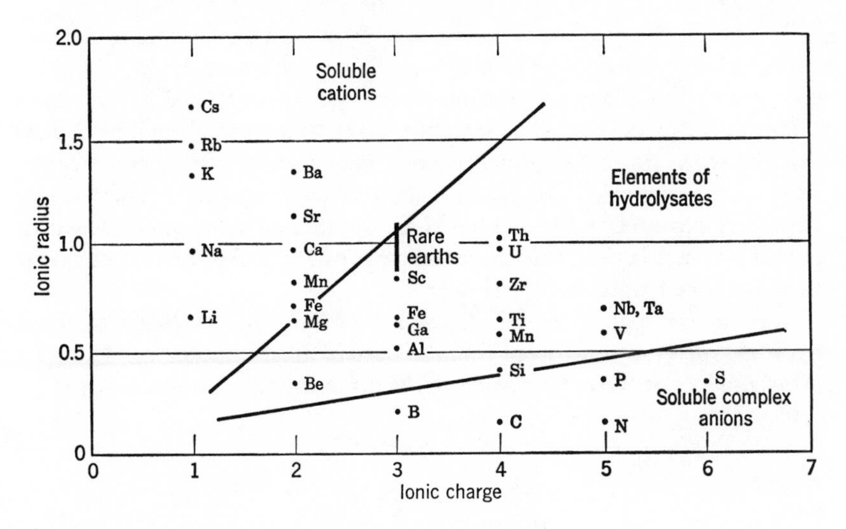

Figure 6.5 Geochemical separation of some important elements on the basis of their ionic potential.

Some specific examples of the significance of ionic potentials in the behavior of certain elements during sedimentary processes may be cited. Iron in the ferrous state is stable in solution ($Z/r = 2.7$), so that precipitation of iron has to be preceded by oxidation to the ferric state, with a much higher potential ($Z/r = 4.7$). Similarly, manganese is stable in solution as manganous ions ($Z/r = 2.5$) and is precipitated in the hydrated quadrivalent form ($Z/r = 6.7$). Thus in many sedimentary rocks products of hydrolysis and of oxidation are associated in the same deposit. Again, a number of less common and rare quadrivalent elements are concentrated in hydrolysate sediments, their ionic potentials falling within the specified range. For instance, not only beryllium and gallium, which are chemically similar to aluminum, but also titanium, zirconium, and niobium may be concentrated in bauxite, the factor of enrichment being often four- to fivefold as compared to the parent material.

HYDROGEN-ION CONCENTRATION

The hydrogen-ion concentration of natural waters is of great significance in chemical reactions accompanying sedimentary processes. In pure water at 20° the hydrogen-ion concentration is 10^{-7} mole/liter. If the concentration is greater than that of pure water at the same temperature, the solution is said to be acid; in the contrary case, alkaline. The neutral point alters with rising temperature in the direction of greater hydrogen-ion concentration. As an inverse measure of hydrogen-ion concentration we use the pH, which is the negative logarithm of this factor; thus the pH of pure water is 7.

The pH of the medium is particularly significant in controlling the precipitation of hydroxides from solution. This is shown in Table 6.3, which gives the pH for beginning precipitation of hydroxides from dilute solutions (about 0.02 *M*) and the pH of some natural environments. It will be noticed that the sea is slightly alkaline, whereas most terrestrial waters are somewhat on the acid side. The pH of many terrestrial waters is controlled by the buffer system $CaCO_3$—CO_2—H_2O; a saturated solution of CO_2 at its partial pressure in the atmosphere has a pH of 5.2, and a solution of calcite in air-saturated water has a pH near 8.

For the transportation and deposition of iron the solubility of ferric hydroxide and the consequent equilibria are of great importance. The solubility product of ferric hydroxide (S) is given by the equation

$$S = [Fe^{3+}][OH^-]^3$$

Hence

$$[Fe^{3+}] = \frac{S}{[OH^-]^3}$$

Table 6.3. The pH of Natural Media and Its Relation to the Precipitation of Hydroxides

pH	Precipitation of hydroxides	Natural media	pH
11			11
	Magnesium		
10		Alkali soils	10
9			9
	Bivalent manganese		
		Sea water	
8			8
7	Zinc	River water	7
6	Copper	Rain water	6
5	Bivalent iron		5
4	Aluminum	Peat water	4
3		Mine waters	3
	Trivalent iron		
2			2
		Acid thermal springs	
1			1

But in water $[OH^-] = K/[H^+]$, where K is the ionization product of water. Therefore,

$$[Fe^{3+}] = \frac{S[H^+]^3}{K^3}$$

At a fixed temperature S and K are constants; hence the concentration of ferric iron in solution is proportional to the *cube* of the hydrogen-ion concentration. For example, at 18° $S = 10^{-38.6}$ and $K = 10^{-14.2}$. Therefore, $S/K^3 = 10^4$, and at pH = 7, $[Fe^{3+}] = 10^{-17}$ mole/liter; at pH = 6, $[Fe^{3+}] = 10^{-14}$ mole/liter. In natural waters iron is present not only as Fe^{3+} but also as Fe^{2+} and $FeOH^{2+}$ ions. Cooper (1937) has determined the total amount of iron in solution at different pH values, with the following results:

pH	Amount
8.5	3.10^{-8} mg/m^3
8	4.10^{-7} mg/m^3
7	4.10^{-5} mg/m^3
6	5.10^{-3} mg/m^3

Thus the solubility of iron at pH 6 is about 10^5 times greater than at pH 8.5. Weakly acid iron-bearing solutions flowing into the sea from neighboring land areas must precipitate most of their iron in the weakly alkaline marine waters. This is borne out by actual figures which show that the average content of iron in river waters is about 1 ppm, whereas in sea water the amount is exceedingly small, about 0.008 ppm.

The pH of the environment is especially significant for the transportation of alumina and silica in solution and their ultimate redeposition. To illustrate this, the solubilities of aluminum hydroxide and silica with respect to pH are plotted on Figure 6.6. At pH < 4 alumina is readily soluble, whereas silica is only slightly soluble. At such pH values alumina would be removed in solution and silica would remain with the parent material. Normal sedimentary environments, however, seldom have pH this low. From pH 5–9 the solubility of silica increases considerably, but alumina is practically insoluble. Under these conditions removal of silica can take place, leaving alumina behind, as has been inferred during the formation of laterites and bauxites. A study of Figure 6.6 also suggests a possible explanation for the formation of either kaolinite or montmorillonite from the same parent materials under different conditions. These clay minerals are probably formed in natural waters carrying both silica and alumina in solution. In acid solutions (pH about 4) the solubilities of silica and alumina are such that relatively much alumina and relatively little silica are present, thus favoring the formation of material of kaolinitic composition (Al_2O_3: SiO_2 = 1:2); in alkaline solutions (pH 8–9) much more silica is present, thus promoting the formation of montmorillonite (Al_2O_3:SiO = 1:4).

OXIDATION-REDUCTION POTENTIALS

Many elements are present in different oxidation states in the earth's crust. The commonest is iron, occurring as the native metal (oxidation state 0), as ferrous compounds (oxidation state 2), and as ferric compounds (oxidation state 3). Similar elements are manganese (2, 3, 4), sulfur (−2, 0, 6), vanadium (3, 4, 5), copper (0, 1, 2), cobalt (2, 3), nitrogen (−3, 0, 5), and many others.

The stability of an element in a particular oxidation state depends upon the energy change involved in adding or removing electrons. A quantitative measure for this energy change is provided by the factor known variously as the "oxidation-reduction potential," "oxidation potential," or "redox potential"; we will refer to it as the oxidation potential. The oxidation potential of any reaction is a relative figure, the reference standard being the reaction

$$H_2 = 2H^+ + 2e \qquad (e = \text{electron})$$

(i.e., the removal of electrons from hydrogen atoms or the oxidation of hydrogen to hydrogen ions). The oxidation potential of this reaction for unit activity (activity is a function of concentration that provides for deviation from the laws of perfect solutions) of the reacting substances is arbitrarily fixed as 0.00 volt, and the scale of oxidation potentials extends on either side of zero. Oxidation potentials are symbolized by $E°$ when the relevant reactions take place under the standard conditions of unit activity of the reacting substances, and by *Eh* when the experimental situation deviates from these conditions.

Table 6.4 lists some reactions in order of decreasing oxidation potential, that is, in order of increasing reducing power, the reduced form of any couple having sufficient energy to reduce the oxidized form of any couple of higher potential. For example, Fe^{2+} reduces Mn^{3+}, and H_2 reduces Fe^{3+}.

The oxidation potential varies with varying concentration of the reacting substances. This variation with concentration is of special importance in reactions involving hydrogen or hydroxyl ions, as many such reactions do (Table 6.4). A variation in pH produces large changes in oxidation potentials involving hydrogen or hydroxyl ions and must be taken into account in applying E^0 values to actual reactions. The influence of pH on the oxidation potentials of some reactions given in Table 6.4 is shown graphically in Figure 6.7.

The range of oxidation potentials of natural environments determines

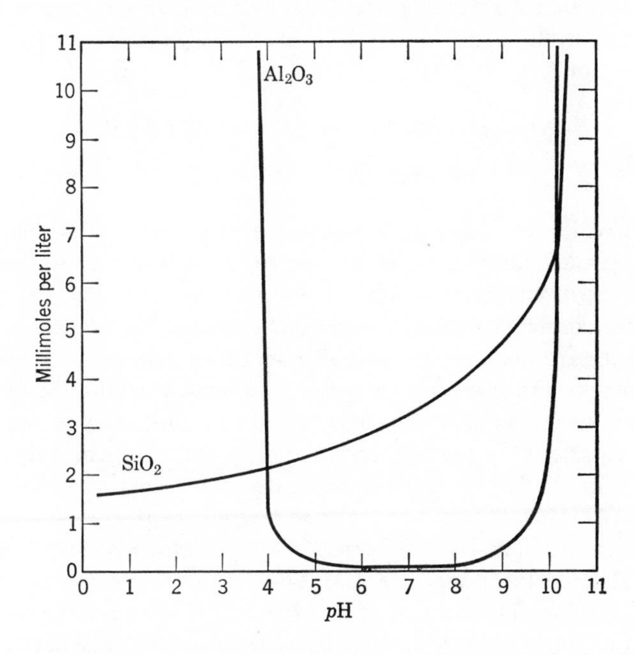

Figure 6.6 The solubility of silica and alumina as a function of pH.

Table 6.4. The Oxidation Potentials of Some Reactions of Geochemical Significance

	E_0 (in volts)
$Co^{2+} = Co^{3+} + e$	1.84
$Ni^{2+} + 2H_2O = NiO_2 + 4H^+ + 2e$	1.75
$Mn^{2+} = Mn^{3+} + e$	1.51
$Pb^2 + 2H_2O = PbO_2 + 4H^+ + 2e$	1.46
$2H_2O = O_2 + 4H^+ + 4e$	1.23
$NH_4^+ + 3H_2O = NO_3^- + 10H^+ + 8e$	0.84
$Fe^{2+} = Fe^{3+} + e$	0.77
$Ni(OH)_2 + 2OH^- = NiO_2 + 2H_2O + 2e$	0.49
$4OH^- = O_2 + 2H_2O + 4e$	0.40
$PbO + 2OH^- = PbO_2 + H_2O + 2e$	0.25
$Co(OH)_2 + OH^- = Co(OH)_3 + e$	0.2
$S^{2-} + 4H_2O = SO_4^{2-} + 8H^+ + 8e$	0.14
$H_2 = 2H^+ + 2e$	0.00
$NH_3 + 9OH^- = NO_3^- + 6H_2O + 8e$	−0.12
$Mn(OH)_2 + OH^- = Mn(OH)_3 + e$	−0.40
$Fe(OH)_2 + OH^- = Fe(OH)_3 + e$	−0.56

the reactions that may take place. Chemical reactions in aqueous media are theoretically limited to those with oxidation potentials between those for the reactions

$$H_2O = \tfrac{1}{2}O_2 + 2H^+ + 2e, \qquad E^0 = 1.23 \text{ volts} \tag{1}$$

$$2H^+ + 2e = H_2, \qquad E^0 = 0.00 \text{ volt} \tag{2}$$

The oxidized form of any couple with a higher potential than that for (1) will theoretically decompose water with the evolution of oxygen. The reduced form of any couple with a lower potential than that for (2) will theoretically decompose water with the evolution of hydrogen. These requirements are not strictly met in practice on account of overvoltage phenomena; that is, it requires a greater potential than the theoretical to produce the evolution of hydrogen or oxygen at a measurable rate. However, the chemistry of sedimentation indicates that the potentials of these two reactions do largely control oxidation and reduction under natural conditions.

Both these reactions involve hydrogen ions, and their potentials are thus strongly affected by changes in pH. The E^0 values given above are for hydrogen-ion concentrations of unity, that is, pH = 0, and the potentials decrease (at 25°) 0.06 volt for each unit increase in pH. The pH of natural waters is variable, ranging from as low as 0 in strongly acid waters of volcanic regions to 10 or more in alkaline areas where sodium carbonate

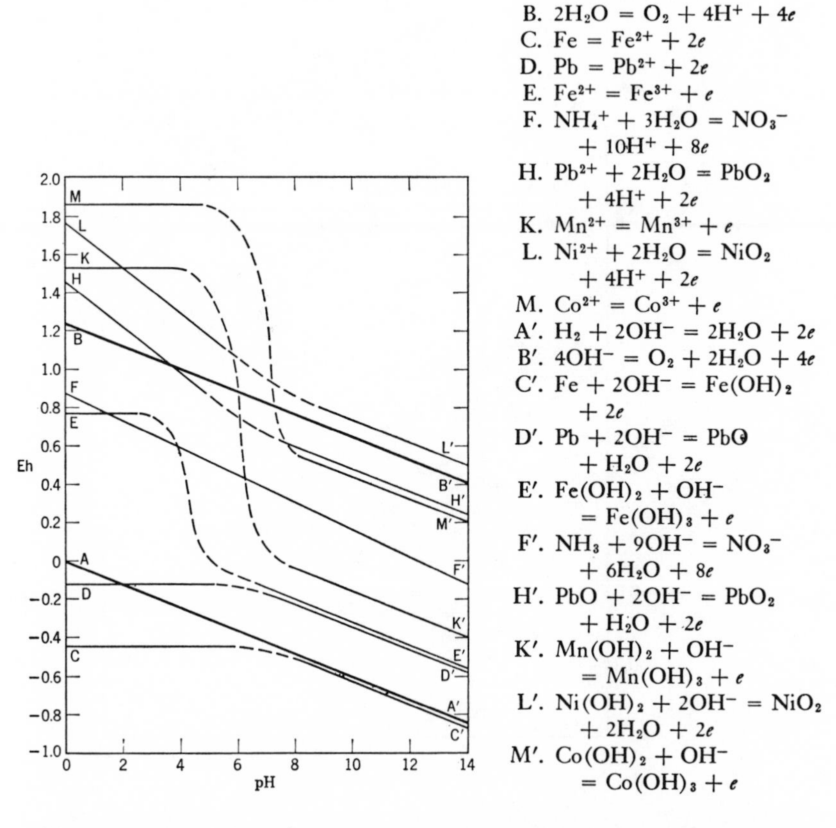

Figure 6.7 Variation of oxidation potential with pH for certain reactions.

is present in solution. As a rule, however, the pH of natural waters lies between 4 and 9, the great majority being within one unit of the figure for pure water (pH = 7). For a pH of 7 the potential of (1) is 0.82 volt and of (2) −0.41 volt. [The potential of 0.82 volt is that of water with pH 7 saturated with oxygen at 760 mm pressure; for oxygen at its partial pressure in the atmosphere (160 mm) this potential is reduced to 0.81 volt.] These figures indicate that the oxidation potentials of natural environments, where the pH is near 7, should lie between −0.41 and 0.82 volt. Measurements of oxidation potentials in natural waters are in agreement with this statement (Figure 6.8) although values as low as −0.5 volt have been recorded in marine-bottom deposits rich in organic matter; hydrogen may actually be generated in such environments.

The solution, transportation, and deposition of elements that may occur

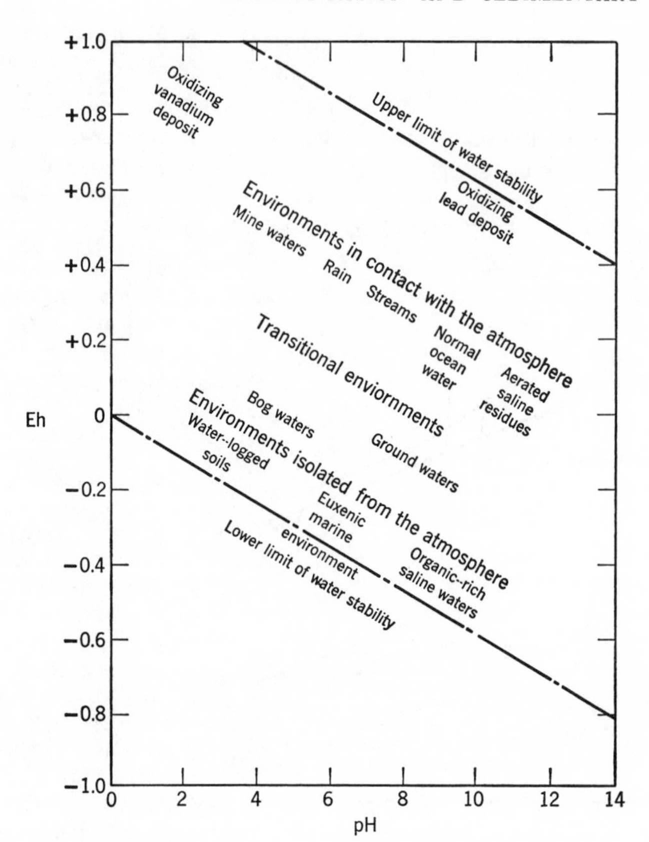

Figure 6.8 Approximate position of some natural environments as characterized by Eh and pH. (From Garrels and Christ, *Solutions, Minerals, and Equilibria*. Courtesy Harper and Row)

in two or more oxidation states are directly and powerfully influenced by the oxidation potential of the environment. For example, the high potentials required to convert bivalent to trivalent cobalt and bivalent lead to PbO_2 in acid solution indicate that the minerals stainierite (CoOOH) and plattnerite (PbO_2) are deposited from alkaline solutions, for which the oxidation potentials are much less (Figure 6.7). Figure 6.7 shows that for many reactions the oxidation potentials decrease rapidly with the increase of pH; generally, therefore, oxidation proceeds more readily the more alkaline the solution. The potentials for some oxidations in alkaline solutions, particularly those that result in the precipitation of almost insoluble compounds, lie far below the potentials for corresponding oxidations in acid solutions. This is especially marked with respect to the oxidation of ferrous to ferric iron; in acid solution the potential is 0.77 volt and is not affected by pH; however, as soon as the pH increases to a figure at which ferric hydroxide is precipitated the oxidation potential drops sharply to a negative figure. Thus ferrous salts are comparatively stable in acid solution, being only slowly oxidized by air, but in solutions sufficiently low in acid for $Fe(OH)_3$ to be precipitated oxidation proceeds rapidly to completion. Deposition of ferrous compounds in nature therefore demands either a very acid environment or one with a very low oxidation potential, on the negative side of zero.

The Eh—pH diagrams provide a useful device for illustrating the stability fields of different minerals in an aqueous environment. These diagrams are discussed in detail by Garrels and Christ (1965), with numerous examples illustrated by them. The system Mn—H_2O under varying conditions of Eh and pH provides an illustrative case (Figure 6.9), the diagram showing clearly why manganese has not been found as the native metal. Under reducing conditions in the geological environment manganese occurs as manganous compounds in acid solutions, and precipitates as $Mn(OH)_2$ when the pH exceeds 8 (for unit activity of Mn; for more dilute solutions the pH required for precipitation is greater, according to the law of mass action). As conditions become more oxidizing, the field of manganous ions shrinks towards more acid conditions; on the alkaline side of the diagram $Mn(OH)_2$ (pyrochroite) is successively replaced by Mn_3O_4 (hausmannite), MnOOH (manganite), and MnO_2 (pyrolusite). Under highly oxidizing conditions MnO_2 is the stable phase over the whole range of pH. The diagram also explains why permanganates do not occur under geological conditions; aqueous solutions of permanganate are unstable and slowly decompose, liberating oxygen and precipitating MnO_2.

The separation of closely related elements in the upper zone of the lithosphere by processes involving solution and redeposition is often brought about by their distinctive properties with respect to oxidation and reduction. Thus the three elements iron, nickel, and cobalt often occur together in primary deposits, yet supergene processes result in their separation. These three elements differ greatly in the potentials required to oxidize them beyond the bivalent state. Iron is readily oxidized to the trivalent state in alkaline and mildly acid environments; cobalt requires a much higher potential even in alkaline solution, and in acid solutions the potential required lies high above that for the release of oxygen from water; nickel does not form trivalent compounds, but a dioxide is known, the formation of which even in alkaline solutions requires potentials somewhat higher than that for the release of oxygen from water. This is reflected in

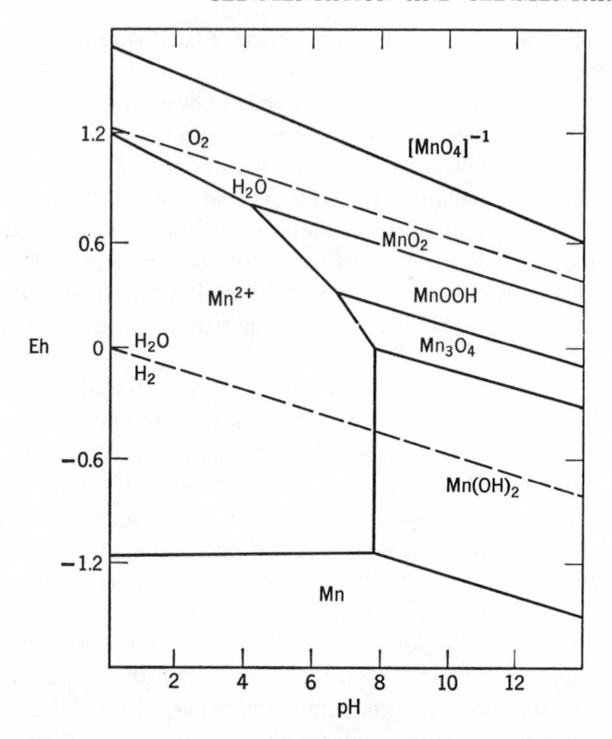

Figure 6.9 Eh-pH diagram for the system Mn-H_2O, for unit activity of Mn. The dashed lines are for the reactions for the decomposition of water (AA^1 and BB^1 of Figure 6.)

natural occurrences: the common form of iron in supergene deposits is hydrated ferric oxide; hydrated cobaltic oxide (stainierite) is found only where conditions have been strongly oxidizing; and the higher oxide of nickel is not known as a mineral. The separation of these three elements by supergene processes is well illustrated where intense weathering of ultrabasic rocks has given rise to lateritic material rich in Fe_2O_3, concentration of the nickel as garnierite and of the cobalt as hydrated cobaltic oxide or as cobaltian wad.

Oxidation processes also result in a similar separation of manganese from iron. Manganese is often present in solid solution in primary minerals containing iron, the ferrous and the manganous ions being mutually replaceable. Supergene processes, however, generally lead to fairly complete separation of iron from manganese, since the potential required to convert iron to the ferric state is much lower than that required to convert manganese to manganese dioxide. The iron readily precipitates as hydrated ferric oxide, whereas the manganese remains in solution longer and is eventually deposited under more oxidizing conditions as comparatively iron-free manganese dioxide.

Oxidation potentials and pH are the basic controls which determine the nature of many sedimentary products. Krumbein and Garrels (1952) have devised an ingenious diagram which illustrates the relation between these factors and the geological materials on which they act (Figure 6.10). In this diagram they develop the concept of the "geochemical fence," a boundary defined by the presence of a particular mineral or material on one side and its absence on the other, in effect, by a certain chemical reaction. A particular geochemical fence may represent a specific pH value or a specific oxidation potential or a combination of both factors. Many geochemical fences can be defined, but Krumbein and Garrels have shown that the most generally useful ones in considering sedimentary processes are the neutral fence, at pH = 7; the limestone fence, at pH = 7.8 (at higher pH calcite is readily deposited, at lower pH it tends to dissolve); the sulfate-sulfide fence, determined by the sulfide-sulfate oxidation potential; the Fe, Mn oxide-carbonate fence, determined by the oxidation potential at which ferrous and manganous compounds (mainly carbonates in the sedimentary environment) oxidize to the higher oxides; and the organic matter fence, below which organic matter is stable and above which it oxidizes to carbon dioxide. As a result, we obtain a classification of sedimentary environments based on the two significant parameters of pH and oxidation potential, and the careful study of the mineralogy of a sedimentary deposit will elucidate the physicochemical conditions under which it developed.

COLLOIDS AND COLLOIDAL PROCESSES

The colloidal state is one of fine subdivision, the size range being approximately 10^{-3} to 10^{-6} mm. Colloidal solutions grade into true solutions on the one hand and into suspensions on the other without any distinct line of demarcation. The degree of dispersion is usually greater than the resolving power of an ordinary microscope but is less than molecular; that is, colloidal particles are generally multimolecular. The colloid particles of the *disperse* phase are separated by the *dispersion* medium; the whole may be referred to as a disperse system. There are a number of different types of disperse or colloidal systems [solid-gas (smokes), liquid-gas (fogs), liquid-liquid (emulsions)], but the important type in sedimentary processes is a solid-liquid system (sols, gels, and pastes) in which the liquid is water. Sols are systems that resemble liquids in their physical properties; thus they flow readily and do not show rigidity. Gels, on the other hand, show some

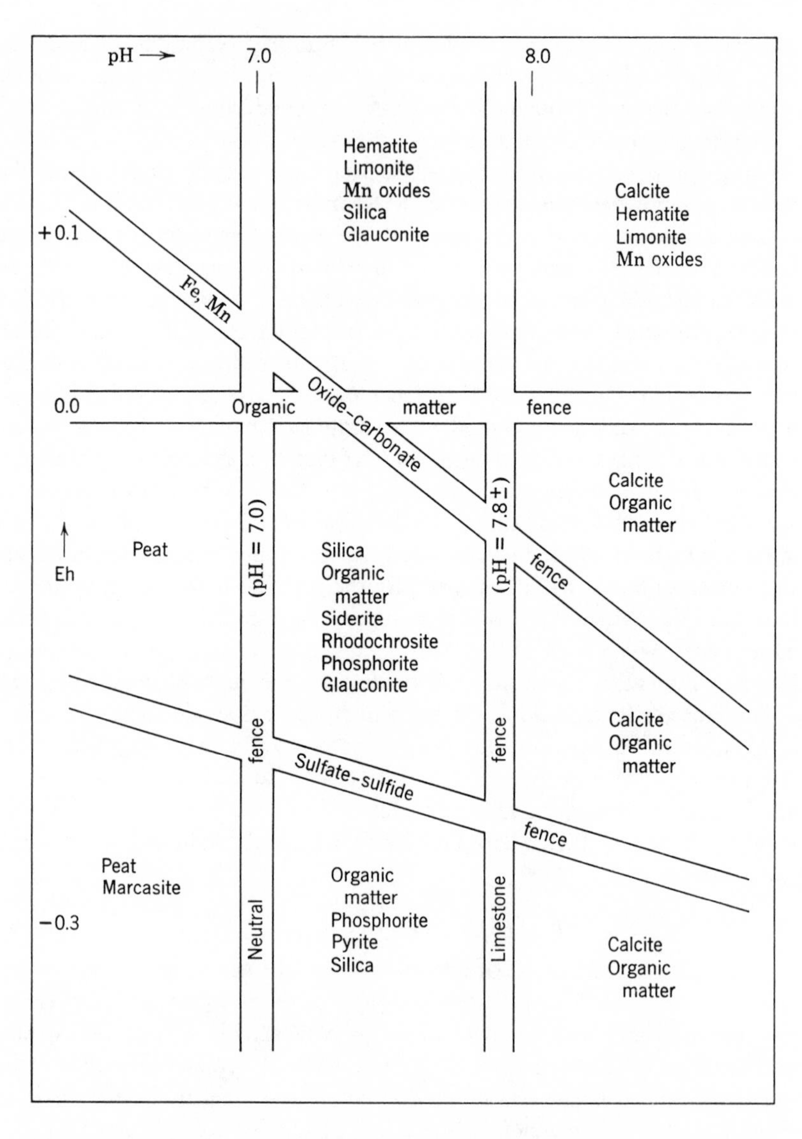

Figure 6.10 Sedimentary associations in relation to environmental limitations imposed by oxidation potential and pH. (After Krumbein and Garrels, *J. Geol.* **60,** 26, 1952)

rigidity. Pastes are systems in which, although the solid part is in the form of discrete particles, the concentration of these particles has been so much increased that they form the bulk of the system, for example, plastic clay.

Sols in water are divided into two types, hydrophilic and hydrophobic. In the first type there is strong interaction between the particles and the water molecules, which serves to stabilize the colloidal solution. In the second type there is no such attraction between the particles and the water molecules, and they are thus much less stable than hydrophilic sols and more easily precipitated. In general the particles in hydrophobic sols are larger than those in hydrophilic sols. Further, when a hydrophobic sol has been precipitated by some physical change it cannot be readily reconverted into a sol by reversing this change, whereas a hydrophilic sol is generally reversible in this respect. No sharp boundary exists between these two types of sols, but the division is useful. Silica is an example of a substance that forms a hydrophilic sol; aluminum hydroxide, on the other hand, forms a hydrophobic sol.

Colloidal particles are electrically charged. This charge may originate in two ways: either by adsorption of ions from the liquid or by the direct ionization of the material of the particle. Some colloids, for example, ferric hydroxide, may acquire either positive or negative charges according to the environment in which they are formed. The charges on some important colloids are as follows:

Positive	Negative
Aluminum hydroxide	Silica
Ferric hydroxide	Ferric hydroxide
Chromic hydroxide	Vanadium pentoxide hydrate
Titanium dioxide hydrate	Manganese dioxide hydrate
Thorium dioxide hydrate	Humus colloids
Zirconium dioxide hydrate	Sulfide sols

Colloidal particles may be either crystalline or amorphous. Most inorganic colloids are crystalline. The establishment of this fact has been an important contribution of X-ray diffraction techniques to the study of colloids.

Colloids may be produced by two procedures; either coarser particles may be broken down to colloidal dimensions, or smaller particles (molecules or ions) may be aggregated into particles of colloidal size. Most naturally formed colloids probably originated in the second way.

The colloidal state is always a metastable one. An increase in the size of the particles will lead to a decrease in the total surface area, and hence a decrease in the free energy of the system, so that all colloids are theoretically unstable, although they may remain unchanged over long periods. The charge on the particles is an important factor in the stability of a colloid, and this generally requires the presence of small concentrations of electrolytes. Large amounts of electrolytes, however, are precipitants of colloids; on this account most colloidal matter is rapidly flocculated by sea water. Colloids show great differences in relative stability. Some are stable under wide variations in chemical and physical environments; others require carefully controlled conditions in order to exist in the colloidal state at all. Obviously only those substances that form rather stable colloids are of importance in geological processes. Unstable colloids can, however, be stabilized by the presence of other substances. Of these the most important are organic compounds. In the geological environment such compounds are generally referred to as humus colloids, for lack of more precise information as to their nature; they are probably albumins. The presence of these humus colloids appears to play an important part in stabilizing the inorganic colloids formed during sedimentary processes, thus enabling transport of such material over much greater distances than would otherwise be possible.

Because silica is the most abundant material in the earth's crust and it readily goes into colloidal solution, the role of colloidal silica in sedimentary processes has been the subject of a great deal of discussion. Much of the geological literature relating to the transportation and precipitation of dissolved silica has been essentially unanimous as to the colloidal state of the silica. However, the development of colorimetric tests that permit discrimination between colloidal silica and silica in ionic solution has shown that this conclusion is based on inadequate evidence. These tests have shown that most of the silica in natural waters is in true solution. The significance of this fact for the dissolution and precipitation of silica during sedimentation has been carefully discussed by Krauskopf (1956). He shows that the origin of chert may be plausibly ascribed to dissolution of remains of siliceous organisms and to reprecipitation of the silica (initially in an amorphous state), but not in general to direct inorganic precipitation.

An important property of colloidal particles is their ability to bind and concentrate certain substances through adsorption. Two types of adsorption are recognized: (*a*) physical or van der Waals' adsorption, and (*b*) chemical adsorption. Both types may act together, and all gradations between extremes exist. Physical adsorption is characterized by low heats of adsorption and by a loose bonding of the adsorbate to the adsorbent. Chemical adsorption, or chemisorption, on the other hand, is characterized by high heats of adsorption and a firm chemical bonding (i.e., by valence bonds) of the adsorbate. It may involve the bonding of a foreign cation or a foreign anion, or both, to open bonds at the surface of the adsorbent; or it may be the exchange or substitution of a foreign cation or anion, or both, for a cation or anion at the surface. The property of base exchange, shown particularly by clays, whereby cations in the clay may be exchanged for other cations present in aqueous solutions in contact with the clay minerals, may be considered as a particular type of adsorption. Some principles governing adsorption may be stated as follows:

1. The amount of adsorption increases as the grain size of the adsorbent decreases, and hence its surface area increases.
2. Adsorption is favored if the adsorbate forms a compound of low solubility with the adsorbent (an example is the adsorption of phosphate ions by ferric hydroxide).
3. The amount of a substance adsorbed from solution increases with its concentration in that solution.
4. Highly charged ions are adsorbed more readily than lower charged ions.

Through adsorption processes many ions may be removed from natural waters. The clay minerals, especially montmorillonite, show a marked adsorptive capacity; the chemical adsorption of potassium ions by montmorillonite may result in the formation of illite. Many complex ions (for example, those containing arsenic and molybdenum) and ions of the heavy metals are also adsorbed and thus removed from solution by natural colloids. Such elements are often enriched in sedimentary iron and manganese ores. Here we are dealing with a systematic "depoisoning" of the hydrosphere (analogous to the "depoisoning" effect of freshly precipitated ferric hydroxide, utilized in medicine); without this phenomenon a number of biologically damaging elements would accumulate in ocean waters. Sufficient amounts of many elements, such as copper, selenium, arsenic, and lead, have potentially been supplied by weathering and erosion during geological time to cause serious poisoning of the ocean had not some process of elimination of these substances been active.

PRODUCTS OF SEDIMENTATION

A cursory examination of sedimentary processes suggests that they would tend to produce an average mixture of the individual components present in the parent material and thus work against any chemical differentiation. This, however, is not the case; weathering, erosion, and sedimentation lead generally to a marked separation of the major elements. As Goldschmidt pointed out, the cycle of matter at the earth's surface can be likened to a chemical analysis and to a quantitative analysis at that. The chemical differentiation that results is remarkable. The steps in this geochemical separation process are as follows:

1. Minerals that are especially resistant to chemical and mechanical breakdown collect as granular material. Of these, the commonest is quartz, and the product is a quartz sand or a sandstone showing a marked enrichment in silicon with respect to the parent material. This may be compared to the separation of silica in the first stage of a rock analysis.

2. Accumulation of the products of chemical breakdown of aluminosilicates, giving a mud consisting essentially of the clay minerals. This results in concentration of aluminum and also of potassium by adsorption. The process corresponds to the second step in a rock analysis, the separation of alumina and other easily hydrolyzed bases.

3. Along with the formation of argillaceous sediments, but often separated in space and time, iron is precipitated as ferric hydroxide. In this process oxidation from the ferrous to the ferric state precedes precipitation by hydrolysis. Concentration of iron is the result, sometimes to the extent of the formation of iron ores.

4. Calcium is precipitated as calcium carbonate either by purely inorganic processes or by the action of organisms. Limestones are formed and calcium thereby concentrated. This may result in almost quantitative separation of calcium, as in a chemical analysis. Limestone can be partly or wholly converted to dolomite by the metasomatic action of magnesium-rich solutions and magnesium thereby precipitated and concentrated together with calcium.

5. The bases that remain in solution collect in the ocean, from which they are removed in quantity only by evaporation, giving rise to salt deposits. The most important of these bases is of course sodium, but lesser amounts of potassium and magnesium also accumulate in sea water.

The chemical breakdown of a rock by weathering can be represented by the following scheme:

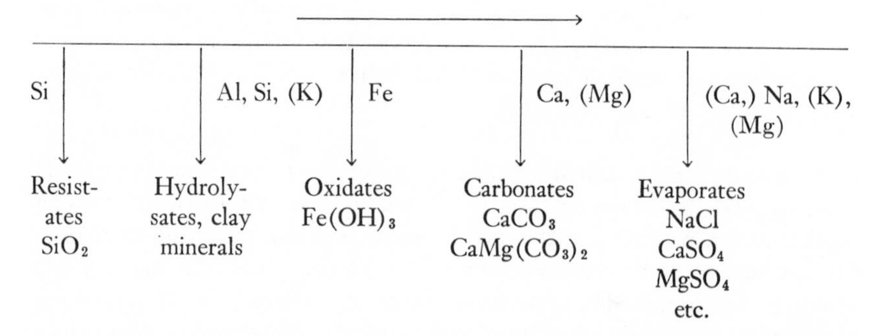

This scheme indicates the course followed by the major elements during sedimentation and gives a useful geochemical classification of sediments into resistates, hydrolysates, oxidates, carbonates, and evaporates. Goldschmidt recognized a further class, the reduzates, which includes coal, oil, sedimentary sulfides, and sedimentary sulfur. Coal and oil are of organic origin and are discussed in the chapter dealing with the geochemistry of the biosphere. Sedimentary sulfides and sulfur also often owe their formation to organic processes, although possibly indirectly.

The fate of the major elements during sedimentation has been fairly well worked out. Silica concentrates in the resistate sediments, alumina in the hydrolysates, iron and manganese in the oxidates, and calcium and magnesium in carbonates. A good part of the sodium remains in solution and eventually accumulates in the ocean; potassium is adsorbed by the clays and may form the minerals illite and glauconite. The fate of the minor elements during sedimentation has not been thoroughly investigated, and much less is known about their behavior under these circumstances than during magmatism. This is in part due to a lesser interest in the chemistry of sedimentary rocks and in part to the lack of guiding principles, such as are provided for igneous rocks by Goldschmidt's empirical rules for capture and admittance of ions by crystal lattices. Sedimentary processes are more complex in this respect, and many factors may play a part in determining the transportation and deposition of an element; they include ionic potential, *p*H and oxidation potential of the medium, colloidal properties, and adsorption. Hence it is not yet possible to make categorical statements regarding the fate of many of the minor elements during sedimentation.

Turekian and Wedepohl have prepared a critical compilation of the available data for the three principal groups of sedimentary rocks—shales, sandstones, and carbonates. Their figures are given in Table 6.5, together with the averages for igneous rocks for comparison. The data indicate that the common sedimentary rocks seldom show marked enrichment in minor and trace elements over the amounts present in igneous rocks—in fact, sandstones and carbonates are usually depleted in these elements. Most trace elements are somewhat more concentrated in shales than in other types of sedimentary rocks, but the amount in ordinary shales seldom exceeds the igneous rock average.

The resistates form the important group of sands and sandstones. Quartz is far and away the commonest and most abundant of residual minerals, and sands and sandstones are sources of silica for industrial uses. Many other minerals may appear in small amounts in sands and sandstones, but most of these can be decomposed and removed by intense weathering. Zircon is one of the most persistent of minerals, and the main ore deposits of zircon are sands from which it can be profitably separated. Magnetite and ilmenite are fairly resistant minerals and accumulate in sands; other industrially important constituents of some sands are rutile, monazite, cassiterite, and, of course, gold and the platinum metals.

Table 6.5. Abundances of the Elements (in parts per million) in the Principal Types of Sedimentary Rocks

Element	Shales	Sandstones	Carbonates	Igneous Rocks
Li	66	15	5	20
Be	3	0.*X*	0.*X*	2.8
B	100	35	20	10
F	740	270	330	625
Na	9,600	3,300	400	28,300
Mg	15,000	7,000	47,000	20,900
Al	80,000	25,000	4,200	81,300
Si	73,000	368,000	24,000	277,200
P	700	170	400	1,050
S	2,400	240	1,200	260
Cl	180	10	150	130
K	26,600	10,700	2,700	25,900
Ca	22,100	39,100	302,300	36,300
Sc	13	1	1	22
Ti	4,600	1,500	400	4,400
V	130	20	20	135
Cr	90	35	11	100
Mn	850	*X*0	1,100	950
Fe	47,200	9,800	3,800	50,000
Co	19	0.3	0.1	25
Ni	68	2	20	75
Cu	45	*X*	4	55
Zn	95	16	20	70
Ga	19	12	4	15
Ge	1.6	0.8	0.2	1.5
As	13	1	1	1.8
Se	0.6	0.05	0.08	0.05
Br	4	1	6.2	2.5
Rb	140	60	3	90
Sr	300	20	610	375
Y	26	40	30	33
Zr	160	220	19	165
Nb	11	0.0*X*	0.3	20
Mo	2.6	0.2	0.4	1.5
Ag	0.07	0.0*X*	0.0*X*	0.07
Cd	0.3	0.0*X*	0.035	0.2
In	0.1	0.0*X*	0.0*X*	0.1
Sn	6.0	0.*X*	0.*X*	2
Sb	1.5	0.0*X*	0.2	0.2
I	2.2	1.7	1.2	0.5
Cs	5	0.*X*	0.*X*	3
Ba	580	*X*0	10	425

Table 6.5. Abundances of the Elements (in parts per million) in the Principal Types of Sedimentary Rocks (Continued)

Element	Shales	Sandstones	Carbonates	Igneous Rocks
La	92	30	*X*	30
Ce	59	92	11.5	60
Pr	5.6	8.8	1.1	8.2
Nd	24	37	4.7	28
Sm	6.4	10	1.3	6.0
Eu	1.0	1.6	0.2	1.2
Gd	6.4	10	1.3	5.4
Tb	1.0	1.6	0.2	0.9
Dy	4.6	7.2	0.9	3.0
Ho	1.2	2.0	0.3	1.2
Er	2.5	4.0	0.5	2.8
Tm	0.2	0.3	0.04	0.5
Yb	2.6	4.0	0.5	3.4
Lu	0.7	1.2	0.2	0.5
Hf	2.8	3.9	0.3	3
Ta	0.8	0.0*X*	0.0*X*	2
W	1.8	1.6	0.6	1.5
Hg	0.4	0.03	0.04	0.08
Tl	1.4	0.8	0.0*X*	0.5
Pb	20	7	9	13
Th	12	1.7	1.7	9.6
U	3.7	0.45	2.2	2.7

Turekian and Wedepohl, *Bull. Geol. Soc. Am.*, **72,** 175, 1961. The figures for igneous rocks are given for comparison. (For some elements only order of magnitude estimates are made; these are indicated by the symbol *X*.

The hydrolysate sediments consist in great part of the clay minerals. Tropical weathering often produces aluminum hydroxides rather than hydrated aluminum silicates, and high-alumina clays and bauxites result. In either process the end product represents a concentration of aluminum over the average amount in the earth's crust. As was indicated in the discussion of ionic potential, many elements, especially those in groups III and IV of the periodic table, may be expected to precipitate in hydrolysate sediments, and the meager data on minor elements in sedimentary rocks bear this out. Shales show concentrations of elements of medium ionic potential and of elements like potassium that are readily adsorbed by colloidal particles; they are sometimes enriched in chalcophile elements, probably precipitated as sulfides by the H_2S often generated in marine muds. The most remarkable hydrolysate sediments from the geochemical viewpoint are the black bituminous shales and the bauxites, both of which have originated under rather special conditions. The black shales were deposited slowly in a strongly reducing marine environment rich in organic matter. Sulfide ions were evidently present, produced by the reduction of sulfate. Analyses of black shale are characterized not only by a considerable content of organic carbon, but generally also by much sulfur, present mainly as FeS_2. Enrichment of the following minor elements has been noted: V, U, As, Sb, Mo, Cu, Ni, Cd, Ag, Au, and metals of the platinum group. Vanadium has been produced commercially from such shales, and they are far more significant than all the primary deposits as a potential source of large amounts of uranium. It has been suggested that the minor elements were accumulated by the vital activity of the organisms now represented by bituminous material, but this is far from certain. Judging from the chalcophile nature of many of these elements, precipitation from solution as sulfides seems a more reasonable explanation. A linear increase of uranium with increasing carbon content has been demonstrated in some of these shales, but this does not necessarily imply that the uranium was present in the organisms that furnished the carbon. The uranium content also shows an excellent correlation with the abundance of colloidal size grades in the sediment; this might suggest that the uranium is present in the clay mineral, which in these black shales is generally illite. The evidence indicates that the concentration of uranium is not the result of biological activity but of later chemical processes probably related, in part at least, to the presence of organic matter in the sediments. Some phosphatic shales, such as those of the Phosphoria formation in Wyoming, Idaho, and Montana, show similar geochemical features, especially in the enrichment in vanadium and uranium; they seem to have been deposited under similar conditions, that is, oxygen-deficient marine environments where organic material was accumulating and the rate of sedimentation was very slow. Bauxites have a different pattern of enrichment and often show a concentration of beryllium, gallium, niobium, and titanium; of these gallium is actually being extracted as a byproduct from the production of aluminum.

The most important oxidate is ferric hydroxide, which, if pure, gives rise to a sedimentary iron ore. Manganese is also deposited as an oxidate sediment in the form of hydrated manganese dioxide, and such deposits or their metamorphosed equivalents are the significant sources of manganese ore. The adsorptive power of precipitated ferric hydroxide and manganese dioxide hydrate is very great; hence many minor elements are found in oxidate sediments. The pattern of enrichment differs; hydrated manganese dioxide, being a negatively charged colloid, adsorbs cations, whereas ferric hydroxide, generally a positively charged colloid, adsorbs anions. These adsorbed elements are sometimes in sufficient amounts to be important commercially, either as profitable byproducts (e.g., Ni in some sedimentary iron ores, W in some manganese ores) or as deleterious impurities (e.g., As in some iron ores). Vanadium, phosphorus, arsenic, antimony, and selenium have been reported in sedimentary iron ores in larger amounts than their average abundance in the crust; Li, K, Ba, B, Ti, Co, Ni, Cu, Mo, As, V, Zn, Pb, and W have been reported in notable concentrations in manganese ore.

The common carbonate sediment is, of course, limestone, which consists essentially of calcite. Calcium carbonate may also be deposited as aragonite, but it is doubtful whether aragonite will persist for any considerable time in a geological formation, since it tends to change to calcite. Whether calcium carbonate was originally deposited as calcite or as aragonite may have significant geochemical consequences; the structure of aragonite permits ready substitution by larger cations, such as strontium and lead, but not the smaller cations, whereas for calcite the reverse is true. Hence the minor elements in a limestone will probably differ in kind and amount according to the nature of the calcium carbonate in the original sediment.

The evaporates are quantitatively unimportant as sediments but are highly significant in the interpretation of geological history. Geochemically they are of especial interest as a type of deposit with a mode of formation which can readily be reproduced in the laboratory. They have been well described by Stewart (1963). As early as 1849 Usiglio made experiments aimed at elucidating the conditions of formation of salt deposits, but his results were unsatisfactory, because he worked with sea water, a highly complex solution with which he failed to get reproducible results. Later the problem was tackled from the other direction by van't Hoff and his coworkers, who began by studying the solubility relations of all the possible compounds that might be produced by the evaporation of sea water. Working initially at 25°, they determined the equilibrium relations in the simple two-component salt-water systems and then extended these researches to multicomponent systems. Similar investigations were made at 83°, and specific reactions involving the appearance or disappearance of individual compounds were studied at the temperature of reaction. Van't Hoff's success in working out phase relations and applying these results to natural occurrences of evaporates (especially the Stassfurt deposits) was one of the first fruits of the application of physicochemical principles (in this case the phase rule) to geological problems.

As sea water evaporates under natural conditions, calcium carbonate is the first solid to separate. The precipitation of calcium carbonate may be followed by that of dolomite, but there is no evidence that extensive deposits

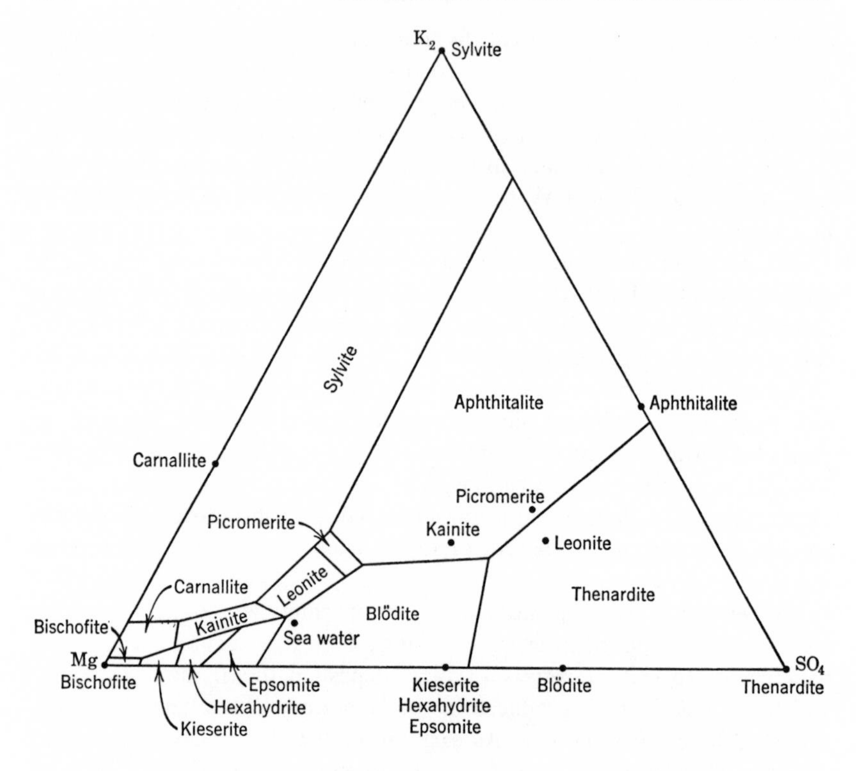

Figure 6.11 Stability fields of evaporate minerals at 25°. (After Stewart, 1963)

of dolomite have been formed in this way. Indeed, evaporation of sea water in a closed basin cannot give rise to thick carbonate deposits—sea water 1000 m deep would give only a few centimeters of limestone.

With continued evaporation calcium sulfate is deposited. Depending on temperature and salinity, either gypsum or anhydrite may be formed. In salt solutions of approximately the composition of sea water at 30° gypsum will begin to separate when the salinity has increased to 3.35 times the normal value; after nearly one half the total amount of calcium sulfate has been deposited anhydrite becomes the stable phase. When the solution has been concentrated to one-tenth of the original bulk, halite starts to separate. Anhydrite and halite then precipitate together until the field of stability of polyhalite, $K_2Ca_2Mg(SO_4)_4 \cdot 2H_2O$, is reached.

Most evaporate deposits contain calcium carbonate, calcium sulfate, and sodium chloride; evidently conditions under which other salts could be deposited have seldom been attained. Only when an evaporating body of sea water has been reduced to 1.54% of the original volume do potassium and magnesium salts begin to crystallize. Important deposits of these salts are worked in Germany, in the Texas–New Mexico area of the United States, in the province of Perm in the U.S.S.R., and in Saskatchewan, Canada.

The further course of crystallization can be discussed in terms of a triangular diagram having corners which represent K_2, Mg, and SO_4 and which shows the compounds in equilibrium with halite and saturated solutions at a particular temperature. Figure 6.11 is such a diagram for 25°. It shows that the continued isothermal evaporation of sea water leads to the crystallization of blödite, $Na_2Mg(SO_4)_2 \cdot 4H_2O$. The composition of the solution follows a path directly away from the point corresponding to blödite and soon crosses the boundary into the field of epsomite, $MgSO_4 \cdot 7H_2O$. Epsomite then begins to crystallize, and the further course depends upon whether the previously separated blödite can react with the solution. If it does, it will be resorbed; under natural conditions, however, it may be crusted over and effectively removed from the system. In either event, the path of crystallization eventually reaches the boundary between the fields of epsomite and kainite, $KMgSO_4Cl \cdot 3H_2O$, and then follows this boundary curve, the two salts crystallizing together. The succession is hexahydrite-kainite, kieserite-kainite, kieserite-carnallite, and finally the three salts kieserite-carnallite-bischofite separate until evaporation is complete. During all this, halite and small amounts of calcium salts are still being deposited. Polyhalite, which began to crystallize almost contemporaneously with halite, ceases to form during the course of separation of kainite.

The accompanying table illustrates the theoretical profile to be expected from the evaporation of sea water at 25°. This theoretical profile shows a

Kieserite, carnallite, bischofite			Bischofite zone
Kieserite, carnallite	Anhydrite		Carnallite zone
Kieserite, kainite			Kainite zone
Hexahydrite, kainite			
Epsomite, kainite		Halite	
Epsomite			Potash-free magnesium sulfate zone
Blödite	Polyhalite		
Polyhalite			Polyhalite zone
Anhydrite			Anhydrite zone
Gypsum			Gypsum zone
Carbonates			Basal limestone and dolomite

general correspondence with the succession in natural salt deposits in most areas. Commonly one finds limestones and dolomites passing up into calcium sulfate and halite, with or without polyhalite. The usual form of calcium sulfate in natural deposits is anhydrite rather than gypsum. The correspondence with the higher parts of the theoretical profile is less marked, however. For example, sylvite, which does not appear in this profile, is the important potassium salt in many deposits. Epsomite and hexahydrite have generally lost water of crystallization and have been converted into kieserite. A bischofite zone is rarely developed, probably because complete evaporation is seldom if ever achieved in nature. Van't Hoff's investigations at higher temperatures showed that most of the mineralogical features of natural deposits can be explained by assuming either a temperature of evaporation considerably higher than 25° or recrystallization at some higher temperatures after burial—a mild thermal metamorphism. The latter suggestion is highly plausible.

TOTAL AMOUNT OF SEDIMENTATION

The total amount of sedimentation during geological time is clearly a figure of great importance for quantitative geochemistry, and a number of attempts have been made to calculate it. Such calculations are generally based upon data regarding the amount and composition of ocean water and the average composition of igneous and sedimentary rocks. If we can assume that the total sodium content of sea water is derived from the weathering and erosion of igneous rocks, this quantity will be proportional to the total amount of sediments. Clarke used the following data:

Sodium content of the ocean	1.14%
Average sodium content of igneous rocks	2.83%
Composition of the 10-mile crust	93% lithosphere, 7% hydrosphere

From these figures the ratio between oceanic sodium and rock sodium is about 1:30. Hence the sodium in the ocean is equivalent to that contained in 1/30 of all the igneous rocks of the earth's crust to a depth of 10 miles. This fraction amounts to 54.8×10^6 cubic miles. However, the calculation is subject to correction, for it rests on the assumption that all the sodium accumulates in the sea. This is not so; the sedimentary rocks contain sodium, even if in much smaller amount than the parent igneous rocks. Clarke put the average sodium content of sedimentary rocks at 0.90%, or 35% of the average in igneous rocks. Thus, of the original sodium, only about 65% has come to rest in the sea. The estimate of the amount of igneous rocks required to give the present sodium content of the ocean must therefore be increased by a factor of 100/65, from 54.8×10^6 cubic miles to 84.3×10^6 cubic miles. An allowance for a 10% increase of volume of sedimentary rocks from the parent igneous material by oxidation, carbonation, and hydration gives a figure of about 93×10^6 cubic miles, or 3.7×10^8 cubic kilometers for the total volume of sediments produced during geological time. This corresponds to a rock shell nearly 2500 feet thick enveloping the whole earth. If the material were confined to the continental platforms (roughly one third of the area of the globe), its thickness would be about 7300 feet.

Goldschmidt used a somewhat different procedure to calculate the total amount of igneous rock weathered during geological time and the amounts of the different sediments. For each square centimeter of the earth's surface, there are 278 kg of sea water; and because sea water contains 1.07% sodium (a later figure than Clarke's 1.14%), the 278 kg contains 2.975 kg of sodium. The average sodium content of igneous rocks is 2.83% and of sedimentary deposits approximately 1%. In the process of weathering a certain amount of the material is leached away, and Goldschmidt estimated that the mass of the sedimentary deposits is 0.97 of the original igneous rocks that gave rise to them.

Let X be the amount of igneous rock eroded per square centimeter of earth's surface. Let Y be the amount of clastic sediments deposited per square centimeter of earth's surface. Then, $Y = 0.97X$.

$$\text{Sodium content of igneous rock per cm}^2 = \frac{2.83}{100} \times X$$

$$\text{Sodium content of clastic rock per cm}^2 = \frac{1}{100} \times Y$$

However, the sodium content of ocean water per square centimeter = 2.975 kg. Therefore,

$$\frac{2.83}{100}X - \frac{Y}{100} = 2.975$$

and $X = 160 \text{ kg/cm}^2$
$Y = 155 \text{ kg/cm}^2$

The value of 160 kg/cm² gives a figure of about 3×10^8 km³ for the total amount of igneous rock weathered during geological time, a value close to that obtained by Clarke.

These calculations overlook, on the one hand, the sodium removed from the ocean during geological time in the form of salt deposits and, on the other, that added by way of volcanic exhalations and in solution in magmatic waters. Both these items are probably small in relation to the sodium cycle as a whole, and they work in opposite directions. In addition, of course, the calculations fail to take into account the sodium that may have been present in the primitive ocean; any such sodium would reduce the amount of weathering necessary to produce the present sodium content of sea water.

Goldschmidt also calculated the quantity of calcium and magnesium carbonate in sedimentary rocks. He estimated the average content of noncarbonate CaO in sandstones and shales to be 0.6% and concluded therefore that CaO in excess of this figure is present as calcium carbonate. Similarly, the average amount of noncarbonate MgO was estimated to be 2.6%. A balance sheet for the cycle of calcium and magnesium in sedimentation was then derived as follows:

In the ocean Ca $0.00042 \times 278 = 0.117$ kg/cm²
Mg $0.00130 \times 278 = 0.361$ kg/cm²

	CaO	MgO
In igneous rocks	5.08%	3.49%
In carbonate-free sediments	0.6%	2.6%
160 kg igneous rocks contain	8.128 kg	5.584 kg
155 kg carbonate-free sediments contain	0.930 kg	4.030 kg
278 kg sea water contain	0.117 kg Ca	0.361 kg Mg
corresponding to	0.164 kg CaO	0.598 kg MgO

Therefore, the following amounts of calcium and magnesium carbonates (per square centimeter) must be present in the sediments:

7.034 kg CaO	0.956 kg MgO
5.519 kg CO_2	1.043 kg CO_2
12.553 kg $CaCO_3$	1.999 kg $MgCO_3$

Assuming that all the magnesium carbonate in sediments is there as dolomite, $CaMg(CO_3)_2$, we obtain these figures: 10.170 kg $CaCO_3$, 4.372 kg $CaMg(CO_3)_2$. Hence the total amount of sedimentary rocks per square centimeter of the earth's surface is 155 kg clastics, 10.2 kg limestone, and 4.4 kg dolomite; the combined CO_2 in sediments is 6.562 kg/cm². From the above figures and the densities of these rock types the total average thickness of sediments can be calculated.

	Mass/cm²	Density	Volume (cm³)	Thickness
Clastics (sandstones and shales)	155 kg	2.65	58,491	585 m
Limestone	10.2 kg	2.7	3,777	38 m
Dolomite	4.4 kg	2.9	1,517	15 m
	169.6 kg			639 m

Clarke's calculation for the total sedimentary rocks of the globe gave an approximate thickness of 2500 feet or 762 meters. The agreement between the figures of Goldschmidt and Clarke is good, considering the assumptions involved in both cases.

Goldschmidt's computations have been extended and refined by more recent workers, but the results obtained are not significantly different. The importance of calculations of this kind lies not in precisely determining the amounts of each sedimentary class but rather in elucidating the general features of the sedimentary cycle. The figures obtained are open to criticism. The results of the geochemical computations are not in agreement with field measurements of the relative percentages of sedimentary rock types. Barth considers the assumptions on which the calculations have been made are unsound. Kuenen has concluded from several lines of evidence that the total amount of sediments is considerably greater than that estimated from the geochemical data. He believes that the major source of error in previous calculations is the figure for the average sodium content of sedimentary rocks, which he considers too low. Kuenen estimates that the total volume of sediments is about 7×10^8 km^3, of which 5×10^8 km^3 is in deep-sea deposits, mainly red clay. He stresses the importance of the deep ocean basins as collecting grounds for vast amounts of sediment throughout geological time and predicts that the average thickness of such sediments is about 3 km. However, this prediction has not been borne out by recent seismic studies of the floor of the Pacific Ocean. These studies indicate that the thickness of sediments on the deep ocean floor is nearly everywhere less than half a kilometer. If this represents the total accumulation throughout geological time, then the rate of accumulation has been much less than estimated by Kuenen.

A number of estimates for the relative amounts of the common sediments are collected in the accompanying table. The lack of agreement is due to

	1	2	3	4	5
Shale	91	80	82	46	57
Sandstone		15	12	32	14
Limestone (and dolomite)	9	5	6	22	29

[1] Goldschmidt, 1933.
[2] Clarke, 1924.
[3] Leith and Mead, 1915.
[4] Leith and Mead, 1915 (708,000 feet of sediments in North America, Europe, and Asia).
[5] Kuenen, *Am. J. Sci.* **239,** 168, 1941 (for the sedimentary rocks of the East Indies).

several causes. The estimates derived from actual measurements of stratigraphic sections are weighted in favor of the proportions of these sediments on the continental shelves, which are not the same as those for the earth as a whole; Kuenen points out that the deposits of the deep sea, which hardly appear in terrestrial outcrops, are by no means insignificant. Since calcareous deposits form a greater proportion of the shelf sediments and red clay those of the deep sea, the sedimentary rocks show more limestone and less argillaceous material than the average for the whole earth. The identification of a formation as a limestone, a sandstone, or a shale is liable to be very gross; shales usually contain considerable sand, sandstones may carry much clay, and the term limestone is applied to many rocks with only 50% or so of carbonate. Hence the disparity between the different estimates is not surprising, and the agreement is more remarkable than the disparity. It is clear, however, that limestones are more prominent in the geological column than one might expect from geochemical calculations; this is certainly significant and confirms that shallow-water environments are the great places of carbonate deposition, whereas much clayey matter is permanently deposited in the ocean deeps.

SELECTED REFERENCES

Barth, T. F. W. (1962). *Theoretical petrology.* Parts II and V.

Bear, F. E. (ed.) (1964). *Chemistry of the soil* (second edition) 515 pp. Reinhold Publishing Corp., New York. A comprehensive monograph by several authors on the different aspects of soil geochemistry; the article on trace elements by R. L. Mitchell is very informative.

Braitsch, O. (1962). *Entstehung und Stoffbestand der Salzlagerstätten.* 232 pp. Springer Verlag, Berlin. A comprehensive discussion of salt deposits and the physical chemistry of their formation.

Clarke, F. W. (1924). *The data of geochemistry.* Chapters 7, 12, and 13.

Cooper, L. H. N. (1937). Some conditions governing the solubility of iron. *Proc. Roy. Soc. (London)* **B124,** 229–307. A presentation of the physicochemical factors controlling the solubility of iron in sea water, natural waters, and in many physiological fluids.

Degens, E. T. (1965). *Geochemistry of sediments.* 342 pp. Prentice-Hall Inc., Englewood Cliffs, N. J. A survey of the applications of low-temperature geochemistry to the study of sediments and sedimentary processes.

Garrels, R. M., and C. L. Christ (1965). *Solutions, minerals, and equilibria.* 450 pp. Harper and Row, New York. A comprehensive account of the principles of solution chemistry and thermodynamics as applied to mineral equilibria in sediments and sedimentary rocks.

Goldich, S. S. (1938). A study in rock weathering. *J. Geol.* **46,** 17–58. A carefully documented account of the chemical and mineralogical changes produced by weathering on a granite gneiss, two diabases, and an amphibolite.

Goldschmidt, V. M. (1954). *Geochemistry.* Chapter 4.

Grim, R. E. (1953). *Clay mineralogy.* 384 pp. McGraw-Hill Book Co., New York. An authoritative work on the structure, composition, properties, occurrence, and mode of origin of the various clay minerals.

Horn, M. K., and J. A. S. Adams (1966). Computer-derived geochemical balances and element abundances. *Geochim. Cosmochim. Acta* **30,** 279–298. A new approach to the determination of elemental abundances in the crust and in major sedimentary rock types, using a computer program.

Imbrie, J., and A. Poldervaart (1959). Mineral compositions calculated from chemical analyses of sedimentary rocks. *J. Sed. Petrol.* **29,** 588–595. This paper provides a carefully reasoned procedure for obtaining a quantitative mineralogical composition from the chemical analysis of a sedimentary rock.

James, H. L. (1966). Chemistry of the iron-rich sedimentary rocks. U. S. Geol. Surv. Prof. Paper 440-W, 61 pp. (*Data of Geochemistry*, (sixth edition), chapter W). A comprehensive account of the composition, distribution, and geochemistry of ironstones and iron-formations.

Keller, W. D. (1955). *The principles of chemical weathering.* 88 pp. Lucas Bros., Columbia, Missouri. An integrated review of the principles of chemical weathering and the significance of weathering in geology and pedology; excellent bibliography.

Krauskopf, K. B. (1955). Sedimentary deposits of rare metals. *Econ. Geol.* 50th Ann. Vol., 413–463. An exhaustive compilation of the data on the occurrence and amounts of rare metals in different kinds of sediments and sedimentary rocks and a discussion of the physicochemical factors involved.

Krauskopf, K. B. (1956). Dissolution and precipitation of silica at low temperatures. *Geochim. Cosmochim. Acta* **10,** 1–26. An account of experimental work on the solubility of silica at low temperatures and the geological applications of this work.

Krumbein, W. C., and R. M. Garrels (1952). Origin and classification of chemical sediments in terms of pH and oxidation-reduction potentials. *J. Geol.* **60,** 1–33. A careful study of the pH and oxidation potentials of the common sedimentary environments and the resultant mineralogy.

Kuenen, Ph. H. (1950). *Marine geology.* 568 pp. John Wiley and Sons, New York. A detailed account of the interpretation of sedimentary rocks in the light of conditions and processes in the marine environment.

Leith, C. K., and W. J. Mead (1915). *Metamorphic geology.* 357 pp. Henry Holt & Company, New York. In spite of the title, the first part of this book gives a careful account with many quantitative data on the chemical and mineralogical changes accompanying rock alteration under surface and near-surface conditions.

Niggli, P. (1952). *Gesteine und Minerallagerstätten, zweiter Band: Exogene Gesteine und Minerallagerstätten.* 557 pp. Verlag Birkhäuser, Basel. A comprehensive account of sedimentation and sedimentary rocks and of the associated mineral deposits.

Pettijohn, F. J. (1957). *Sedimentary rocks* (second edition). 718 pp. Harper & Brothers, New York. An excellent source of information on all aspects of sedimentary rocks, especially useful on account of the carefully chosen groups of analyses of different rock types.

Rankama, K., and Th. G. Sahama (1950). *Geochemistry.* Pp. 189–242.

Stewart, F. H. (1963). Marine evaporites. U. S. Geol. Surv. Prof. Paper 440-Y, 53 pp. [*Data of Geochemistry* (sixth edition), chapter Y]. A brief but comprehensive account, both physiochemical and geological, of marine salt deposits.

Yoder, H. S., and H. P. Eugster (1955). Synthetic and natural muscovites. *Geochim. Cosmochim. Acta* **8,** 225–280. An extensive study of the synthesis and stability of muscovite and the application of the data to the natural occurrence of this mineral; illite is shown to consist largely of clay-size muscovite or mixed-layer muscovite-montmorillonite.

CHAPTER SEVEN

THE HYDROSPHERE

THE NATURE OF THE HYDROSPHERE

The hydrosphere is the discontinuous shell of water—fresh, salt, and solid—at the surface of the earth. It comprises the oceans with their connected seas and gulfs, the lakes, the waters of the rivers and streams, ground water, and snow and ice. The oceans are clearly of first magnitude, for they cover an area of 361×10^6 km^2, or 70.8% of the earth's surface. From the mean depth of 3800 m, the volume of the ocean waters is 1372×10^6 km^3. At the surface the density of sea water of normal salinity at 0° is 1.028, and it increases with depth on account of the compressibility of water under increased pressure. Clarke accepts a mean density of 1.03, which when multiplied by the above volume gives 1413×10^{21} g for the total mass of the oceans.

It is more difficult to arrive at an accurate measure of the water in the other parts of the hydrosphere. Goldschmidt estimated that there are 273 liters of water in all its forms for every square centimeter of the earth's surface, made up as follows:

	Liters	Kilograms
Sea water	268.45	278.11
Fresh water	0.1	0.1
Continental ice	4.5	4.5
Water vapor	0.003	0.003

On the basis of these figures, the mass of fresh water and of continental ice works out at 0.5×10^{21} g and 22.8×10^{21} g, respectively. (Improved knowledge of the Antarctic ice has raised the figure for continental ice to approximately 33×10^{21} g.) Sea water thus comprises about 98% of the mass of the hydrosphere, and its composition can therefore be taken without serious error as giving an average composition of the hydrosphere, since the small amounts of fresh water and continental ice cannot affect the results significantly.

THE COMPOSITION OF SEA WATER

Two arbitrarily defined quantities, the chlorinity and the salinity, are commonly used in discussing the composition of sea water. The *chlorinity* is determined by the precipitation of the halides with a silver salt and is essentially the total amount in grams of chloride, bromide, and iodide contained in 1 kg of sea water, assuming that the bromide and iodide have been replaced by chloride. The *salinity* is also a defined quantity, slightly less than the total weight of dissolved solids per kilogram, and can be calculated from the chlorinity or determined from a measurement of the density. Both chlorinity and salinity are customarily expressed in grams per kilogram (g/kg) of sea water or parts per thousand (‰). In the open ocean the salinity averages about 35‰ but rises to as much as 40‰ in the Red Sea and the Persian Gulf, where evaporation is high and inflow and precipitation low. However, in all samples of sea water the relative proportions of the major ions are practically constant, and so the determination of one constituent provides a measure of the others. Because of the homogeneity of sea water the most accurate data of geochemistry are those regarding the ocean.

Precise knowledge of the average composition of sea water was provided by Dittmar, who in 1884 made careful analyses of 77 water samples, representative of all oceans and taken both from the surface and from the depths. These samples had been collected on the voyage around the world of H.M.S. *Challenger* (1872–1876). He determined the halides, sulfate, magnesium, calcium, and potassium. On composite samples he found the ratio of bromine to chlorine and estimated the carbonate. Sodium was calculated by difference from the sums of the chemical equivalents of the negative and positive ions. This procedure was followed because he was unable to achieve satisfactory direct determinations for sodium. Dittmar's work showed that there are no significant regional differences in the relative composition of sea water, and his average values can be used to represent the ratios between the major dissolved constituents. Since 1884 much research has been devoted to the chemical composition of sea water, and great advances have been made in the science of analytical chemistry. Nevertheless, the figures obtained by Dittmar agree closely with those accepted today as the best available, a remarkable tribute to his work.

The figures for the major constituents are given in Table 7.1 and are referred to a chlorinity of 19‰, which is taken as the standard concentration of sea water. Because of the constancy of the relative proportions of the

Table 7.1. The Major Dissolved Constituents of Sea Water

Ion	Cl = 19‰	Per Cent
Cl	18.980	55.05
Br	0.065	0.19
SO_4	2.649	7.68
HCO_3	0.140	0.41
F	0.001	0.00
H_3BO_3	0.026	0.07
Mg	1.272	3.69
Ca	0.400	1.16
Sr	0.008	0.03
K	0.380	1.10
Na	10.556	30.61
Total	34.477	99.99

major constituents any one may be used as a measure of the others, and chlorinity is that most readily determined. The complexity of sea water makes it impossible by direct chemical analysis to determine the total quantity of dissolved solids in a given sample. Furthermore, reproducible results cannot be obtained by evaporating sea water and weighing the residue, because some of the components, particularly chloride, are lost in the last stages of drying. Hence the use of indirect methods based on the chlorinity factor.

So far we have been considering only the major constituents. However, about seventy elements have been identified in sea water, and others are certainly present, although not detectable by the analytical methods used. Table 7.2 gives the available data on the amounts of the minor and trace elements. It should be realized that the quality of these data is not always satisfactory, because some figures are based on the analysis of a single sample and many of the samples were taken from inshore surface waters, which for minor and trace elements may not be representative of the ocean as a whole. The concentrations of some elements fluctuate from place to place as a result of biological activity. These fluctuations are often related to the depth at which the sample was taken, biological activity being greatest in surface and near-surface waters. Calcium may be relatively diminished in surface layers through abstraction by organisms. Silica is also removed from surface waters in this way, and its contents in sea water generally shows a regular increase with depth. The distribution of phosphorus is greatly affected by organic agencies; this element is markedly enriched in the deeper parts of the ocean as a result of the dissolution of dead organic matter.

Sea water also contains gases in solution. Since the atmosphere and the

Table 7.2. Elements Present in Solution in Sea Water Excluding Dissolved Gases

Element	Abundance, mg/l	Principal Species	Residence Time, Years
Li	0.17	Li^+	2.0×10^7
Be	0.0000006		1.5×10^2
B	4.6	$B(OH)_3$; $B(OH)_2O^-$	
C	28	HCO_3^-; H_2CO_3; CO_3^{2-}; organic compounds	
N	0.5	NO_3^-; NO_2^-; NH_4^+; N_2 (g); organic compounds	
F	1.3	F^-	
Na	10,500	Na^+	2.6×10^8
Mg	1,350	Mg^{2+}; $MgSO_4$	4.5×10^7
Al	0.01		1.0×10^2
Si	3.0	$Si(OH)_4$; $Si(OH)_3O^-$	8.0×10^3
P	0.07	HPO_4^{2-}; $H^2PO_4^-$; PO_4^{3-}; H_3PO_4	
S	885	SO_4^{2-}	
Cl	19,000	Cl^-	
K	380	K^+	1.1×10^7
Ca	400	Ca^{2+}; $CaSO_4$	8.0×10^6
Sc	0.00004		5.6×10^3
Ti	0.001		1.6×10^2
V	0.002	$VO_2(OH)_3^{2-}$	1.0×10^4
Cr	0.00005		3.5×10^2
Mn	0.002	Mn^{2+}; $MnSO_4$	1.4×10^3
Fe	0.01	$Fe(OH)_3$ (s)	1.4×10^2
Co	0.0001	Co^{2+}; $CoSO_4$	1.8×10^4
Ni	0.002	Ni^{2+}; $NiSO_4$	1.8×10^4
Cu	0.003	Cu^{2+}; $CuSO_4$	5.0×10^4
Zn	0.01	Zn^{2+}; $ZnSO_4$	1.8×10^5
Ga	0.00003		1.4×10^3
Ge	0.00007	$Ge(OH)_4$; $Ge(OH)_3O^-$	7.0×10^3
As	0.003	$HAsO_4^{2-}$; $H_2AsO_4^-$; H_3AsO_4; H_3AsO_3	
Se	0.0004	SeO_4^{2-}	
Br	65	Br^-	
Rb	0.12	Rb^+	2.7×10^5
Sr	8.0	Sr^{2+}; $SrSO_4$	1.9×10^7
Y	0.0003		7.5×10^3
Zr			
Nb	0.00001		3.0×10^2
Mo	0.01	MoO_4^{2-}	5.0×10^5
Ru			
Rh			

Table 7.2. Elements Present in Solution in Sea Water Excluding Dissolved Gases (Continued)

Element	Abundance, mg/l	Principal Species	Residence Time, Years
Pd			
Ag	0.00004	$AgCl_2^-$; $AgCl_3^{2-}$	2.1×10^6
Cd	0.00011	Cd^{2+}; $CdCl^+$	5.0×10^5
In	<0.02		5.0×10^5
Sn	0.0008		5.0×10^5
Sb	0.0005		3.5×10^5
Te			
I	0.06	IO_3^-; I^-	
Cs	0.0005	Cs^+	4.0×10^4
Ba	0.03	Ba^{2+}; $BaSO_4$	8.4×10^4
La	0.0000029		0.44×10^3
Ce	0.0000013		0.08×10^3
Pr	0.00000064		0.32×10^3
Nd	0.0000023		0.27×10^3
Sm	0.00000042		0.18×10^3
Eu	0.00000011		0.30×10^3
Gd	0.00000060		0.26×10^3
Tb			
Dy	0.00000073		0.46×10^3
Ho	0.00000022		0.53×10^3
Er	0.00000061		0.69×10^3
Tm	0.00000013		1.8×10^3
Yb	0.00000052		0.53×10^3
Lu	0.00000012		0.45×10^3
Hf			
Ta			
W	0.0001	WO_4^{2-}	1.0×10^3
Re			
Os			
Ir			
Pt			
Au	0.000004	$AuCl_4^-$	5.6×10^5
Hg	0.00003	$HgCl_3^-$; $HgCl_4^{2-}$	4.2×10^4
Tl	<0.00001	Tl^+	
Pb	0.00003	Pb^{2+}; $PbSO_4$	
Bi	0.00002		4.5×10^5
Th	0.00005		3.5×10^2
U	0.003	$UO_2(CO_3)_3^{4-}$	5.0×10^5

Largely from E. D. Goldberg, *The Sea*, **2**, 4, courtesy of Interscience, New York

*Table 7.3. Dissolved Gases in Sea Water**

	Concentration (ml/l)
Oxygen	0–9
Nitrogen	8.4–14.5
Total carbon dioxide	34–56
Argon (residue after removal of N)	0.2–0.4
Helium and neon	1.7×10^{-4}
Hydrogen sulfide	0–22 or more

* Sea water of 19‰ chlorinity at 0° in equilibrium with normal dry atmosphere will contain 8.08 ml/l oxygen and 14.40 ml/l nitrogen.

ocean are in contact, a relation must exist between the amount of the gases in solution and their partial pressures in the atmosphere. The surface waters are in equilibrium or near equilibrium with the oxygen and nitrogen of the air. It is generally assumed that nitrogen dissolved in sea water does not enter into chemical reactions; hence its concentration is not subject to appreciable variation. Oxygen on the other hand plays an active part in metabolism and in the decay of organic matter, and its percentage varies considerably from place to place. The atmosphere also regulates the carbon dioxide content of surface waters, but the relationship is complex because carbon dioxide is present in sea water in four distinct forms—free carbon dioxide, carbonate ions, bicarbonate ions, and undissociated H_2CO_3. The carbon dioxide content of sea water, which is the most important factor controlling the solubility of $CaCO_3$, is also dependent on the nature and amount of biological activity. Ammonia, argon, helium, and neon have been recorded in sea water. Hydrogen sulfide is often locally present (probably in part as sulfide ions rather than free gas), and may be widespread in stagnant bottom waters. Table 7.3 gives some data on dissolved gases in sea water.

THE COMPOSITION OF TERRESTRIAL WATERS

Although the total amount of terrestrial waters is insignificant in comparison with the mass of the hydrosphere, these waters are important geochemically, because they are responsible for most of the weathering and erosion of the land masses. A knowledge of their amount and composition is clearly essential to an understanding of the evolution of the ocean. The ultimate source of most terrestrial waters is rain, although some magmatic water is undoubtedly added through thermal springs. Part of the total rainfall runs directly off into streams, part is taken up as ground water and may ultimately reappear as springs, part is retained by the formation of hydrated compounds, and part is returned to the atmosphere. Because of the solvent power of water the runoff from the land is never pure H_2O but always contains dissolved material. The amount of this dissolved material, however, differs greatly from time to time and from place to place.

It has been estimated that the total annual rainfall on the land areas of the earth amounts to 123.4×10^{18} g, of which 27.4×10^{18} g drains off to the sea. Clarke estimates that this water carries 27.35×10^{14} g of dissolved material, which gives an average salinity for river water of about 100 ppm. Actual salinities are very variable; analyses of river water show salinities from 13 to 9185 ppm, but figures greater than 1000 ppm are uncommon. Conway (1942) has critically examined Clarke's data and shows that the concentrations of carbonate, calcium, and magnesium rise rapidly with salinity until limiting values are reached at about 200 ppm; higher salinities are largely due to increases in sodium, sulfate, and chloride. Conway points out that waters of salinity up to 50 ppm drain areas consisting mainly of igneous or metamorphic rocks, whereas for salinities of 50–200 ppm the drainage is largely from sedimentary rocks; higher figures indicate human contamination on a large scale or drainage of arid regions where saline soils are common.

Clarke computed an average composition of the dissolved matter in river water by weighting the composition of the water from different river systems in proportion to the total supply of dissolved material. Revised figures based on more recent data have been provided by Livingstone (1963), and these are given in Table 7.4. These figures are for the major dissolved constituents, those present in amounts of the order of 1 ppm or greater. Comparatively little information on minor elements is available, but Durum and Haffty have made numerous determinations for the large rivers of North

Table 7.4. Composition of Dissolved Solids in River and Sea Water

	River Water		Sea Water	River Water less Cyclic Salts
	ppm	%	%	%
HCO_3	58.5	48.6	0.41	54.6
SO_4	11.2	9.3	7.68	9.6
Cl	7.8	6.5	55.04	0.0
NO_3	1	0.8	—	0.9
Ca	15	12.5	1.15	13.9
Mg	4.1	3.4	3.69	3.4
Na	6.3	5.3	30.62	2.5
K	2.3	2.0	1.10	2.1
Fe	0.7	0.6	—	0.7
SiO_2	13.1	11.0	—	12.3
Sr, H_3BO_3, Br	—	—	0.31	—
	120.0	100.0	100.	100.0

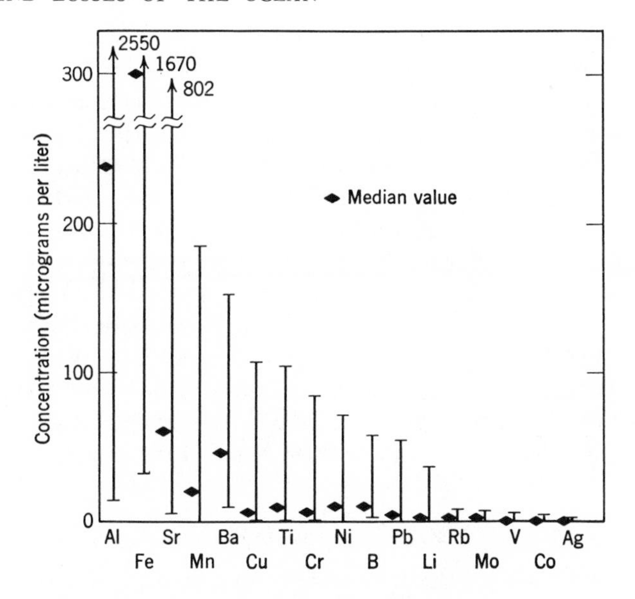

Figure 7.1 Minor and trace elements in river waters of North America. (Durum and Haffty, *Geochim. et Cosmochim. Acta* **27,** 3, 1963)

America (Figure 7.1). The figures are in micrograms per liter, that is, in parts per billion.

GAINS AND LOSSES OF THE OCEAN

Comparison shows that river water and sea water are opposites in chemical character. In sea water Na > Mg > Ca and Cl > SO_4 > CO_3; in average river water Ca > Na > Mg and CO_3 > SO_4 > Cl. In addition, as Conway has convincingly shown, over 90% of the chloride and a good deal of the sulfate in river waters is cyclic, being derived ultimately from the ocean via the atmosphere. Hence the average composition must be adjusted for these cyclic salts in discussing its effect on oceanic composition. This correction has been made in Table 7.4 by assuming that all the chloride in river water is cyclic and modifying the amounts of the other elements in the proportions in which they occur in sea water. The assumption is somewhat arbitrary, but the resulting figures undoubtedly give a truer picture of the net contribution to the sea of material in solution.

Evidently factors other than the supply of dissolved material in runoff from the lands regulate the composition of sea water. Many reactions take place in the sea to alter the balance of dissolved substances. Adsorption and base exchange by particles of sediment remove some ions from solution; other ions react with sedimentary material to form new minerals, as for example glauconite and phillipsite. Biological activity is responsible for the extraction of much of the dissolved material from the land. This is particularly true for calcium carbonate, which forms the hard parts of marine organisms, and for silica, which is used by diatoms and radiolaria.

It is not certainly known whether the addition of dissolved matter has brought about progressive changes in the relative composition of sea water during geological time. In any event, such changes must be exceedingly slow. The total quantity of dissolved solids in the ocean, assuming an average salinity of 35‰ and a mass of 1413×10^{21} g, is 49.5×10^{21} g. The total amount of material in solution contributed annually by runoff from the land (27.35×10^{14} g) is only an infinitesimal fraction, 5.5×10^{-8}, of this amount; it is, nevertheless, significant in terms of the length of geological time.

THE BALANCE OF DISSOLVED MATTER IN SEA WATER

It is generally believed that the ocean is, to a large degree, in a steady-state condition, whereby the amount of an element introduced per unit time is balanced by an equal amount deposited in the sediments. Accepting this hypothesis of a steady-state condition, we can then define a residence time in years for each element, being the total amount of that element in solution in the oceans divided by the amount introduced by the rivers per year. Lacking adequate data for many minor and trace elements in average river water, we can assume hopefully that the amounts introduced will be proportional to their crustal abundances. The residence times for many elements are given in Table 7.2.

Although many independent assumptions are involved in these calculations, both the absolute and relative values for most of the residence times seem reasonable. Sodium has the longest residence time, within an order of magnitude of the age of the oceans. This long residence time reflects a lack of reactivity of sodium in the marine environment; it is not readily incorporated in the common sedimentary minerals, nor is it removed by biological reactions. The abundant cations K, Ca, and Mg have residence times of the order of 10^7 years. Elements of low abundance in sea water have short residence times, especially those which might potentially be available in considerable amounts. The very low residence time of manganese, 1400 years, is evidently related to its removal from solution by oxidation to the tetravalent state and precipitation as the manganese dioxide nodules that are abundant over wide areas of the ocean floor. Silicon and aluminum are among the elements with the shortest residence times. Much silicon is removed from solution by the activity of organisms; aluminum apparently has no biological function, but is probably rapidly removed from solution as newly precipitated clay minerals.

The significance of residence times for practical geochemical problems is illustrated in predicting the behavior of artificial radionuclides introduced into the ocean by nuclear explosions or as wastes from reactors. The radioisotopes of strontium and cesium, both hazardous to living organisms, have residence times which are very high, relative to their half-lives. Hence the radionuclides of these elements, when introduced into the ocean, are not rapidly precipitated; they remain in solution and are available to marine organisms.

Goldschmidt made an interesting study of the balance between supply and removal of several elements in sea water. The basis of his comparison was the total amounts of different elements supplied by weathering, disintegration, and sedimentation during geological time. According to Goldschmidt, about 160 kg of igneous rocks have been weathered for each square centimeter of the earth's surface. Since there are 278 kg of ocean water per square centimeter, approximately 600 g of rock have been weathered for each kilogram of water in the oceans. The 600 g of igneous rock have therefore been the potential source of the dissolved matter supplied by weathering to 1 kg of sea water. Of course, only a part of this 600 g has actually dissolved and remained in solution. Goldschmidt drew up a "balance sheet" relating the potential supply of several elements in 600 g of igneous rock to the amounts present in solution in 1 kg of sea water. The "percentage in solution" was obtained by dividing the amount of each element in sea water by the potential supply. The most striking feature is that the concentrations of a few elements are far greater than can have been supplied by weathering. These are the common anions of sea water—chloride, sulfate, borate, and bromide. Either they were present in great amounts in the primeval ocean or they have been supplied largely from volcanic gases and thermal springs throughout geological time. Judging by the data available on amounts and composition of igneous exhalations, the latter alternative is adequate to explain at least the major part of these "superabundant" elements in sea water.

In the chapter on sedimentation and sedimentary processes (Chapter 6) we saw the importance of ionic potential as a measure of behavior of an element in aqueous solution. Comparison of the percentages of different elements in solution in sea water shows that other factors besides ionic potential must be significant in the ocean. Thus, of the alkali elements, all of which have low ionic potentials, only sodium remains to any great extent in the sea; little potassium is present, and rubidium and cesium have been even more effectively removed, evidently as a consequence of the strong adsorption of their ions by colloidal particles of hydrolysate sediments. The alkaline earth elements are also largely removed from sea water. Biological activity is, of course, responsible for the extraction of much of the calcium. Strontium remains in sea water to four times the extent of calcium. The very small amounts of trivalent, quadrivalent, and quinquivalent elements in sea water evidently reflect their high ionic potentials, causing them to be almost quantitatively precipitated. In this connection, it is interesting to note that some early spectrographic work by Sir William Crookes showed the presence of minute amounts of rare earths in the calcium carbonate of marine organisms, an observation later confirmed by Goldschmidt and his coworkers. The substitution of a rare-earth metal for calcium is perhaps accompanied by the replacement of boron for carbon in the acid radical.

A remarkable feature is the very low percentages of certain poisonous metals and metalloids in sea water, including some, such as selenium and arsenic, which form soluble complex anions and therefore might be expected to accumulate. The quantities of these elements potentially supplied to the ocean during geological time are so great that a serious poisoning action would have resulted had not some process been active in eliminating them.

Krauskopf (1956) has made an extensive study of the factors controlling the concentrations of Zn, Cu, Pb, Bi, Cd, Ni, Co, Hg, Ag, Cr, Mo, W, and V in sea water. He investigated in detail four processes for the removal of these elements: precipitation of insoluble compounds with ions normally present in sea water; precipitation by sulfide ion in local regions of low oxidation potential; adsorption by materials such as ferrous sulfide, hydrated ferric oxide, hydrated manganese dioxide, and clay; and removal by the metabolic action of organisms. He showed both by calculation and experiment that sea water is greatly undersaturated in all these ions, that is, precipitation of insoluble compounds cannot be responsible for the observed concentrations. Local precipitation of sulfides is a possible control mechanism for some of the elements but is probably not the chief control because the concentrations are unrelated to sulfide solubilities. Adsorption is a possible mechanism for all elements except V, W, Ni, Co, and Cr; if Cr is assumed to be removed by local reduction and precipitation of the hydroxide and the other four elements by biological processes, the existing concentrations can be fairly adequately accounted for. Adsorption processes supplemented by biological removal also furnish an explanation for the distribution of minor and trace metals in marine sedimentary rocks.

With few exceptions, therefore, all the elements have been potentially available in much larger amounts than are actually present in sea water. The solubility of its compounds, physicochemical factors such as adsorption and coprecipitation, and biological activity control the removal of an element from solution. Probably some ions in river water are precipitated as insoluble compounds as soon as they enter the sea because of the difference in chemical environment. Some elements are removed from solution by complex reactions between ions in solution and solid material, which may be particles in suspension or bottom sediments. Such processes are termed halmyrolysis.

An interesting example of halmyrolysis is the accumulation of radioactive material in the upper layers of deep-sea sediments. The breakdown of U^{238} in sea water results in the production of Th^{230} (ionium), which is precipitated almost quantitatively and accumulates in the bottom sediments, giving them abnormally high radioactivity. The half-life of ionium is 83,000 years; hence the amount of ionium falls off with the age of the sediment; that is, the ionium content is greatest in the surface layers and decreases below. It is thus possible to calculate the age of different layers in deep-sea cores. Some valuable results on rates of deposition of deep-sea sediments have been obtained by this method; it has been shown that red clay in the central Pacific accumulates at a rate of about 1 mm in 1000 years, whereas the red clay in the Atlantic generally shows a deposition rate considerably more rapid.

Biological activity is undoubtedly of great importance in controlling the concentrations of many of the elements in the sea, not only the abundant ones, such as calcium and carbon, but also rarer ones, such as copper, vanadium, and phosphorus. For example, the concentration of copper and iron is extremely low in sea water, but one or the other is an essential constituent of the blood of many marine animals. Organisms remove from solution elements that would not otherwise precipitate, and, if conditions are such that some of the organic material becomes a permanent part of the sediments, this process must play an important part in controlling sea-water composition.

For many elements in solution in the oceans there is probably a dynamic equilibrium between supply and removal. Under such conditions the amount extracted from the sea by various processes balances the amounts returned by runoff from the land. The ocean may thereby act as a self-balancing mechanism in which most of the elements have reached an equilibrium concentration.

THE HISTORY AND EVOLUTION OF THE OCEAN

A review of the literature reveals little agreement concerning the history and evolution of the ocean. Some geologists have suggested that there was probably no deep sea until the end of the Palaeozoic, implying that the earth had an insignificant hydrosphere for the greater part of its history and has since acquired its present volume of sea water at a rapid rate. Twenhofel has proposed similar, but less extreme, views, believing that much of the growth of the oceans has occurred since Devonian times. Kuenen, on the other hand, considers that the volume of the oceans has remained almost constant since well back in the Precambrian. These ideas can best be illustrated by graphs in which the suggested volume of the oceans is plotted against geological time, as in Figure 7.2. Whether or not the volume of the ocean has increased, the trend in ideas is towards the view that its composition has not altered greatly throughout geological time; continual addition of water and dissolved matter may have occurred, but the various reactions that take place in the sea have served to maintain comparative constancy of composition. This position has been vigorously maintained by Rubey (1951). By a careful analysis of a great deal of data he shows that the composition of the oceans and the atmosphere can have varied little since Archean times.

Conway (1943) has discussed the chemical evolution of the oceans in terms of the following hypotheses:

1. The water condensed from the primeval atmosphere and the chloride has been added throughout geological time (constant volume-volcanic chloride hypothesis).
2. Both water and chloride are the result of initial condensation (constant volume–constant chloride hypothesis).
3. Both components have appeared gradually (volcanic ocean–volcanic chloride hypothesis).
4. The chloride was present as metallic chloride on the earth's surface, and the water has been added throughout geological time by volcanic activity (volcanic ocean–constant chloride hypothesis).

Conway appears to favor the first hypothesis as being most consistent with the data available. He considered the fourth hypothesis most unlikely and

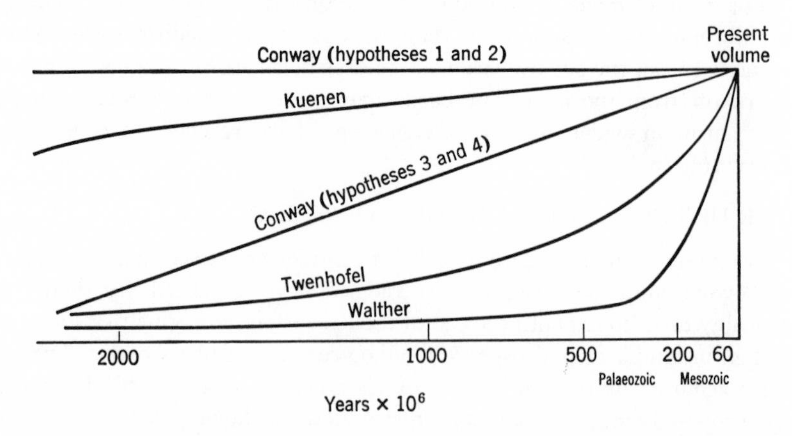

Figure 7.2 Variation in the amount of ocean waters during geological time, according to the views of Walther, Twenhofel, Kuenen, and Conway.

gave it little consideration, although he was able to show that, after the Precambrian, oceanic evolution would be much the same on both the fourth and the second hypotheses. The third hypothesis leads to results intermediate between those of the first and second.

The uniformitarian view that the composition of the ocean has been similar to that at the present day as far back as the Archean is borne out by the available data. We know that magmas contain water vapor and gases such as carbon dioxide, hydrogen chloride, hydrogen fluoride, and hydrogen sulfide, and that, therefore, these have been continually added to the hydrosphere. Much of the water, and some of the other gases also, may be cyclical. On the whole, it appears that no one of Conway's theses can be accepted as presenting the full story. Current ideas on the origin of the earth suggest the formation of an ocean as soon as the surface was cool enough for water to condense. The amount and salinity of the primitive ocean are unknown. Throughout geological history the oceans have grown by the addition of primary magmatic water. This growth may well have been considerably more rapid in early geological time, when a thinner crust and a greater production of radiogenic heat probably resulted in more violent igneous activity. Such a view of the growth of the oceans is essentially that of Kuenen, as expressed in Figure 7.2. The nature of sedimentary rocks and fossils indicates that the content of dissolved substances has probably remained much the same, approximate constancy of composition being maintained by the complex chemical reactions outlined in the previous section.

Whether igneous activity has provided sufficient water to account for all or the greater part of the hydrosphere is still a matter of dispute. For example, Verhoogen (1946) concludes that the total lava extruded since the beginning of the Cambrian cannot exceed a volume of 3×10^{22} cm^3 or, adopting an average density of 3.3, a mass of 1×10^{23} g. Even if this lava were accompanied by as much as 10% of its mass of water, not more than 0.7% of the ocean can have been added since the beginning of the Palaeozoic or less than 5% during the whole of geological time. However, Verhoogen's estimate does not appear to include submarine volcanic activity, which is probably as widespread as terrestrial volcanism and would thus increase his figures by a factor of three. In addition, much magmatic water undoubtedly reaches the surface in thermal springs. Rubey discusses this problem in detail and estimates that the hot-spring discharge from the continents and ocean floor combined is about 66×10^{15} g H_2O per year. This rate, continued over a period of 3×10^9 years, would yield an amount of 2.0×10^{26} g H_2O. Rubey estimates the total quantity of water in the atmosphere, hydrosphere, and biosphere and that buried in ancient sedimentary rocks as $16{,}700 \times 10^{20}$ g, which is 0.8% of this amount. In other words, if hot springs are delivering to the surface an average of only 0.8% of juvenile water, derived from a primary magmatic source (the remaining 99.2% being heated ground water), the entire volume of the ocean could be accounted for in the course of geological time.

If the oceans have increased in volume throughout geological time, we are faced with the problem of correlating this with the relationship between continents and ocean basins. The doctrine of the permanence of continents and ocean basins is one for which there is a great deal of geological evidence. This evidence is, however, largely based on the record of the fossiliferous rocks, which represent only the latest sixth of geological time. If the amount of sea water has increased continually, then the ocean basins must have increased in area or in depth or both. The relationship of continents and ocean basins during Precambrian time is very poorly known. Since for the most part the oldest rocks are concentrated in the central shield areas, with younger rocks around their edges, it has been suggested that the present continents have grown laterally throughout geological time. This implies a considerable reduction in the area occupied by the ocean, and the sea floor must necessarily have become deeper in relation to the continental level. Revelle (1955) has suggested that a considerable portion, perhaps as much as one fourth of all the water in the oceans, has appeared since the late Mesozoic, that is, in the last thirtieth or fortieth of geological time. This hypothesis is based on the presence of coral atolls and flattopped sea mounts in the Pacific Ocean, which indicate a subsidence of the sea floor in relation to sea level. Remains of reef corals and shallow-water mollusks of Middle Cretaceous age have been dredged from the flattopped summits of the submerged mid-Pacific mountains, which now lie at depths of 5000 to 6000 feet. However, the question as to whether this represents a world-wide rise of sea level of this amount or a regional depression of the sea floor remains unanswered.

SELECTED REFERENCES

Clarke, F. W. (1924). *The data of geochemistry.* Chapters 3, 4, 5, and 6.

Conway, E. J. (1942). Mean geochemical data in relation to oceanic evolution. *Proc. Roy. Irish Acad.* **B48,** 119–159.

Conway, E. J. (1943). The chemical evolution of the ocean. *Proc. Roy. Irish Acad.* **B48,** 161–212.

The two papers above present a careful analysis of the data of marine geochemistry, based largely on Clarke's figures, and the possible deductions therefrom regarding the development of the ocean during geological time.

Goldschmidt, V. M. (1954). *Geochemistry.* Chapter 4.

Harvey, H. W. (1955). *The chemistry and fertility of sea waters.* 224 pp. Cambridge University Press, Cambridge, England. An interesting and well-documented account of the chemistry and the biochemistry of the ocean.

Hess, H. H. (1962). History of ocean basins. *Geol. Soc. Amer.*, Buddington Vol., 599–620. A comprehensive discussion of the geophysical and geochemical data pertinent to the origin and evolution of the continents and oceans.

Krauskopf, K. B. (1956). Factors controlling the concentrations of thirteen rare metals in sea water. *Geochim. et Cosmochim. Acta* **9,** 1–32. A critical discussion of the factors controlling the concentration of minor and rare elements in the ocean, supported by laboratory experiments.

Kuenen, Ph. H. (1950). *Marine geology.* 568 pp. John Wiley and Sons, New York. A valuable source of information and ideas on the ocean, written from the geological viewpoint.

Livingstone, D. A. (1963). *Chemical composition of rivers and lakes.* U. S. Geol. Surv. Prof. Paper 440-G, 64 pp. [*Data of Geochemistry* (sixth edition), chapter G]. An excellent source of data on the composition of natural waters, with estimates for the rate of chemical denudation for the different continents.

Rankama, K., and Th. G. Sahama (1950). *Geochemistry.* Chapter 6.

Revelle, R. (1955). On the history of the oceans. *J. Marine Res.* **14,** 446–461. A discussion of the evolution of the oceans, with special reference to the evidence for a general sinking of the ocean floor.

Riley, J. P., and G. Skirrow (eds.) (1965). *Chemical oceanography.* 1452 pp. (2 vols.). Academic Press, New York. A series of articles on all aspects of the geochemistry of sea water and marine sediments.

Rubey, W. W. (1951). Geologic history of sea water. *Bull. Geol. Soc. Amer.* **62,** 1111–1147. A critical discussion of the evolution of the oceans, based on a careful evaluation of the geochemical data.

Sears, M. (ed.) (1961). *Oceanography.* 654 pp. Publication 67 of the Amer. Assoc. Adv. Sci., Washington, D. C. A compendium of 30 invited lectures given at the International Oceanographic Congress, 1959, many of which are of geochemical significance.

Sverdrup, H. U., M. W. Johnson, and R. H. Fleming (1942). *The oceans, their physics, chemistry, and general biology.* 1087 pp. Prentice-Hall, New York. The standard work on the subject. Chapters 6, 7, and 20 are especially concerned with the geochemistry of the ocean.

Verhoogen, J. (1946). Volcanic heat. *Am. J. Sci.* **244,** 745–751. An examination of the fundamentals of igneous activity in which the reasoning is based as far as possible on the thermodynamics involved; makes a tentative evaluation of the amount of lava extruded during geological time and the amount of water thereby added to the hydrosphere.

CHAPTER EIGHT

THE ATMOSPHERE

THE COMPOSITION OF THE ATMOSPHERE

The atmosphere is rather simple in composition, being made up almost entirely of three elements: nitrogen, oxygen, and argon. Other elements and compounds are present in amounts that although small are nevertheless significant in terms of the geochemistry of the atmosphere. The average composition is set out in Table 8.1. By virtue of Avogadro's law, the volume concentrations give directly the relative number of the different molecules and atoms. Convection currents maintain constancy of the proportions of the different components at elevations below 60 km, but above that level gravitational separation according to molecular weight may begin, although such separation does not become significant until above 100 km. Auroral

Table 8.1. The Average Composition of the Atmosphere

Gas	Composition by Volume (ppm)	Composition by Weight (ppm)	Total Mass ($\times 10^{20}$ g)
N_2	780,900	755,100	38.648
O_2	209,500	231,500	11.841
Ar	9,300	12,800	0.655
CO_2	300	460	0.0233
Ne	18	12.5	0.000636
He	5.2	0.72	0.000037
CH_4	1.5	0.94	0.000043
Kr	1	2.9	0.000146
N_2O	0.5	0.8	0.000040
H_2	0.5	0.035	0.000002
O_3[1]	0.4	0.7	0.000035
Xe	0.08	0.36	0.000018

[1] Variable, increases with height.

observations indicate that even at high altitudes oxygen and nitrogen predominate. The atmosphere gradually thins out into interplanetary space; above an altitude of about 600 km the atoms and molecules describe free elliptical orbits in the earth's gravitational field. Although from the viewpoint of composition the atmosphere can be looked upon as becoming uniformly more diffuse upward, it does show some structural features of remarkable physical significance. The lower part of the atmosphere, in which convection is prominent, is known as the troposphere; above this is the stratosphere, so named because it appears to be stratified in a number of layers between which no strong vertical circulation seems to exist. The boundary between the troposphere and the stratosphere varies both with the latitude and with the season, but can be placed at a height of about 10–15 km. In the troposphere the temperature decreases with increasing height, whereas in the stratosphere the temperature is independent of altitude. Above about 80 km the stratosphere passes into the ionosphere; in these regions the atmosphere is rendered conducting by ionization induced by ultraviolet radiation from the sun. At least three apparent layers are recognized in the ionosphere, the E (Heaviside) layer and the F_1 and F_2 (Appleton) layers; they are characterized by their specific absorptive and reflective properties towards radio waves.

The minor constituents of the atmosphere are highly important. Carbon dioxide, whose concentration is only a few hundredths of one per cent, provides the raw material for plant life. Ozone, most of which is present in a diffuse layer (the ozonosphere) within the stratosphere, plays a vital part in that it is responsible for absorption of ultraviolet radiation; if this absorption did not take place the excessive ultraviolet radiation at the earth's surface would be fatal for most forms of life. The ozonosphere is also of some consequence in reducing the escape of terrestrial heat by radiation into space.

The mass of the atmosphere is given by the product of the average height of the mercury barometer in centimeters, the density of mercury, and the area of the earth in square centimeters. The result thus obtained by several authorities is about 50×10^{20} g; the latest and most careful estimate gives 51.17×10^{20} g for the total mass of the dry atmosphere, that is, without any water vapor.

THE EVOLUTION OF THE ATMOSPHERE

The evolution of the atmosphere can be discussed in terms of the following factors:

1. The composition of the primeval atmosphere.
2. Additions during geological time.
3. Losses during geological time.

Although the quantitative results produced by these factors are difficult to evaluate, considerable data have been accumulated. These data are discussed briefly in the following sections.

THE COMPOSITION OF THE PRIMEVAL ATMOSPHERE

Our ideas as to the composition of the primeval atmosphere are conditioned largely by the mode of origin that we ascribe to the earth. On the planetesimal hypothesis, the particles aggregating to form the earth had no atmosphere associated with them; the primeval atmosphere originated from the gases occluded or combined within the planetesimals and released by the heat and chemical reactions accompanying and following aggregation.

There seems little question now that the constituents of the atmosphere have been largely, if not entirely, exhaled from the interior of the earth. The strongest evidence for this lies in a comparison between terrestrial and cosmic abundances of the elements. Table 8.2 provides such a comparison in terms of a deficiency factor, based on the assumption that silicon, as an element forming stable non-volatile compounds, has the same relative abundance both in the earth and throughout the universe. In the last column of Table 8.2 the deficiency factor is expressed on a logarithmic scale in which zero represents comparable abundance in the earth and throughout the

Table 8.2. The Abundances of Some Elements in the Earth and in the Universe as a Whole

(atoms/10,000 atoms Si)

	Atomic Number	Whole Earth (*a*)	Cosmos (*b*)	Deficiency Factor log (*b*/*a*)
H	1	84	3.5×10^8	6.6
He	2	3.5×10^{-7}	3.5×10^7	14
C	6	71	80,000	4.0
N	7	0.21	160,000	5.9
O	8	35,000	220,000	0.8
F	9	2.7	90	1.5
Ne	10	1.2×10^{-6}	50,000	10.6
Na	11	460	462	0
Mg	12	8,900	8,870	0
Al	13	940	882	0
Si	14	10,000	10,000	0
P	15	100	130	0.1
S	16	1,000	3,500	0.5
Cl	17	32	170	0.7
Ar	18	5.9×10^{-4}	1,200	6.3
Kr	36	6×10^{-8}	0.87	7.2
Xe	54	5×10^{-9}	0.015	6.5

universe, that is, neither enrichment nor deficiency. The truth of the assumption that silicon is an element that has the same relative abundance in the earth and in the universe is supported by the fact that comparable elements, such as sodium, magnesium, and aluminum, also give deficiency factors of zero; that is, their terrestrial and cosmic abundances are of the same order. The degree of precision of these deficiency factors is not great, but qualitatively they are undoubtedly significant. The much greater deficiency factor of helium than hydrogen can be ascribed to the retention of hydrogen by chemical combination with other elements; helium, being an inert gas, cannot be retained in this way. A similar relation is evident when the deficiency factors of oxygen, nitrogen, and neon are compared; clearly, chemical activity has played a significant role in the retention of the gaseous elements. The deficiency of these lighter constituents in the earth is explicable on either the nebular or the planetesimal hypothesis; however, as Table 8.2 shows, the heavy inert gases are also deficient. Krypton and xenon are about a million times less abundant in the earth than their immediate neighbors in the periodic table. Since their nuclear properties are not strikingly dissimilar to those of their neighbors, it is reasonable to expect these rare gases to be of the same order of abundance. If the earth condensed directly from solar matter, krypton and xenon, unlike the lighter elements, should have been completely retained because of their high atomic weight. Their deficiency therefore is an important point in favor of some form of the planetesimal hypothesis.

The evolution of the atmosphere on an earth formed by planetesimal accretion has been carefully discussed by Holland (1962). He distinguishes three stages (Table 8.3). The accretion of the earth was an exothermic process, and this, plus radiogenic heat, would soon result in volcanism. Holland points out that the chemistry of volcanic gases at the first stage in the evolution of the atmosphere was probably very different than it is today. Before the formation of the core, metallic iron would be present in the upper part of the mantle (if the planetesimals were similar in composition to ordinary chondrites, the metal would be part of the accreting material; if they were like carbonaceous chondrites, nickel-iron would be formed as the material heated up). Volcanic gases in equilibrium with free iron would consist largely of H_2, H_2O, and CO, with minor amounts of N_2, CO_2, and H_2S. Hydrogen was probably the dominant constituent. As the gases cooled the CO and CO_2 reacted with H_2 to form CH_4. Nitrogen may have reacted to form NH_3, provided the rate of escape of hydrogen from the planet was sufficiently slow to permit the existence of an appreciable hydrogen pressure. If the earth's surface was cool enough and nearly all the water condensed, the main constituents of the atmosphere would then be CH_4 and H_2.

The duration of the first stage was determined by the time that elapsed

Table 8.3. Summary of Data on the Probable Chemical Composition of the Atmosphere during Stages 1, 2, and 3

	Stage 1	Stage 2	Stage 3
Major components	CH_4	N_2	N_2
$P > 10^{-2}$ atm	H_2 (?)		O_2
Minor components	H_2 (?)		Ar
$10^{-2} < P < 10^{-4}$ atm	H_2O	H_2O	H_2O
	N_2	CO_2	CO_2
	H_2S	Ar	
	NH_3		
	Ar		
Trace components	He	Ne	Ne
$10^{-4} < P < 10^{-6}$ atm		He	He
		CH_4	CH_4
		NH_3 (?)	Kr
		SO_2 (?)	
		H_2S (?)	

between the accretion of the earth and the formation of a core by the gravitational separation of the metallic iron from the mantle. Holland suggests that this did not exceed 500 million years. The removal of the iron phase left a mantle essentially like that existing today, and subsequent volcanism was accompanied by gases much more highly oxidized than those of the first stage, and probably similar in composition to the Hawaiian volcanic gases today. Water was the dominant component, and CO_2, CO, H_2, SO_2, and N_2 minor constituents. The atmosphere during the second stage contained largely N_2 with minor amounts of CO_2 and H_2O. Photochemical dissociation of water vapor in the upper atmosphere would produce oxygen and hydrogen, with the hydrogen escaping into outer space. However, the free oxygen would not accumulate in the atmosphere, but would be used up in oxidizing the more reduced constituents of the volcanic gases.

The second stage came to an end when oxygen production exceeded oxygen use. The problem of the time of appearance of free oxygen and its mode of formation is an intriguing one. Increasing knowledge and understanding of Precambrian rocks has contributed greatly to the solution of this problem. Cloud (1965) has summarized the combined evidence of paleontology, stratigraphy, and geochemistry. The evidence indicates that green plant photosynthesis existed at least 1.7 to 2.0 billion years ago and that atmospheric oxygen first began to be available in relatively large quantities probably about 1.2 billion years ago. From the time of the first available photosynthetic oxygen until about 1.2 billion years ago ferrous

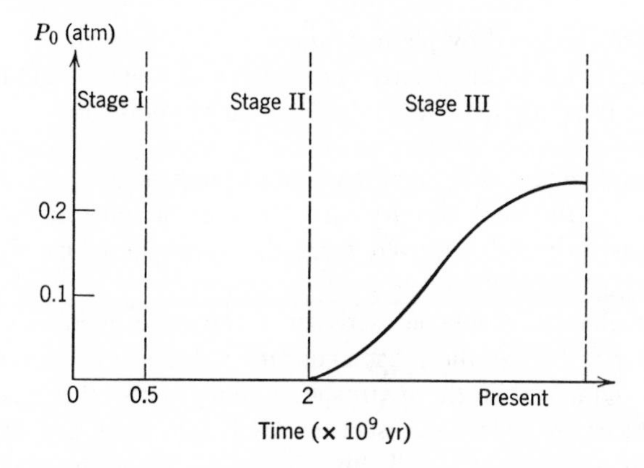

Figure 8.1 Possible development of oxygen partial pressure in the atmosphere over time, according to Holland.

iron, carried into large water bodies or the ocean by reducing surface waters, may have served as a vast oxygen sump, retarding the evolution of free oxygen to the atmosphere. Cloud remarks that the existence and facies of the unique Precambrian iron formations of the Lake Superior-type possibly reflect the onset and nature of early photosynthesis. Typical Precambrian iron formation rocks range in age from about 2.7 to 1.7 billion years before the present. The oldest fossils, of possibly algal origin, are apparently somewhat older than 2.6 billion years. Cloud suggests that the appearance of oxygen-generating organisms in the hydrosphere was necessary to trigger the deposition of this type of iron formation, and that deposition ceased at some point of compensation or evolution between 1.7 and 1.2 billion years ago. As early as about 1.2 billion years ago there was enough oxygen in the atmosphere to produce extensive red beds (nonmarine sedimentary rocks colored by ferric oxides). By the beginning of Paleozoic time, about 600 million years ago, enough oxygen had accumulated in the atmosphere to permit the evolution of the Metazoa (multicelled animals). Thus the buildup of oxygen in the atmosphere was quite gradual and did not reach its present concentration until quite late in earth history (Figure 8.1).

ATMOSPHERIC ADDITIONS DURING GEOLOGICAL TIME

The following additions have been made to the atmosphere during geological time:

1. Gases released by the crystallization of magmas.
2. Oxygen produced by photochemical dissociation of water vapor.

3. Oxygen produced by photosynthesis.
4. Helium from the radioactive breakdown of uranium and thorium.
5. Argon from the radioactive breakdown of potassium.

Carbon dioxide and a very small amount of methane are contributed to the atmosphere by the vital activity and decay of organisms, but, since this material was ultimately derived from the atmosphere, its significance is discussed in a later section.

The contribution of igneous activity to the atmosphere is considerable. Water vapor is by far the most abundant volcanic gas, but it is rapidly condensed and added to the hydrosphere along with other readily soluble gases produced by volcanism, such as HCl, HF, H_2S, and SO_2. Next to water, carbon dioxide is probably the most important contribution of igneous activity to the atmosphere. Some of this CO_2 is presumably secondary, having been picked up from surrounding rocks during the ascent of the magma, but the greater part is probably juvenile, brought up from the depths as a new contribution to the atmosphere and hydrosphere. Igneous activity is evidently the source of the CO_2 needed to replace that locked up as carbonate minerals and free carbon during geological time. Rubey (1951) has carefully evaluated the contribution of igneous activity to the atmosphere and hydrosphere, and his results are given in Table 8.4. The excess volatiles unaccounted for by rock weathering must be the amounts present in the primeval atmosphere and hydrosphere, plus that added by igneous activity during geological time. From Rubey's data it seems reasonable to assume that the latter quantity far exceeds that originally present.

Table 8.4. Estimated Quantities (in Units of 10^{20} Grams) of Volatile Materials Now at or Near the Earth's Surface

	H_2O	Total C as CO_2	Cl	N	S	H, B, Br, Ar, F, etc.
In present atmosphere, hydrosphere, and biosphere	14,600	1.5	276	39	13	1.7
Buried in ancient sedimentary rocks	2,100	920	30	4.0	15	15
Total	16,700	921	306	43	28	16.7
Supplied by weathering of crystalline rocks	130	11	5	0.6	6	3.5
Excess volatiles unaccounted for by rock weathering	16,600	910	300	42	22	13

Some nitrogen has also been added to the atmosphere from magmatic sources during geological time. It is usual to ascribe the nitrogen recorded in volcanic gases to atmospheric contamination, and most of it probably originates thus. However, analyses by Rayleigh (1939) and others indicate nitrogen as a constituent of igneous rocks and so presumably of magmatic gases. Rayleigh found that the nitrogen content of igneous rocks of all compositions, from ultrabasic to acidic, is remarkably constant, averaging about 0.04 cm^3 N_2/g, or 0.005 wt %. The nitrogen is largely present in combination because an appreciable amount is driven off as ammonia when the rocks are heated with soda lime. Chemical combination of the nitrogen is borne out by the nitrogen:argon ratio, which is 2000 in igneous rocks as against 120 in the atmosphere.

A balance sheet for the production and consumption of atmospheric oxygen has been provided by Holland (Table 8.5). The photochemical dissociation of water vapor in the upper atmosphere, with the subsequent escape of hydrogen into outer space, is a process which has been adding to the earth's supply of free oxygen throughout its history. However, the estimated amount produced in this way is insignificant compared with that contributed by photosynthesis.

The total amount of oxygen released by photosynthesis can be calculated if we know the amount of organic carbon which has been "fossilized" in the sedimentary rocks. Rubey estimates that the total organic carbon in the sedimentary rocks is the equivalent of 250×10^{20} g CO_2, which would correspond to the photosynthetic liberation of 181×10^{20} g oxygen. No

Table 8.5. Production and Use of Oxygen

Total estimated production	
By photosynthesis in excess of decay	181×10^{20} gm
By photodissociation of water vapor followed by hydrogen escape	1
Total	182×10^{20} gm
Total estimated use	
Oxidation of ferrous iron to ferric iron during weathering	14
Oxidation of S^{-2} to SO_4^{-2} during weathering	12
Oxidation of volcanic gases	
CO to CO_2	10
SO_2 to SO_3	11
H_2 to H_2O	<150
Total	$<197 \times 10^{20}$ gm
Free in atmosphere	$\sim 12 \times 10^{20}$ gm

reasonable explanation of the organic carbon of sediments other than a biological origin has been proposed, so we may assume 181×10^{20} g oxygen has been liberated by photosynthesis and not reconverted to carbon dioxide by decay or respiration.

The rate of production of helium is 1.16×10^{-7} cm^3/g of uranium per year and 2.43×10^{-8} cm^3 per gram of thorium per year. A simple calculation based on the amounts of uranium and thorium in the earth and an age of 4×10^9 years shows that the quantity of helium liberated by radioactive decay during geological time is many times that now present in the atmosphere. This is explained by the loss of helium from the earth into interplanetary space, which is discussed in the next section.

Argon has three isotopes, Ar^{36}, Ar^{38}, and Ar^{40}, of which Ar^{40} is by far the most abundant (the percentages of Ar^{36}, Ar^{38}, and Ar^{40} are 0.307, 0.061, and 99.632). The amount of argon on the earth is anomalously high when compared with that of the other inert gases. This is evidently due to the production of Ar^{40} by the radioactive decay of K^{40} throughout geological time. Much of the argon thereby produced has probably remained occluded in the potassium minerals where it originated, but sufficient has been liberated to the atmosphere to give it a comparatively high argon content.

ATMOSPHERIC LOSSES DURING GEOLOGICAL TIME

Atmospheric losses during geological time can be summarized under the following heads:

1. Loss of oxygen by oxidation of hydrogen to water, ferrous to ferric iron, sulfur compounds to sulfates, manganese compounds to manganese dioxide, and similar reactions.
2. Loss of carbon dioxide by the formation of coal, petroleum, and disseminated carbon from the death and burial of organisms.
3. Loss of carbon dioxide by the formation of calcium and magnesium carbonates.
4. Loss of nitrogen by the formation of oxides of nitrogen in the air and by the action of nitrifying bacteria in the soil.
5. Loss of hydrogen and helium by escape from the earth's gravitational field.

Removal of oxygen is observable wherever weathering is taking place. Ferrous iron oxidizes readily to ferric iron under most surface conditions, as is evidenced by the yellow or red colors produced by ferric oxides in many sediments. Clarke's figures for the average composition of igneous rocks show $FeO > Fe_2O_3$, whereas for sedimentary rocks the relationship is reversed. Iron is the principal consumer of oxygen during weathering, but manganese is converted from the bivalent to the quadrivalent state, and oxidation converts sulfides into sulfates or free sulfuric acid. From

Table 8.5, however, it is clear that the principal consumption of free oxygen has been through the oxidation of volcanic gases, especially hydrogen.

The loss of carbon dioxide from the atmosphere by deposition as carbonate and organic carbon in sedimentary rocks was estimated by Rubey as totaling 920×10^{20} g. More recently, Wickman (1956) has published some revised figures. He places the amount of carbonate carbon per square centimeter of the earth's surface at 2420 ± 560 g and of organic carbon at 700 ± 200 g. Taking the figure of 3100 g/cm^2 for the total amount of carbon transferred from the atmosphere to the sedimentary rocks, this is equal to a total of 158×10^{20} g carbon, or 580×10^{20} g CO_2. The latter figure is of the same order of magnitude as Rubey's estimate but considerably lower. The figures show clearly that the amount of carbon dioxide deposited in sedimentary rocks far exceeds the amount in the present atmosphere, hydrosphere, and biosphere (about 1.5×10^{20} g), and thus indicate that large amounts of carbon dioxide must have been released from magmatic sources throughout geological time to maintain organic activity. Wickman's figures show, in addition, that far more carbon dioxide has been removed as limestone and dolomite than as coal or other organic carbon.

Nitrogen is removed from the air by both organic and inorganic processes. The organic processes include nitrogen-fixing microorganisms living in the root nodules of certain plants, soil organisms (both aerobes and anaerobes), and some blue-green algae. Inorganic processes produce oxides of nitrogen by electrical discharges and photochemical reactions in the atmosphere. The amount of organically fixed nitrogen greatly exceeds that fixed by inorganic processes; Hutchinson (1954) has carefully analysed the geochemical cycle of nitrogen and estimates biological fixation at 0.008–0.07 mg N_2/cm^2 per year for the land surfaces of the earth, nonbiological fixation at not more than 0.0035 mg N_2/cm^2 per year. Much of this nitrogen is eventually returned to the atmosphere by the decay of organic matter. Some, however, remains in the sediments and in a few places may be sufficiently concentrated to form nitrogenous deposits, such as the Chilean nitrates; lower concentrations of nitrogen are found in guano deposits. Hutchinson gives a figure of 67–108 g/cm^2 of the earth's surface for the amount of fossil nitrogen in sediments; of this fossil nitrogen he estimates that about 8 g/cm^2 have been derived from the weathering of igneous rocks, so that 60–100 g/cm^2 have been abstracted from the atmosphere. Assuming a figure of 80 g/cm^2, this amounts to a total of 4.8×10^{20} g nitrogen removed from the atmosphere during geological time, a very small amount compared to that of carbon dioxide.

The loss of gases by escape from the earth depends on the strength of its gravitational field and the mean square velocity of the gas molecules. The mean square velocity is the velocity whose square is equal to the mean of the squares of the individual velocities; it varies inversely as the square root of the molecular weight of the gas and directly as the square root of the absolute temperature. At 0° the mean square velocity is 1.84 km/sec for hydrogen, 1.31 for helium, 0.62 for water vapor, 0.49 for nitrogen, 0.46 for oxygen, and 0.39 for carbon dioxide. At 100° these velocities are increased by 17%. For any body there is a so-called velocity of escape which may be calculated from the formula

$$V^2 = \frac{2GM}{R}$$

in which V is the velocity, G is the gravitational constant, M is the mass of the body, and R its radius. The gravitational constant is 6.67×10^{-8}, the mass of the earth 5.97×10^{27} g and its radius 6.37×10^{8} cm; therefore, the velocity of escape is 11.2 km/sec. Even if the mean velocity of the molecules is considerably less than the velocity of escape, an atmosphere will gradually be lost by the escape of fast-moving molecules from its extreme upper region, where the free paths of the molecules are so long that they stand a chance of getting away without being stopped by collisions. Jeans showed that an atmosphere will be stable throughout geological time ($> 10^9$ years) if the mean square velocity of the molecules is less than one fifth the velocity of escape. On this basis no loss of gases into interplanetary space should have taken place since the earth cooled to its present temperature, because even for hydrogen the mean square velocity is considerably less than the velocity of escape. The situation is, however, not quite so simple as it may appear. First, hydrogen and helium atoms colliding with metastable oxygen atoms in the upper atmosphere acquire sufficient momentum for their velocities to exceed the escape velocity. In addition, temperatures in the upper atmosphere are considerably higher than at the earth's surface (recent estimates suggest temperatures between 200 and 1200° at altitudes of 120–700 km, as a result of absorption of ultraviolet radiation) and this sufficiently increases the mean square velocities of hydrogen and helium to permit their escape. The fact that no more than one tenth the amount of helium supplied by radioactive disintegration during geological time is present in the atmosphere shows that such escape has occurred.

CONSTANCY OF ATMOSPHERIC COMPOSITION

Geological evidence indicates a uniformity in climatic and biological conditions since early Paleozoic times which could not have existed had the atmosphere been subject to marked changes in composition. Nevertheless, minor changes in atmospheric composition have been suggested as possibly responsible for climatic variations. For example, the onset of glacial periods has been ascribed to a decrease in carbon dioxide, since this gas is a selective absorbent for solar radiation. On the other hand, water vapor performs a similar function, and its presence in the atmosphere in large amounts far outweighs the climatic significance of CO_2. The present attitude is that only a small part of climatic variations can be attributed to varying CO_2 content of the atmosphere.

Eskola made the interesting suggestion of a correlation between periods of rapid organic evolution and periods during which carbon dioxide was available in greatest amounts. He points out that rapid organic evolution has coincided with worldwide orogeny. During orogenic periods igneous activity is at a maximum; hence the addition of carbon dioxide to the atmosphere also reaches a maximum. The increased supply of carbon dioxide would stimulate organic activity and be at least a contributory factor in the evolution of higher forms of life.

Considerable interest has been aroused by the prospect of a marked increase of carbon dioxide in the atmosphere as a result of the greater use of fossil fuels in recent years. The latest figure for the amount of coal mined annually is 1.63×10^9 tons; assuming an average of 78% C, this would produce 4.67×10^{15} g CO_2. For oil the annual production figure is 3.4×10^9 barrels (1 barrel = 160 liters); assuming a density of 0.9 and an average of 85% C, this would produce 1.53×10^{15} g CO_2. The total CO_2 produced by the burning of the annual production of coal and oil is 6.2×10^{15} g, or about $\frac{1}{300}$ of the amount in the atmosphere today. This might suggest that at the present rate of consumption of fossil fuels atmospheric carbon dioxide will be doubled in 300 years. However, in this connection the importance of the hydrosphere as a reservoir of carbon dioxide should be emphasized; its significance has been discussed by Revelle and Suess (1957). Sea water contains 20 g CO_2/cm^2 of the earth's surface, as against 0.4 g/cm^2 in the atmosphere. Oceanic and atmospheric carbon dioxide are interdependent, the former being a function of the partial pressure of CO_2 in the atmosphere. Thus to double the partial pressure of carbon dioxide in the atmosphere would require the addition of much more than is now present therein, because most of that added would be absorbed by the ocean; similarly, to decrease the carbon dioxide in the atmosphere by one half would require the removal of many times the present content. It is apparent that the oceans, by controlling the amount of atmospheric CO_2, play a vital part in maintaining stable conditions suitable for organic life on the earth. The dense carbon dioxide atmosphere of Venus may well be due to a total lack of a hydrosphere to act as a regulating medium.

VARIABLE CONSTITUENTS OF THE ATMOSPHERE

So far we have been discussing the invariable constituents of the atmosphere—those that are present in all samples of the air and in amounts that do not vary appreciably from time to time or from place to place. Besides these constituents there are others that must be taken into account. Of these, the most important is water vapor, which may be present in amounts varying from 0.02 to 4% by weight. The content of water vapor depends on a number of factors, of which temperature is the most significant; thus the annual average values for different latitudes are 0°, 2.63 vol. %; 50° N, 0.92 vol. %; 70° N, 0.22 vol. %. Water vapor in the atmosphere plays an important part in regulating climatic conditions. It absorbs heat radiation and acts like the glass roof in a greenhouse. This absorption of heat brings about a closer approach to uniformity in the temperatures of different latitudes than would exist on a dry earth.

Sulfur compounds are present in the atmosphere in variable quantities and may be considered as contaminants rather than normal constituents. Hydrogen sulfide is a product of putrefaction and is also given off, together with sulfur dioxide, by volcanoes. An important local source of sulfur compounds in the air is the combustion of coal, and as is to be expected the highest concentrations of such compounds are found in industrial areas. Sulfurous gases are temporary constituents of air insofar as they are readily soluble in water and are washed out by rain. The total amount of sulfur thereby circulated through the atmosphere is by no means insignificant; careful measurements over a period of years at Rothamsted, England, showed that the equivalent of 17.26 lb of SO_3 was annually precipitated on each acre of land at the agriculture experimental station there.

Similar measurements have been made in many localities to evaluate the amount of combined nitrogen brought to the surface of the earth in solution in rain water. The figures vary from 1 to 10 lb per acre per year, mainly in the form of ammonium nitrate. A large part of this has been added to the atmosphere by the decomposition of organic matter at the earth's surface, but some is due to the inorganic production of oxides of nitrogen. Nitrous oxide is produced in the soil by bacterial activity, and the small concentration of this gas in the atmosphere can be regarded as a steady state between the biological production and the direct or indirect decomposition of the gas in the upper atmosphere. As mentioned previously, Hutchinson estimates that the maximum nonbiological fixation of nitrogen amounts to about 0.0035 mg/cm^2 of surface per year, corresponding to about one seventh of the mean annual precipitation of nitrate nitrogen in temperate localities.

Of the temporary constituents of the atmosphere, sodium chloride is very important. Figures for the amount of this substance precipitated upon the land by rain range from 20 to 200 lb per acre annually. Most if not all represents salt raised by vapor from the ocean. As would be expected, sodium chloride in the air is greatest near the sea and rapidly diminishes away from the coast. The atmospheric circulation of salt has received much attention, since cyclic salt of this kind is largely responsible for the sodium chloride of inland waters.

The other halogens—fluorine, bromine, and iodine—have also been detected in the atmosphere. Fluorine is probably an industrial contaminant liberated by the burning of fuel and the calcination of material containing fluorine or released in the manufacture of phosphate fertilizers from rock phosphate, which always contains some fluorine (in the form of apatite). In places this results in an accumulation on surrounding grasslands which may be harmful to animals. Bromine and iodine appear to be universally present in the atmosphere in very small and variable amounts; analyses of rain water show 0.03–0.002 mg/liter of bromine and 0.002–0.0002 mg/liter of iodine. The ocean is probably the major source of these elements, although some will be contributed by industrial gases. The ratio of Cl:Br:I is 100:0.34:0.00021 in sea water and about 100:5:0.5 in rain water; this indicates a moderate enrichment of bromine and an enormous enrichment of iodine in atmospheric precipitation. Iodine is readily liberated from its compounds, and because it has an appreciable vapor pressure at ordinary temperatures it can easily exist free in the atmosphere.

The influence of the internal combustion engine on atmospheric geochemistry has become increasingly marked in recent years, especially in large cities. The automobile is a primary source of carbon monoxide; atmospheric carbon monoxide begins to be hazardous to humans at concentrations of about 100 ppm, if this level is maintained for several hours. Such concentrations may be reached occasionally in areas of heavy traffic. When combustion takes place at high pressures, as in the cylinders of an internal combustion engine, nitric oxide is produced, and this is oxidized in the atmosphere to nitrogen dioxide, a considerably more toxic gas. A further source of atmospheric pollution is the lead compounds used as antiknock agents in gasolines. Since leaded gasolines were introduced in 1923 more than 2.6 million tons of lead have been converted into lead alkyls and burned in automobiles. The present annual production of lead alkyls is equivalent to about 300,000 tons of lead. The average cumulative lead contamination in the northern hemisphere is about 10 mg/m^2 from gasoline burning alone. In highly industrialized and motorized areas the amount of lead voided into the atmosphere is many times greater than the average.

SELECTED REFERENCES

Abelson, P. H. (1966). Chemical events on the primitive earth. *Proc. Nat. Acad. Sci.*, **55**, 1365–1372. A review of the nature of the primitive atmosphere and ocean in the light of geological and geophysical information.

Brancazio, P. J., and A. G. W. Cameron (eds.) (1964). *The origin and evolution of atmospheres and oceans*. 314 pp. John Wiley and Sons, New York. The proceedings of a conference on the processes responsible for the production and subsequent development of planetary atmospheres and oceans; also includes a reprint of Rubey's article "Geologic history of sea water" (1951).

Cloud, P. E. (1965). Significance of the Gunflint (Precambrian) microflora. *Science* **148,** 27–35. A critical evaluation of the evidence from geochemistry, paleontology, and stratigraphy bearing on the evolution of the atmosphere.

Holland, H. D. (1962). Model for the evolution of the earth's atmosphere. *Geol. Soc. Am.*, Buddington Vol., 447–477. A carefully reasoned account of the processes by which the present atmosphere was developed from a primordial reducing atmosphere.

Hutchinson, G. E. (1954). The biogeochemistry of the terrestrial atmosphere. *The earth as a planet*, G. P. Kuiper (ed.) Chapter 8 (pp. 371–433). University of Chicago Press, Chicago. An exhaustive review of the factors which influence the composition of the atmosphere, with evaluations of their quantitative effects.

Kuiper, G. P., and others (1952). *The atmospheres of the earth and planets*. 434 pp. University of Chicago Press, Chicago. A collection of fourteen papers presented at a symposium on planetary atmospheres held at the University of Chicago in 1947. Although most of the book deals with the upper atmosphere of the earth, it contains two general papers by Kuiper, one by R. T. Chamberlin on geological evidence regarding the evolution of the earth's atmosphere, and one by Brown on the formation of the earth's atmosphere.

Rankama, K., and Th. G. Sahama (1950). *Geochemistry*. Chapter 7.

Rayleigh, Lord (1939). Nitrogen, argon, and neon in the earth's crust, with applications to cosmology. *Proc. Roy. Soc. (London)* **A170,** 451–464. Nitrogen, argon, and neon occur in measurable amounts in igneous rocks; most of the nitrogen is present in chemical combination.

Revelle, R., and H. E. Suess (1957). Carbon dioxide exchange between atmosphere and ocean. *Tellus* **9,** 18–32. A quantitative discussion of the carbon dioxide cycle, with special reference to the possibility of increased atmospheric carbon dioxide as a result of the burning of coal and oil.

Rubey, W. W. (1951). Geologic history of sea water. *Bull. Geol. Soc. Amer.* **62,** 1111–1147. This paper contains a great amount of data on the gases contributed to the atmosphere and the hydrosphere by magmatic activity and discusses the geological implications.

Rubey, W. W. (1955). Development of the hydrosphere and atmosphere, with special reference to probable composition of the early atmosphere. *Geol. Soc. Amer. Spec. Paper* **62,** 631–650. The different hypotheses for the origin of the earth's atmosphere and hydrosphere are considered in the light of physicochemical principles and geological evidence.

Urey, H. C. (1952). *The planets*. 245 pp. Yale University Press, New Haven, Conn. One topic discussed in this book is the origin and composition of the earth's primitive atmosphere.

Wickman, F. E. (1956). The cycle of carbon and the stable carbon isotopes. *Geochim. Cosmochim. Acta* **9,** 136–153. The amounts of carbonate carbon and organic carbon in sedimentary rocks is estimated from measurements of C^{12}/C^{13} fractionation.

CHAPTER NINE

THE BIOSPHERE

THE NATURE OF THE BIOSPHERE

The concept of the biosphere was introduced by Lamarck, and the term is used in two senses, one denoting that part of the earth capable of sustaining life and the other the sum total of living matter—plants, animals, and microorganisms. In the first sense it appears to comprise a narrow zone at the surface of the hydrosphere and the lithosphere, and this indeed is the region in which life is most prolific, thanks to the favorable conjunction of water, air, and radiant energy. However, the zone of life is more extensive than this: insects and spores have been collected at high altitudes, living organisms have been dredged from the ocean bottom at great depths, and bacteria have been found in oil-well brines from within the upper crustal layers.

The biosphere evidently developed later than the other geochemical spheres, since life can hardly have begun before surface conditions were much as they are today, presumably with an atmosphere and a hydrosphere. How much later is largely a matter of speculation. Well-preserved fossils first appear in Cambrian formations, but organisms complex enough to leave fossil remains must surely have been preceded by a long evolutionary cycle. Even in very ancient Archean formations slates and schists containing free carbon are found, and it is reasonable to assume that this free carbon is the residue of some form of life. Structures believed to be those of primitive algae have been recognized in limestones of the Bulawayan system of Southern Rhodesia, one of the oldest Precambrian systems, with an age of about 2.6 billion years. On the north shore of Lake Superior well-preserved microfossils identified as bacteria and blue-green algae are found in cherts of the Gunflint formation, believed to be about 1.9 billion years old. It thus seems probable that life began about three billion years ago, but evolved comparatively slowly until the development of green plant photosynthesis, apparently about 1.5 to 2 billion years ago. Once the atmospheric oxygen had risen to sufficiently high levels for respirational metabolism (the "Pasteur point," approximately one per cent of present atmospheric oxygen), metazoan organisms could develop. This was perhaps about one billion years ago. The fact that Metazoa are not unequivocally recorded as fossils before the Cambrian may mean that there was not sufficient atmospheric oxygen for metazoan metabolism until that time, or that the evolution of such organisms required a long chain of events that could not begin until adequate free oxygen was available.

In the material sense the biosphere is made up of a group of complex organic compounds. Life as we know it consists of a relatively small number of organic molecules—alcohols, fatty acids, amino acids, and purines—out of which are built up the more complex molecules of the carbohydrates, proteins, fats, and nucleic acids. Life has one common material characteristic, the presence of protein molecules, and one common physicochemical process, the stepwise formation of organic substances carried out practically isothermally. Living matter can in general be divided into the two great classes of plants or producers, animals or consumers. Because animals are completely dependent on vegetable matter for their nourishment, they cannot be considered as independent organisms but rather as parasitic on the plants that constitute their food. The assimilation of carbon dioxide by means of chlorophyll and light is of dominant importance for life on this planet. The process can be expressed by the following equation:

$$6CO_2 + 6H_2O = C_6H_{12}O_6 + 6O_2$$

For the reaction as written $\Delta G = 688{,}000$ cal and $\Delta H = 673{,}000$ cal. Because ΔG is positive, energy must be supplied if the reaction is to go from left to right; this energy is provided by solar radiation. In this way the radiant energy of sunlight is converted into the chemical energy of organic compounds. The reverse reaction is one of biological oxidation, or respiration, and the energy thus liberated may appear as heat or as work. A fairly close balance presumably exists between photosynthesis and respiration, although over the whole of geological time respiration has been exceeded by photosynthesis, the energy derived therefrom being stored in coal and petroleum.

THE MASS OF THE BIOSPHERE

Looking on the biosphere as the sum total of living matter, we may attempt to determine its mass, just as we have determined the mass of other geochemical zones. Here, however, we are faced with many difficulties, difficulties that were not present in other calculations of this kind. The matter of the biosphere is not uniformly distributed, as is, for example, that of the atmosphere. It is also in a constant state of change, and the cycle of changes is very rapid. The life span of any organism is minute in comparison to geological time, and the life cycles of different organisms are immensely varied—compare, say, the life of a redwood tree with that of a protozoon.

In spite of all these difficulties quantitative work in biogeochemistry demands some idea of the total amount of matter in the biosphere. Borchert (1951) estimated that the total amount of carbon in living organisms is 2.8×10^{17} g. The total annual productivity is also an important figure in discussions of the biosphere. Hutchinson (1954) has analyzed this problem and accepts the data of Riley as the most reliable. Riley expressed his results in terms of the total annual productivity of organic carbon per square kilometer of different environments. He arrived at a figure of about 160 (metric) tons/km^2 as the average for terrestrial environments (forest, cultivated lands, steppe, and desert) and 340 tons/km^2 for open oceanic waters. The total annual production of organic carbon thus determined is $20 \pm 5 \times 10^9$ tons for terrestrial environments and $126 \pm 82 \times 10^9$ tons for marine environments, a grand total for the earth of $146 \pm 87 \times 10^9$ tons.

Examination of these figures should induce a properly humble opinion of the land as the predominant domain, even in the realm of life. The ocean is twice as fertile as the land per square kilometer, and because of its much greater area it is quantitatively predominant in the amount of living matter it supports. Admittedly, life on land is practically limited to the surface, whereas life in the ocean ranges to depths of several kilometers; nevertheless, the concentration of marine life is near-surface, within the range of radiant energy—the euphotic zone. The great bulk of life in the sea consists of minute, free-floating organisms, the phytoplankton and the zooplankton, which occur in all oceans from the surface to the depth of effective light penetration (roughly 40 to 150 m). The enormous number of individuals compensates for their small size. The comparative uniformity in fertility of the marine environment contrasts strongly with the variability on the land. Large areas of the land, for example, arid tracts and polar regions, are almost unproductive, whereas no such areas occur in the seas. The abundance of life in polar seas contrasts particularly with the sterility of the land in those regions.

The mass of the biosphere is insignificant in comparison to that of other geochemical zones. According to Rankama and Sahama, the relative weights of the hydrosphere, the atmosphere, and the biosphere are expressed by the following figures:

Hydrosphere	69,100
Atmosphere	300
Biosphere	1

In spite of the negligible mass of the biosphere, however, it is a zone of great chemical activity, and its geochemical effects are of considerable significance. Its significance can best be realized in terms of the turnover of material taking place therein. If the mass of the biosphere has been approximately constant over the last 500 million years and if the average life cycle were one year, then the total amount of matter that has passed through the biosphere in that time is comparable with the total mass of the earth. Of course, most organisms appear to us as inconspicuous blobs of organic matter which have a certain life cycle, reproduce, die, and leave their surroundings very much as they were before. This picture is, however, a false one. We saw in the previous chapter how much of the free oxygen of the atmosphere has been produced by the chemical activity of plants. An individual organism secreting calcium carbonate does not change the character of the earth much, but during geological time such organisms have taken calcium in enormous amounts from a dilute solution in sea water and thereby formed most of the carbonate rocks. An even more remarkable separation is the deposition of silica by radiolaria and diatoms from sea water which contains as little as 0.02–4 g/ton of silicon. Millions of square kilometers of the ocean floor are covered with siliceous sediments, and cherts representing deposits of such organisms are common in the geological column. The organically deposited carbon in sediments has been extracted from an atmosphere which throughout geological time has probably never contained more than a few hundredths of a per cent of carbon dioxide. In addition, every organism has some metal compounds in its structure, and the biosphere is thus of considerable significance in the migration and concentration of many elements.

THE COMPOSITION OF THE BIOSPHERE

If an estimate of the mass of the biosphere is difficult, an estimate of its average composition is even more so. Living matter varies greatly in composition, the analytical data are rather meager, and our knowledge of the relative amounts of different organisms in the biosphere little more than guesses. To a greater or less extent water is the principal component of all organic matter; wood contains about 50% and vertebrates about 66%, whereas marine invertebrates may be more than 99% H_2O. Webb and Fearon (1937) discussed the ultimate composition of biological material, and Table 9.1 is taken from their paper, with some additions. The metabolic role of the invariable elements is fairly well understood. They can be roughly classified into energy elements (C, O, H, N), macronutrients (P, Ca, Mg, K, S, Na, Cl—Na is essential for animals, but not for plants), and micronutrients (Fe, Cu, Mn, Zn for both animals and plants, B, Mo, and Si for plants only, and Co and I for animals only). A few of the variable elements,

*Table 9.1. Distribution of Elements as Percentage Body Weight of Organisms**

Invariable			Variable		
Primary 60–1	Secondary 1–0.05	Microconstituents <0.05	Secondary	Microconstituents	Contaminants
H	Na	B	Ti	Li	He
C	Mg	Fe	V	Be	A
N	S	Si	Br	Al	Se
O	Cl	Mn		Cr	Au
P	K	Cu		F	Hg
	Ca	I		Ni	Bi
		Co		Ge	Tl
		Mo		As	
		Zn		Rb	
				Sr	
				Ag	
				Cd	
				Sn	
				Cs	
				Ba	
				Pb	
				Ra	

* Webb and Fearon, 1937, with additions. Webb and Fearon remark that for some of the elements the number of organisms analyzed by reliable methods is insufficient to enable to one to decide whether an element is strictly variable or invariable, and, furthermore, that the quantitative classification, though convenient, is necessarily arbitrary.

such as vanadium and bromine, may play an important part in the metabolism of some organisms, but for many of them no biological function is known. Many trace elements, whether or not they are essential, are noxious above quite low levels. It has been suggested that radioactive elements may play a part in cell division and the makeup of genes.

In all, over sixty elements have now been recorded in analyses of organisms, but many of these are "ultramicro" constituents. Great difficulty is experienced in cleaning organisms for analysis, since inorganic matter on the surface and in the digestive system is not easy to remove, and many published data are therefore suspect. However, if analytical techniques were sufficiently refined, presumably every element could be detected in uncontaminated organic matter, if only to the extent of a few thousand atoms. The statement that an element is "present" in one species and "absent" from another may give unwarranted prominence to comparatively small differences in concentration which happen to span the limits of sensitivity of the analytical method used. For an element to have biogeochemical significance it must either be shown to participate in the life of the organism or else to be present consistently in concentrations greater than that in the environment.

When the data on the occurrence of the different elements in organisms are plotted on the periodic table (Figure 9.1), certain relations are apparent. The invariable elements (except iodine and molybdenum) are all of low atomic number, belonging to the first four horizontal rows in the table.

The probability that an element plays an important part in metabolism depends upon its ability to participate in the aqueous colloidal systems constituting the bodies of living organisms. Common elements are more likely to occur than rare ones, elements usually soluble in aqueous systems more than those highly insoluble. Hutchinson (1943) showed that ionic potential is significant in this. Table 9.2 presents his data comparing the amounts of some elements in igneous rocks and in plant material. It is evident that elements of comparable abundance are very unequally taken up by plants. When the ratio between the percentage of an element in plant material and the average in rocks is plotted on Goldschmidt's ionic potential diagram

Table 9.2. The Relative Amounts of Different Elements in Igneous Rocks and in Organisms

	(1) Percentage in Igneous Rocks	(2) Percentage in Living Terrestrial Plants	Ratio (2):(1)
Li	0.0065	0.00001	0.0015
Na	2.83	0.02	0.007
K	2.59	0.3	0.12
Rb	0.031	0.0002	0.0066
Mg	2.09	0.07	0.034
Ca	3.63	0.5	0.14
Sr	0.015	0.002	0.13
Ba	0.025	0.003	0.12
Rare earths	0.0148	0.00004	0.0027
Al	8.13	0.002	0.00025
Ti	0.64	0.0002	0.00032
Si	27.72	0.15	0.0055
B	0.0003	0.005	1.7
P	0.08	0.07	0.88
S	0.052	0.05	0.96
V	0.015	0.00002	0.0013
Mo	0.0015	0.00005	0.033

After Hutchinson, 1943.

H ≡																	He
Li −	*Be* −											B −	C ≡	N ≡	O ≡	*F* −	Ne
Na ≡	Mg ≡											*Al* −	Si −	P ≡	S =	Cl =	A
K =	Ca =	*Sc* −	*Ti* =	*V* =	*Cr* −	Mn −	Fe −	Co −	*Ni* −	Cu −	Zn −	*Ga* −	*Ge* −	*As* −	*Se* −	*Br* =	Kr
Rb −	*Sr* −	*Y* −	Zr	Nb	Mo −		Ru	Rh	Pd	*Ag* −	*Cd* −	In	*Sn* −	Sb	Te	I −	Xe
Cs −	*Ba* −	*La-Lu* −	Hf	Ta	W	Re	Os	Ir	Pt	Au	Hg	Tl	*Pb* −	Bi			
	Ra −		Th		U												

Invariable {Primary (1–60%) ≡; Secondary (0.05–1%) =; Microconstituents (<0.05%) −}
Variable {*Secondary* =; *Microconstituents* −}

Figure 9.1 Elements classified according to their distribution as percentage body weight of organisms. (After Webb and Fearon, 1937, with some more recent data)

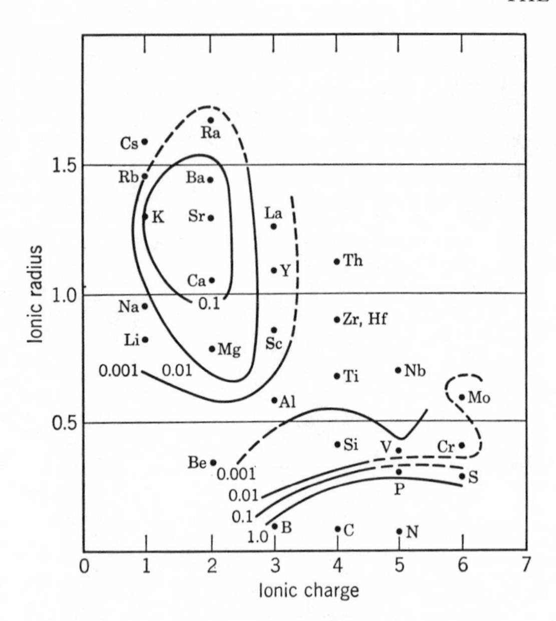

Figure 9.2 Concentration ratio of elements in terrestrial plants and in the crust, in relation to ionic potential. (Hutchinson, 1943, p. 337)

(Figure 9.2), the correlation is readily seen. Elements of low ionic potential, which form soluble cations, and those of high ionic potential, which form soluble anions, are readily assimilated; the elements of intermediate ionic potential, which form insoluble hydrolysates, are unavailable and generally play no part in metabolism. Since plants sustain herbivorous animals and, ultimately, carnivorous animals as well, the same features apply to these organisms.

Some data on chemical composition of specific organisms are reproduced in Table 9.3. If we compare these figures with those for the average abundances of the same elements in the earth's crust, we find that a few are decidedly enriched in organisms. Hydrogen and oxygen head the list, of course, since organisms are so largely water, but carbon, nitrogen, phosphorus, and sulfur are also concentrated by the vital activities of animals and plants. Considered in terms of the ash, instead of the living material, several more elements are enriched—Na, K, Br, and I in a marine invertebrate such as *Calanus finmarchicus;* Ca, K, Cl, Mg, and B in alfalfa; P, Na, K, Cl, Br, and I in man. On the average, however, plants and animals concentrate relatively few of the elements. Exceptions are known, especially for plants, which are often capable of taking up large amounts of an element

Table 9.3. Elementary Composition (Weight Per Cent) of Different Organisms

Calanus finmarchicus (a copepod)*			Alfalfa (lucerne)†			Man‡		
O	79.99		O	77.90		O	62.81	
H	10.21		C	11.34		C	19.37	
C	6.10	n. 10^{0}	H	8.72	n. 10^{0}	H	9.31	n. 10^{0}
N	1.52		N	8.25		N	5.14	
Cl	1.05					Ca	1.38	
Na	5.4	n. 10^{-1}	P	7.06	n. 10^{-1}	S	6.4	n. 10^{-1}
K	2.9		Ca	5.80		P	6.3	
S	1.4		K	1.70		Na	2.6	
P	1.3		S	1.037		K	2.2	
						Cl	1.8	
Ca	4	n. 10^{-2}	Mg	8.2	n. 10^{-2}	Mg	4	n. 10^{-2}
Mg	3		Cl	7.0				
Fe	7	n. 10^{-3}	Si	9.3	n. 10^{-3}	Fe	5	n. 10^{-3}
Si	7		Fe	2.7		Si	4	
B	1.5		Al	2.5		Zn	2.5	
Br	9	n. 10^{-4}	B	7.0	n. 10^{-4}	Rb§	9	n. 10^{-4}
I	2		Rb	4.6		Cu	4	
Cu	6		Mn	3.6		Sr	4	
			Zn	3.5		Br	2	
			Cu	2.5		Sn	2	
			F	1.5		Mn	1	
			Mo‖	1		I	1	
V	4	n. 10^{-5}	Ti	9	n. 10^{-5}	Al	5	n. 10^{-5}
Mo	2.4		Ni	5		Pb	5	
Ti	2		Br	5		Ba	3	
			Ti	4.6		Mo§	2	
			V	1.6		B	2	
Co	6	n. 10^{-6}	I	2.5	n. 10^{-6}	As	5	n. 10^{-6}
			Co	2		Co	4	
						Li§	3	
						V§	2.6	
						Ni	2.5	

As, Sn, Pb, Sr, Ba n. 10^{-n}

* Vernadsky, *Z. Krist. Mineral. Petrog.*, Abt. B, *Mineral Petrog. Mitt.* **44,** 191, 1933.
† Water, 75.1%; org., 22.45%; ash, 2.45%; from Bertrand, 1950, p. 442.
‡ Water, 60%; org., 35.7%; ash, 4.3%; from Bertrand, 1950, p. 442.
§ Mean figure for mammals.
‖ Normally in plants, 3.10^{-5}.

if it is readily available in the soil. A classic example is afforded by some of the plants associated with seleniferous soils of the western Great Plains of the United States; selenium contents as high as 1.5% have been reported in some specimens of locoweed (*Astragalus racemosus*). The ability of certain plants to reflect the presence of unusual concentrations of a particular element in the soil and subsoil has been applied to the search for hidden ore bodies (one method of geochemical prospecting; see, for example, Cannon, 1960).

Generally, however, the most remarkable examples of chemical concentration through biological processes are provided by marine organisms. The ability of many marine invertebrates to extract calcium carbonate and of diatoms and radiolaria to extract silica has already been mentioned. Bromine is concentrated as dibromoindigo, the classical Tyrian purple dye, in some species of *Murex*. Vanadium is concentrated by ascidians and a few other organisms. Some sponges and corals are rich in iodine, but marine plants are particularly effective in extracting this element, so much so that certain seaweeds were once harvested to provide commercial iodine. Oysters commonly build up a concentration of copper two hundred times that in sea water; along with many other marine animals their respiratory fluid is hemocyanin, a close relative of hemoglobin, but with copper instead of iron as an essential constituent.

THE BIOGENIC DEPOSITS

We have been considering the composition of living organisms as a background to the understanding of deposits in the geological column resulting from biological activity. Such deposits may be called bioliths and were divided by Grabau into two groups, the acaustobioliths, or noncombustibles, and the cautobioliths, or combustibles. The most important acaustobiolith is limestone, which has been discussed in the chapter on sedimentary rocks. Siliceous deposits of organic origin are much less widespread. The biological origin of nonclastic siliceous rocks in the geological column is difficult to prove, since the silica in organisms (diatoms, radiolaria, and sponges) is present as opal, which is easily reconstituted by diagenesis with the obliteration of organic structures. However, sea water is not saturated with SiO_2, and therefore the initial deposition of silica from solution in sea water must be biogenic. Chert is a general term for sedimentary nonclastic siliceous material, and it may occur dispersed as nodules, generally in limestone, and sometimes as the major part of whole geological formations hundreds and occasionally thousands of feet thick. Bramlette (1946) has given a careful account of one of these great chert formations, the Monterey of California, and summarized the work of previous investigators on this and similar deposits. The Monterey chert lends itself to the elucidation of the mode of formation of these rocks, since it is geologically young (Miocene) and little altered. Its original nature is therefore more certainly deciphered than that of older formations in which diagenesis and mild metamorphism have caused solution and redeposition of the silica with the obliteration of primary structures. Bramlette concludes that the silica was deposited originally as organic remains, mainly diatoms. In effect, the Monterey cherts and probably many other such deposits are primarily organic in origin.

Microorganisms probably play a considerable role in the formation of many biogenic deposits. The evidence for this is carefully presented by Beerstecher (1954). Bacteria and algae are capable of great chemical activity. Several types precipitate hydrated ferric oxide, utilizing the energy of a reaction of the type

$$4FeCO_3 + O_2 + 6H_2O = 4Fe(OH)_3 + 4CO_2 + 40{,}000 \text{ cal}$$

Some geologists consider that these organisms have been responsible for the deposition of extensive iron ores. Gruner has photographed structures that he considers are the remains of bacteria and algae in ferruginous cherts of the Precambrian iron formations of the Mesabi Range; however, these identifications have been criticized and the structures ascribed to inorganic processes.

Because of the delicacy of microorganisms they seldom leave any recognizable trace behind them, so the positive identification of fossil bacteria is exceedingly difficult. However, the demonstration of a bacterial origin for deposits in the geological column may be obtained indirectly. A case in point is the occurrence of sulfur deposits in sedimentary rocks. These have been investigated by Feely and Kulp (1957). Anaerobic bacteria can reduce sulfates to hydrogen sulfide, and the formation of these sulfur deposits has been ascribed to their activity. This hypothesis has been greatly strengthened by work on the sulfur isotopes. Bacterial reduction of sulfate to sulfide results in a higher S^{32}/S^{34} ratio in the sulfide than the sulfate from which it originated. Sulfide and sulfur of the deposits in salt domes of Louisiana and Texas have higher S^{32}/S^{34} ratios than the associated sulfate, indicating enrichment in S^{32} during their formation. It is believed that bacterial reduction of calcium sulfate produced hydrogen sulfide, which then reacted with more calcium sulfate to produce the sulfur of these deposits.

In this chapter we are mainly concerned with the caustobioliths, of which there are two principal divisions, coal and petroleum. Geochemically, these materials represent concentrations of carbon compounds produced by organic activity. The origin of coal and petroleum presents the following problems: the nature and composition of the parent organisms; the mode of accumulation of the organic material; and the reactions whereby it was transformed into the end products.

THE ORIGIN OF COAL

Geological evidence provides ample justification for the belief that coal has been formed from terrestrial plant remains. The chief constituents of plant material are cellulose and lignin, although a great variety of minor components—proteins, essential oils, organic acids and their salts, tannins, etc.—enter into the composition of vegetable matter. The empirical formula of cellulose is $C_6H_{10}O_5$ and that of lignin approximately $C_{12}H_{18}O_9$, so that they do not differ greatly in the relative proportions of carbon, hydrogen, and oxygen. The structure of lignin is, however, aromatic—that is, made up of ring-like groups of carbon atoms, the benzene ring being a familiar example—in contrast to the aliphatic nature of cellulose. On this account some authorities believe that lignin is the chief mother substance of coal, since the breakdown products of coal are largely aromatic. In addition, lignin is resistant to attack by microorganisms, whereas cellulose is readily decomposed into carbon dioxide, methane, and aliphatic acids. However, it has been shown that cellulose can be converted into aromatic compounds under conditions similar to those believed to exist during coal formation. The story is, of course, more involved than the conversion of lignin or cellulose into coal substances. Neither cellulose nor lignin contains nitrogen, yet nitrogen compounds are important in coal. Some of these may be derived from minor constituents, such as plant proteins, but the nitrogen content of coal is relatively high compared with that of most plants. It has been suggested, therefore, that bacteria play a major part in the first stages of the formation of coal, since the nitrogen content of bacteria runs up to 13%.

The transformation of vegetable matter to coal involves two stages, one largely biochemical and the other metamorphic. The biochemical stage is essentially part of the conversion of vegetable matter into peat, during which microorganisms are active in reconstituting the organic matter. Peat formation involves the following processes: (a) rapid decomposition of water-soluble substances; (b) slow decomposition of the cellulose compounds, the nature of the peat depending to a large extent on the completeness of decomposition of these compounds; (c) gradual accumulation of resistant constituents—lignin, resins, and waxes; (d) accumulation of the cell substance of the microorganisms, which accounts for the increase in nitrogen content. The biochemical stage is eventually brought to an end when conditions become unsuitable for bacterial activity, either by burial under inorganic sediments or by the development of toxic conditions within the organic matter. Beyond that stage metamorphism, the action of heat and pressure, is responsible for coalification.

The progressive change from wood through peat to coal can readily be followed by chemical analyses. Table 9.4 gives the average compositions of wood, peat, and three successively higher ranks of coal—lignite, bituminous coal, and anthracite. Rank is a measure of the degree of metamorphism a coal has undergone. Various suggestions have been made regarding the geological factors that have determined the rank of a coal. Some of these are (*a*) the length of time since burial (Palaeozoic coals are generally high-rank, Tertiary coals low-rank), but many exceptions are known, and it is now believed that time has no direct influence on rank; (*b*) the action of heat from earth movements or from igneous intrusions; (*c*) pressure resulting from compression during folding and faulting; and (*d*) increased pressure and temperature resulting from depth of burial. Of these the latter is most readily evaluated and has been formulated into a rule (Hilt's rule): *In a series of coal seams the fixed carbon increases and the volatile matter decreases with depth.* Hilt's rule has been applied to many coal basins, and the decrease in volatile matter is generally of the order of 0.2–0.8% per 100 feet of descent. Suggate has analysed the depth-volatile relationship in a number of coal-fields and has demonstrated (Figure 9.3) a general pattern of depth-volatile distribution for British coals from low-rank bituminous (42% volatile matter) to anthracite (5% volatile matter). He presents good evidence that depth of burial is the cause of rank increase, except in areas of contact metamorphism. For the transformation of peat into low-rank bituminous coal he estimates a depth of burial of 7500 feet and for transformation into anthracite with 5% volatile matter a depth of burial of about 19,000 feet. These data can be usefully applied in reverse to determine original depth of burial of coal seams whose present position is the result of tectonic movements or erosion.

Chemically, the passage from wood to anthracite is mainly an increase in carbon coupled with a decrease in oxygen. Hydrogen also decreases, but much less rapidly. The process is essentially one of reduction. The H:O

Table 9.4. The Average Composition of Fuels

	C	H	N	O
Wood	49.65	6.23	0.92	43.20
Peat	55.44	6.28	1.72	36.56
Lignite	72.95	5.24	1.31	20.50
Bituminous coal	84.24	5.55	1.52	8.69
Anthracite	93.50	2.81	0.97	2.72

From Clarke, *The Data of Geochemistry*, p. 773.

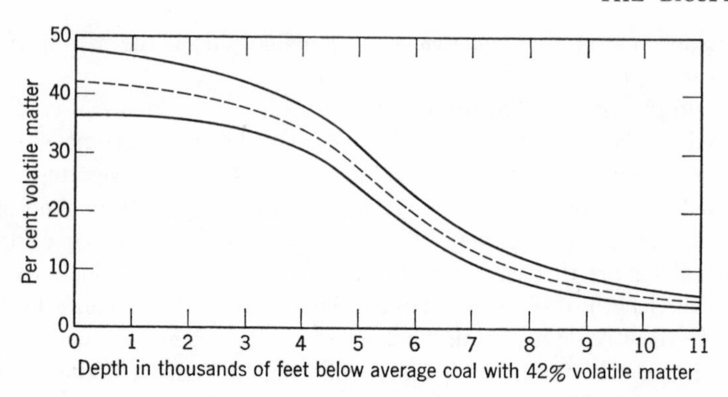

Figure 9.3 Depth-volatile relationship for British bituminous and anthracite coals, calculated on a dry, ash-free basis. (After Suggate, *Geol. Mag.* **93,** 212, 1956)

ratio (weight per cent), which is 1:8 in cellulose and 1:7 in wood, increases to 1:1 in anthracite. The changes in composition from peat to anthracite in terms of the changes in carbon and hydrogen are shown diagrammatically in Figure 9.4.

The processes outlined above give an empirical account of the transformation of plant material to coal. Little, however, is known of the chemi-

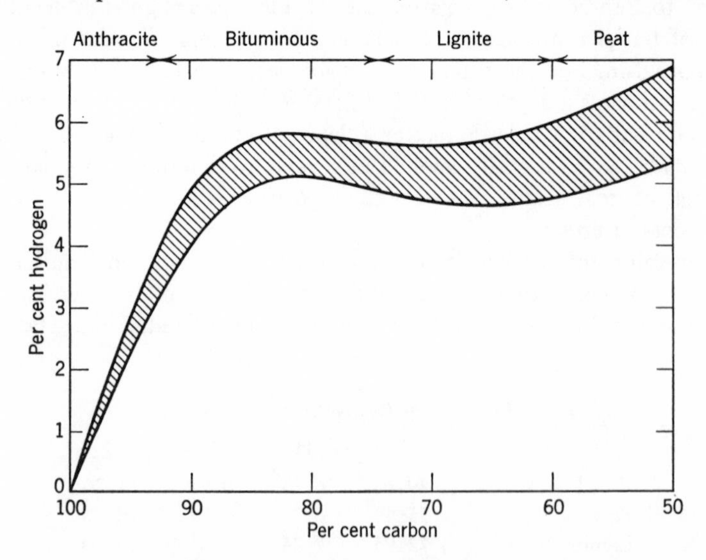

Figure 9.4 Compositional changes, calculated on a dry, ash-free basis in the series from peat to anthracite, in terms of the content of carbon and hydrogen the-so-called coal band.

cal reactions involved. Van Krevelen (1961) has provided a useful summary of the present state of knowledge in this important field. Research on the chemical constitution of coal has indicated that the coal material consists essentially of complex organic substances of high molecular weight. Study of the breakdown products of coal shows that these compounds are largely aromatic and that the degree of aromatization increases with rank, until ultimately complete aromatization to graphite may be reached. The oxygen is evidently present as OH and COOH groups and is expelled during rank increase as water and carbon dioxide. The rapid drop in hydrogen content from bituminous coal to anthracite is a reflection of the expulsion of methane at this stage of metamorphism. Nitrogen is present in coal as amino groups or substituting carbon in the ring structures. The evidence of the chemical reactions of coal suggests that the carbon in the ring systems is largely saturated, that is, the rings are naphthenic, and that six-membered rings predominate.

Maximum alteration of coaly material produces graphite; this alteration is analogous with the metamorphism of rocks, although the organic compounds in coal react more readily to increasing temperatures and pressures than do minerals. Coal composition is in fact an extremely sensitive indicator of the degree of alteration not only of its own material but also of the enclosing rocks. However, even at the anthracite stage the enclosing rocks are usually indurated but would not be described as metamorphic. In a few areas, such as the Narragansett basin of Rhode Island (Quinn and Glass, 1958), it has been possible to follow in the field and in the laboratory the transition from coal metamorphism to rock metamorphism. The change in the coal is indicated by an increasing degree of graphitization, as shown by the appearance of graphite reflections in an X-ray diffraction pattern. The anthracite is practically amorphous to X-rays, and the enclosing rocks show only a minor degree of recrystallization. With increasing metamorphism the rocks grade successively into chlorite schists and then into biotite schists. The meta-anthracite in the chlorite schist gives fairly well-defined graphite reflections, and in the biotite schist it is a well-crystallized graphite.

THE ORIGIN OF PETROLEUM

General agreement exists that the source material of petroleum was biological and was laid down in a sedimentary environment, although the theory of the inorganic origin of petroleum dies hard, and a supposedly serious book published in 1950 ascribed oil to a cosmic rain. There is some uncertainty whether nonmarine sediments are to be entirely excluded as petroleum source beds, but the predominance of marine sediments in oil-bearing formations supports the belief that the parent material of most petroleum was deposited in a marine environment.

Important clues to the origin of petroleum may be derived from a knowledge of its composition. Crude oil shows a remarkable constancy in elementary composition. Generally it has a carbon content of 83–87%, a hydrogen content of 11–14%, and other elements, mainly oxygen, nitrogen, and sulfur, up to a maximum of about 5%. Over 99% of many crude oils consists of carbon and hydrogen only. The ash content is uniformly low, from 0.001 to 0.05%.

In contrast to this uniformity and simplicity of elementary composition, crude oil shows a great variability and complexity with respect to the compounds which may be present. It consists of a variable mixture of a large number of different hydrocarbons which may be grouped into three homologous series: (*a*) paraffins, with the general formula C_nH_{2n+2}; (*b*) cycloparaffins or naphthenes, ring-type hydrocarbons with the general formula C_nH_{2n}; and benzenoid hydrocarbons, with the general formula C_nH_{2n-6}. The proportions of these major types vary considerably from one crude to another, and crudes are distinguished as paraffinic, naphthenic, or benzenoid according to the predominant type.

If crude petroleum is assumed to be a mixture in thermodynamic equilibrium, it should be possible to calculate its composition, provided adequate thermodynamic data on the many possible hydrocarbons are available. Much work on the free energies of hydrocarbons has been carried out at the U. S. Bureau of Standards and elsewhere in recent years. Some of the major results of this work are summed up in the following statements:

1. Paraffins are comparatively the most stable hydrocarbons at lower temperatures.
2. In an homologous series the stability increases as the number of carbon atoms decreases. Methane is the most stable hydrocarbon.
3. The stability of the naphthenes does not differ greatly from that of the paraffins.
4. In the temperature range most directly concerned with oil genesis benzenoid hydrocarbons are less stable than the corresponding saturated hydrocarbons.

Because of the wide variation in amounts of the different hydrocarbons in crudes of approximately the same bulk composition it is evident that crude oil is not a thermodynamic equilibrium mixture. The composition of a crude probably depends largely on the relative reaction rates of the competing reactions which have been responsible for its formation. However, oils from older strata as a rule contain more paraffins and more volatile (i.e., low molecular weight) hydrocarbons than those from younger formations. This suggests that increasing age is reflected in a closer approach to thermodynamic equilibrium.

Other clues as to origin provided by the composition of crude oil are the presence of heat-sensitive compounds, such as porphyrins and complex nitrogen-substituted hydrocarbons, which are decomposed at temperatures around 200°. Evidently temperatures of 200° are not exceeded in petroleum formation, and probably the maximum temperature has generally been much lower. This is supported by an overall correlation between the occurrence of oil and the rank of coal within the same general area; oil cannot be expected when the rank of coal exceeds a certain value, best expressed by the fixed carbon content (fixed carbon in excess of 70% usually signifies absence of petroleum). Oil is clearly more sensitive to metamorphism and is readily destroyed by conditions which merely cause an increase of rank in the coal.

The nature of the original organic matter from which crude oil has been formed is a subject of considerable controversy. It is conceivable that practically any organism, animal or plant, may have contributed to the formation of petroleum. Since, however, plankton is the most abundant organic material in the sea and forms the food of higher organisms, it is reasonable to consider its composition as representative of the composition of the raw material from which petroleum has been formed. The data on the chemical composition of plankton are summarized by Vinogradov (1953). Plankton, like higher organisms, contains carbohydrates, proteins, and fats, and small amounts of hydrocarbons have been found in diatoms. On death planktonic organisms are either eaten or sink to the bottom to become part of the marine sediments. Oceanographic investigations have revealed the environment of deposition whereby organic matter may be preserved for burial and ultimate transformation into petroleum.

Reducing conditions are clearly necessary, since oxidation rapidly decomposes organic matter into carbon dioxide and water. The environment must also be inimical to carrion-eating animals for which dead organisms serve as food. These conditions are present on the ocean floor in places where the bottom waters are stagnant, especially in basins and troughs. In such areas organic matter sinking to the bottom is not destroyed in the usual manner but is decomposed by the action of anaerobic bacteria. Putrefaction takes place, and the product is a black mud known as sapropel. A present-day example is the Black Sea, where water circulation is restricted and bottom sediments with as much as 35% organic matter have been recorded (the average marine sediment contains about 2.5% organic matter).

Sapropel is believed to be the parent substance of petroleum, just as peat is the parent substance of coal. The conversion of sapropel to petroleum probably involves both biochemical and inorganic processes, although the relative part each of these plays is subject to much difference of opinion.

Good evidence exists for the importance of bacterial action in converting carbohydrates and proteins into compounds which may change into hydrocarbons. Chemically, the conversion of organic matter to hydrocarbons can be represented schematically by the equation

$$\underset{\text{Carbohydrate}}{(CH_2O)_n} = \underset{\text{Carbon dioxide}}{xCO_2} + \underset{\text{Hydrocarbons}}{yCH_z}$$

that is, a reciprocal oxidation and reduction in which part of the organic material is completely oxidized to CO_2 and part completely reduced to hydrocarbons. However, this is certainly an oversimplification. Probably biochemical processes first produce fatty acids, which are later converted into hydrocarbons by a series of reactions involving fission, condensation, cyclation, and dehydration. These reactions may be promoted by catalysts, in which connection the significance of clay deserves to be emphasized. Clay particles are strong adsorbents and thereby bring foreign molecules into close contact. In addition, such molecules are not fixed at random but in definite positions relative not only to the clay, but also to each other, where they may be able to interact and form new compounds.

A controversial point of much significance in discussing the origin of oil has been the time factor. Since little oil has been produced from rocks younger than Pliocene (and that may have migrated from older rocks), the widely accepted view has been that the formation of oil is a slow process. This view was seemingly confirmed by the failure of past investigators to detect liquid hydrocarbons in recent sediments. However, the situation has been completely changed by the work of Smith (1954) and others, who have succeeded in extracting paraffin, naphthene, and benzenoid hydrocarbons from recent marine sediments. The quantity of free hydrocarbons varied up to 11,700 parts per million of dried sediment. That this material has formed in place and is not petroleum migrating from older formations is proved by carbon-14 dating, which has indicated an apparent age of about 10,000 years for the hydrocarbons. Extrapolation of the data for sediments of the Gulf of Mexico off Louisiana and the Santa Cruz Basin off California gives estimates varying from 4,500,000 to 10,400,000 barrels of crude oil per cubic mile of sediments in these areas. The problem of the ultimate source of petroleum thus appears solved; however, the hydrocarbons in recent sediments still differ significantly from the assemblages in crude oil. For example, the *n*-paraffin hydrocarbons extracted from recent sediments show a strong predominance of molecules with an odd number of carbon atoms, whereas no such preference is noted in crude oils. Hydrocarbons of low molecular weight are lacking in recent sediments, but they are abundant in many crude oils. Evidently the organic matter deposited with the sediments must undergo further evolution during the generation and accumulation of petroleum.

THE CONCENTRATION OF MINOR ELEMENTS IN BIOGENIC DEPOSITS

In terms of the average composition of the earth's crust coal and petroleum are greatly enriched in carbon and hydrogen and sometimes in nitrogen and sulfur. In addition, the biogenic deposits often show remarkable concentrations of rare elements. The classic example was Goldschmidt's discovery of 1.6% GeO_2 in the ash of a coal from the Newcastle district in England (a figure that has been surpassed by the finding of up to 7.5% germanium in the ash of lignite from the District of Columbia). Goldschmidt followed up this discovery by a systematic study of the geochemistry of coal ash. Table 9.5 is a summary of some of his results. The figures in Table 9.5 are for the average content of "rich" ashes, and individual samples, such as the Newcastle ash with 1.6% GeO_2, may show much higher enrichment factors. Additional data are available in a paper by Swaine, in which he compares the trace element contents in New South Wales coal ash with those recorded in the literature (Figure 9.5). The remarkable feature of these results is the heterogeneous group of elements found in coal ash. These elements differ widely in their geochemical behavior. Both chalcophile and lithophile elements are present, although the chalcophile predominate. Ionic or atomic radii are apparently not significant, for the association in the coal ashes embraces elements with small, medium, and large radii.

Table 9.5. Rare Elements in Coal Ash

Element	Average Content in Coal Ash (g/ton)	Average Content in Earth's Crust (g/ton)	Factor of Enrichment
B	600	10	60
Ge	500	1.5	330
As	500	2	250
Bi	20	0.2	100
Be	45	2.8	16
Co	300	25	12
Ni	700	75	9
Cd	5	0.2	25
Pb	100	13	8
Ag	2	0.1	20
Sc	60	22	3
Ga	100	15	7
Mo	50	1.5	30
U	400	2.7	150

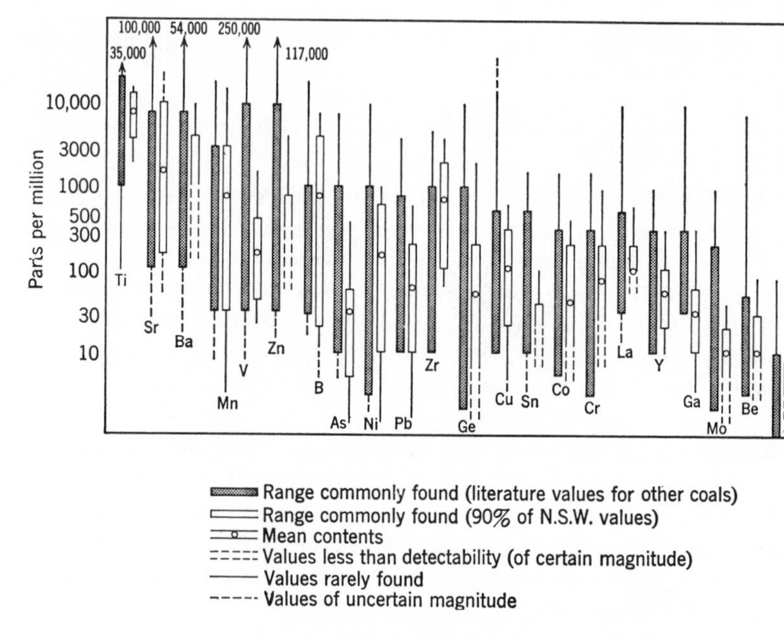

Figure 9.5 Range of trace element contents of coal ash; solid bars are for data from the literature, open bars are for determinations on New South Wales coals. (Swaine, *Technical Communication* 45, Division of Coal Research, C.S.I.R.O., Australia, 1962)

Several possibilities may be suggested for the enrichment of a rare element in coal ash:

1. The element was accumulated by the vital processes of the plants that formed the parent material of the coal (possibly followed by selective decay whereby the rare element was further concentrated).
2. The element was precipitated from ground water by adsorption or chemical reaction during coalification.
3. The element formed part of the mineral matter deposited along with the organic material.

The third possibility can be discounted because there is no reason to expect mineral matter deposited with organic material to be especially rich in rare elements. In addition, Goldschmidt's analyses show clearly that the rare elements are not present in the mineral matter, since the concentration of these elements is usually greatest in low-ash coals and least in those with high-ash content. The first possibility may be responsible for some instances of enrichment of trace elements, although in general plants show little capacity to accumulate elements from the soil. Goldschmidt favored a concentration during decay of the plant remains whereby the more soluble elements were leached out, leaving others retained either as insoluble compounds or as metal-organic complexes. The second possibility, however, seems to provide particularly favorable circumstances for concentration of rare elements. The low oxidation potential of the medium would reduce sulfur compounds to H_2S, and thereby lead to the precipitation of chalcophile elements, many of which are markedly enriched in coal ashes. Another significant process is probably chelation of metallic ions by complex organic molecules; experiments have shown that humus has considerable capacity for adsorbing a variety of cations from solution and retaining them in stable combination. This process has been held responsible for the enrichment of uranium in some lignites, in which the uranium has been leached from overlying strata and is especially concentrated in the upper part of the lignite.

Compared with coal, petroleum (including asphaltites and bitumens) is far more specific in the concentration of rare elements. Vanadium shows a strong affinity for petroleum, and over 70% V_2O_5 has been recorded from petroleum ash. Some shipping companies have found it profitable to buy fuel oil from a particular locality, since the ashes can be sold as vanadium ore. In the famous Minasragra deposit in Peru the vanadium occurs with asphaltite. According to Goldschmidt, the typical elements associated with petroleum and bitumen are vanadium, molybdenum, and nickel. He believed that they are present as organometallic compounds which migrate with the hydrocarbons, and he pointed out that these elements are effective catalysts for the synthesis of hydrocarbons; hence they may have been active in facilitating the formation of petroleum from organic remains. These elements may have been extracted from sea water by organisms that utilized them in the form of metal-organic porphyrin compounds. The porphyrin compounds are exceedingly stable and have been recognized in shales, asphalts, and petroleum dating back to the Palaeozoic; they are evidently able to withstand the ordinary processes of diagenesis.

As discussed in the chapter on sedimentary rocks, the black bituminous shales also show unusual concentrations of minor elements. An economically important example is the Mansfeld "Kupferschiefer" of Germany, which is a bituminous shale with a considerable copper content. It is worked as a copper ore and is markedly enriched in As, Ag, Zn, Cd, Pb, V, Mo, Sb, Bi, Au, and the platinum metals. The increased interest in uranium in recent years has revealed that this element, too, is often enriched in the black shales, contents of up to 400 g/ton being not uncommon.

THE GEOCHEMICAL CYCLE OF CARBON

Carbon, although not one of the more abundant elements in the earth, plays an important role, perhaps the most important one, in geochemistry, because carbon compounds are essential for every known form of life. The geochemistry of carbon is closely linked with that of the other essential elements of organisms, especially hydrogen, oxygen, nitrogen, and sulfur. Various aspects of the carbon cycle have been studied for over a century. In 1933 Goldschmidt developed the concept of the carbon cycle in both its biological and geological aspects and made quantitative estimates of the amount of carbon in the different parts of the cycle. Some of these estimates have been revised by later workers, especially Borchert (1951) and Wickman (1956); figure 9.6 is a synthesis of the available data. Borchert gives a figure of 5500 g/cm² for the total amount of carbon in the earth's crust (mean thickness 16 km); the excess over the amount in carbonate rocks and as organic carbon represents the carbon in igneous and metamorphic rocks. Of the 700 g as organic carbon Borchert estimates that about 1 g is present as extractable coal and petroleum; the remainder is disseminated in carbonaceous and bituminous sedimentary rocks.

Much work has been done in recent years on variations in the relative

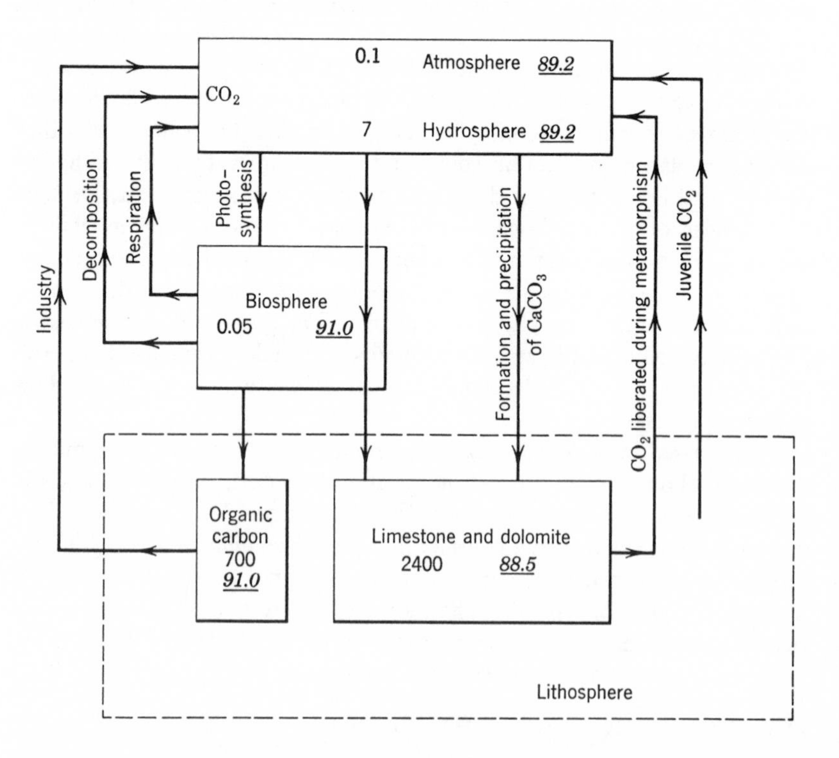

Figure 9.6 The circulation of carbon in nature; figures are amounts of carbon in g/cm² of the earth's surface, and (in italics) the average C^{12}/C^{13} ratio.

abundances of the two stable isotopes of carbon. Craig (1953) and Wickman (1956) have made many measurements of the C^{12}/C^{13} ratio in materials from different geochemical environments and have found small but regular variations (Figure 9.7). The ratio varies from about 88.3 to 91.4 and shows consistent values for specific types of material. Carbonate rocks are richest in C^{13}, with an average ratio of 88.55, and fossil organic carbon (from coal, bituminous shale, and petroleum) is richest in C^{12}, with an average ratio of 91.00. In 1941 Wickman, in one of the earliest applications of isotopic variations in geology, used the data then available on C^{12}/C^{13} ratios in an ingenious calculation to determine the amount of organic carbon in sedimentary rocks. This basis of his calculation is that the differences in C^{12}/C^{13} ratios between carbonate carbon and organic carbon represent fractionation of the two isotopes from original crustal carbon of uniform isotopic composition. If the total amount of carbonate carbon is A, organic carbon, B, their isotope ratios, x and y, and the isotope ratio of original crustal carbon, z, then

$$A/B = (y - z)/(z - x)$$

Wickman (1956) has discussed the concept of "crustal carbon" at some length and shows that the C^{12}/C^{13} ratio for diamonds, for carbonates of probable magmatic origin, and for a number of graphites all give consistent values for this ratio around an average of 89.11, which he accepts as the

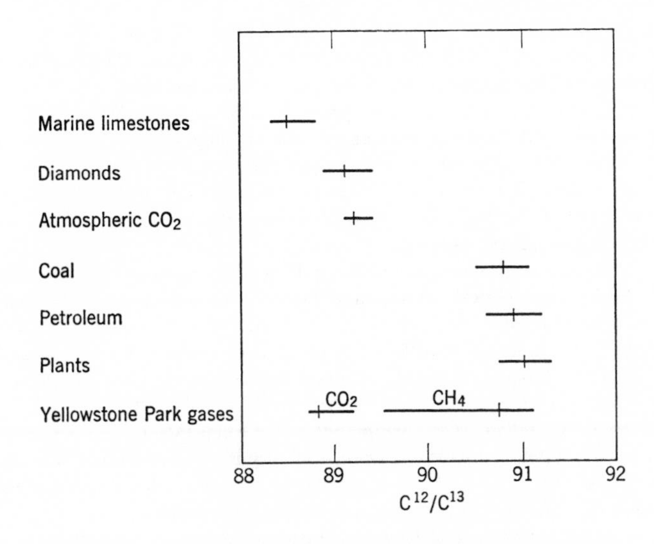

Figure 9.7 Variations in C^{12}/C^{13} ratios in different natural materials; the bar gives the range of values, and the cross-bar the average figure. (After Craig and Wickman)

figure for z. Then, using the figures given above for the C^{12}/C^{13} ratio in carbonate rocks and in fossil organic carbon and his earlier estimate of 2420 ± 560 g/cm^2 of the earth's surface for A, he calculates that B is 700 ± 200 g/cm^2. This corresponds to a content of $0.41 \pm 0.13\%$ C in the average shale, a figure consistent with direct analytical data.

SELECTED REFERENCES

Beerstecher, E. (1954). *Petroleum microbiology.* 375 pp. Elsevier Press, New York. This book is not only an exhaustive account of the role of bacteria in the origin and utilization of petroleum, but also includes an excellent chapter on the role of microorganisms in geological processes.

Bertrand, D. (1950). Survey of contemporary knowledge of biogeochemistry: 2. The biogeochemistry of vanadium. *Bull. Am. Museum Nat. Hist.* **94,** 403–456. Although concerned with a specific element, this paper has a good deal of material of general significance.

Borchert, H. (1951). Zur Geochemie des Kohlenstoffs. *Geochim. et Cosmochim. Acta* **2,** 62–75. A detailed discussion of the carbon cycle, with quantitative data for the amounts of carbon in various parts of the cycle.

Bramlette, M. N. (1946). The Monterey formation of California and the origin of its siliceous rocks. *U. S. Geol. Survey, Profess. Paper* 212. 57 pp. A detailed description of a thick and extensive non-clastic siliceous deposit and a discussion of the origin of this and similar formations in which the author presents good evidence for the organic deposition of much of the silica.

Breger, I. A. (ed.) (1964). *Organic geochemistry.* 658 pp. Pergamon Press, New York. Contains 15 chapters by different authors covering the origin, classification, and geochemistry of the major biogenic deposits; excellent bibliographies.

Cannon, H. L. (1960). Botanical prospecting for ore deposits. *Science* **132,** 591–598. A useful review of this subject, with an extensive bibliography.

Clarke, F. W. (1924). *The data of geochemistry.* Chapters 16 and 17.

Colombo, U., and G. D. Hobson (eds.) (1964). *Advances in organic geochemistry.* 488 pp. Pergamon Press, New York. A symposium covering recent developments in the geochemistry of biogenic deposits.

Craig, H. (1953). The geochemistry of the stable carbon isotopes. *Geochim. Cosmochim. Acta* **3,** 53–92. A survey of the variation of the $C^{12}/^{13}$ ratio in nature, based on measurements on several hundred samples of carbon from different geological environments.

Feely, H. W., and J. L. Kulp (1957). Origin of Gulf Coast salt-dome sulfur deposits. *Bull. Am. Assn. Petroleum Geol.* **41,** 1802–1853. These studies provide good evidence for the origin of native sulfur by the bacterial reduction of sulfate.

Goldschmidt, V. M. (1954). *Geochemistry.* Part II, Group IV.

Hedberg, H. D. (1964). Geologic aspects of origin of petroleum. *Bull. Am. Assn. Petroleum Geol.* **48,** 1755–1803. A critical review of the information from different approaches on the origin of petroleum; extensive bibliography.

Hutchinson, G. E. (1943). The biogeochemistry of aluminum and of certain related elements. *Quart. Rev. Biol.* **18,** 1–29, 129–153, 242–262, 331–363. A discussion of biological aspects of geochemistry and an account of the factors involved in the accumulation of elements through the vital activities of plants and animals.

———. (1954). The biogeochemistry of the terrestrial atmosphere. Chapter 8 (pp. 371–433) of *The earth as a planet;* ed. by G. P. Kuiper. University of Chicago Press, Chicago. A critical review of the factors which influence the composition of the atmosphere, with special reference to biological processes.

Quinn, A. W., and H. D. Glass (1958). Rank of coal and metamorphic grade of rocks of the Narragansett basin of Rhode Island. *Econ. Geol.* **53,** 563–574. An account of the chemical and mineralogical changes accompanying the transformation of coal into graphite.

Rankama, K., and Th. G. Sahama (1950). *Geochemistry.* Chapter 8.

Rutten, M. G. (1962). *The geological aspects of the origin of life on earth.* 146 pp. Elsevier Press, New York. A brief account of the hypotheses for the origin of life, and its possible evolution during Precambrian times.

Smith, P. V. (1954). Studies on origin of petroleum: occurrence of hydrocarbons in recent sediments. *Bull. Am. Assoc. Petroleum Geol.* **38,** 377–404. A classic paper describing the research which provided the first proof of the formation of liquid hydrocarbons in recent marine sediments.

Van Krevelen, D. W. (1961). *Coal.* 514 pp. Elsevier Press, New York. A comprehensive account, with special reference to the origin, metamorphism, and chemical constitution of coal.

Vinogradov, A. P. (1953). *The elementary chemical composition of marine organisms.* 647 pp. Sears Foundation for Marine Research, Yale University, New Haven. An exhaustive compilation of all available data on the composition of marine organisms, originally published in Russian in three parts in 1935, 1937, and 1944.

Webb, D. A., and W. R. Fearon (1937). Studies on the ultimate composition of biological material. Part 1. Aims, scope and methods. *Sci. Proc. Roy. Dublin Soc.* **21,** 487–504. A comprehensive investigation of the elements present in a number of different organisms.

Wickman, F. E. (1956). The cycle of carbon and the stable carbon isotopes. *Geochim. Cosmochim. Acta* **9,** 136–153. Gives data on the isotopic composition of carbon from different geological environments and uses these data to calculate the amount of organic carbon in sedimentary rocks.

CHAPTER TEN

METAMORPHISM AND METAMORPHIC ROCKS

METAMORPHISM AS A GEOCHEMICAL PROCESS

Metamorphism may be defined as the sum of the processes that, working below the zone of weathering, cause the recrystallization of rock material. During metamorphism the rocks remain essentially solid; if remelting takes place a magma is produced, and metamorphism has passed into magmatism. Metamorphism is induced in solid rocks as a result of pronounced changes in temperature, pressure, and chemical environment. These changes affect the physical and chemical stability of a mineral assemblage, and metamorphism results from the effort to establish a new equilibrium. In this way the constituents of a rock are changed to minerals that are more stable under the new conditions, and these minerals may arrange themselves with the production of structures that are more suited to the new environment. Metamorphism accordingly results in the partial or complete recrystallization of a rock, with the production of new structures and new minerals.

It is not feasible to define a sharp boundary between diagenesis—the complex of processes which converts a newly deposited sediment into an indurated rock—and the onset of metamorphism. As was discussed in the preceding chapter, the alteration of peat to anthracite and ultimately to graphite is essentially a metamorphic process, but the enclosing sedimentary rocks show little change except induration until the carbonaceous matter has been considerably graphitized. Nearly all salt deposits show evidence of partial or complete recrystallization following burial, because they contain numerous minerals which are highly sensitive to the moderate increase in temperature accompanying burial to the depth of a few thousand feet. Many tuffaceous sediments, especially those that contained large amounts of volcanic glass, have been partly or completely recrystallized to zeolitic rocks, whereas interbedded lava flows are likely to show little alteration. Clearly, the onset of metamorphism cannot be defined in terms of a fixed temperature and pressure, and depends also on the mineralogical composition and fabric of the rock—even rocks of practically the same chemical composition, say a granite and a rhyolite ash, may react very differently.

The three factors of heat, pressure, and chemically active fluids are the impelling forces in metamorphism. The heat may be due to the general increase of temperature with depth or to contiguous magmas. Pressures may be resolved into two kinds: hydrostatic or uniform pressure, which leads to change in volume; and directed pressure or shear, which leads to change of shape or distortion. Uniform pressure results in the production of granular, nonoriented structures; shear results in the production of parallel or banded structures. Uniform pressure affects chemical equilibria by promoting a volume decrease, that is, the formation of minerals of higher density. The action of chemically active fluids is a most important factor in metamorphism, because even when they do not add or subtract material from the rocks they promote reaction by solution and redeposition. When they add or subtract material the process is called *metasomatism*. Probably some degree of metasomatism accompanies most metamorphism. Water is the principal chemically active fluid, and it is aided by carbon dioxide, boric acid, hydrofluoric and hydrochloric acids, and other substances, often of magmatic origin.

THE CHEMICAL COMPOSITION OF METAMORPHIC ROCKS

The bulk chemical composition of metamorphic rocks is exceedingly variable. It may correspond to that of any of the igneous and sedimentary rocks, and metasomatism may produce a composition different again from these. It is, however, often possible to determine the nature of the original rock from its chemical composition even after it has been totally recrystallized and the original structure completely destroyed. We have seen how sedimentary rocks can be more extreme in their composition than igneous rocks, and this feature aids in the elucidation of their metamorphosed equivalents. Some of the chemical criteria that may be used to establish an originally sedimentary origin of metamorphic rocks are (*a*) an excess of alumina, which will appear as C when the norm is calculated (if C exceeds 5%, a sedimentary origin may be suspected; if C is greater than 10%, a sedimentary origin is almost certain); (*b*) $K_2O > Na_2O$ combined with $MgO > CaO$ (this feature is characteristic of argillaceous rocks, especially those containing appreciable amounts of illite and montmorillonite); and (*c*) a very high SiO_2 content (say greater than 80%, or more than 50% Q in the norm), which suggests that the rock was originally a sandstone or a chert.

The general tendency of metamorphism is to smooth out variations and to produce rocks of more or less uniform mineral composition over wide areas. This is exemplified by regions of monotonous Precambrian gneisses, consisting almost entirely of quartz, feldspar, hornblende, muscovite, and biotite. Recrystallization during metamorphism may bring about a segregation of certain minerals into lenses and bands, thereby producing chemical segregation on a small scale; this is known as metamorphic differentiation.

The bulk chemical composition of a rock may remain constant during metamorphism (isochemical metamorphism), or it may change through the introduction and/or removal of material (allochemical metamorphism). Even isochemical metamorphism involves some transference of material, although it may be over very small distances. For example, Hutton and Turner have described metacrysts of manganese garnet which have grown in a chlorite schist, even though the average manganese content of the parent rock is only 0.1%. Manganese has been transferred from the surrounding rock and concentrated in the garnet metacrysts, but the distance of migration is quite small, of the order of a millimeter. Even the change from one polymorph to another in response to changing physical conditions involves the rearrangement of atoms and hence their movement within the crystal lattice. When we speak of transference of material during metamorphism, we generally have in mind processes that result in marked composition changes of large rock masses, but it is well to remember that the same processes are active on a small scale in essentially isochemical metamorphism.

Allochemical metamorphism or metasomatism raises the question as to how the introduction or removal of material has taken place. Three ways may be suggested: transportation in a gas phase; transportation by liquids; and transportation involving neither of these. The last process may be visualized as a migration of atoms or ions along crystal boundaries or even through solids, and its significance in metamorphism is a subject of acute controversy. That such a process is possible is universally accepted. The controversy, like so many others in geology, rages around the magnitude of the effects this form of material transfer has actually produced in the rocks. Some workers ascribe to it the alteration of enormous volumes of material; specifically, it has been stated (Perrin and Roubault, 1949) that granite masses of batholitic dimensions are the result of allochemical metamorphism of preexisting rocks without the intervention of liquid or gas phases. Other workers state that migration of ions in and through solids is an insignificant factor in metamorphism at best, being completely overshadowed by other types of transportation, and that in any case the requirements for extensive migration of this kind are unlikely to be realized under the usual conditions of metamorphism.

On the whole, the evidence favors the latter view. Measurements of diffusion in silicates indicate that the rate of migration of ions in solids of this kind is much too slow to produce extensive changes, even in the time available during cycles of metamorphism. These experimental data are complemented by deductions based on energy considerations. Diffusion in and through solids is conditioned by the kinetic energy of the ions and the presence of defects in the crystal lattices. Rising temperatures will promote such diffusion by increasing the kinetic energy of the ions and the degree of disorder. An indication of the temperatures necessary for significant diffusion within a crystal structure is given by the Tammann temperature. The Tammann point is the temperature at which the mobility of ions in the structure rapidly increases; its value is approximately $0.5T$, where T is the melting point of the substance (in degrees absolute). For the common silicate minerals the Tammann temperatures range from 420° for albite to 720° for quartz. Thus at temperatures in this range (medium to high-grade metamorphism) we can expect a considerable mobility of the ions in rock-forming minerals. Metamorphic reactions could take place in the absence of a fluid phase, but the presence of water will greatly increase the speed of such reactions and lower the minimum temperature required. A laboratory example is the formation of forsterite from MgO and quartz; when the dry solids were heated together at 1300° only about 10% reacted in two hours, whereas at 600°, in the presence of water vapor under pressure, reaction was virtually complete in the same time. Water and other volatile substances are practically omnipresent, at least in small amounts, in all rocks, and are also set free in large amounts both by igneous activity and by progressive metamorphism, thereby providing a universal and effective medium for the transport of material. Field observations and laboratory experiments indicate that metasomatism is essentially the result of introduction or removal of material in a fluid phase.

THE MINERALOGY OF METAMORPHIC ROCKS

Since the chemical composition of metamorphic rocks is so varied, it is to be expected that their mineralogy will be correspondingly diverse. In addition, metamorphic rocks are formed under a wide range of temperatures and pressures, and even if no change in bulk composition takes place a mineral assemblage stable under certain (P, T) conditions may be replaced by a totally different assemblage under other (P, T) conditions. In these circumstances it is noteworthy that the mineralogy of metamorphic rocks is not more complex; this is largely due to the stability of some common minerals over considerable ranges of bulk composition and of physical conditions.

A list of some of the silicate minerals of metamorphic rocks is given in Table 10.1, along with information regarding their relative compositions. The figures in Table 10.1 are derived from the ideal formulas of the minerals,

Table 10.1. The Composition of Minerals of Metamorphic Rocks in Atomic Proportions, Based on a Common Content of 24(O,OH) Ions

Mineral	Formula	Si	Al	Mg	Ca	Fe	Na	K
Quartz	SiO_2	12						
Andalusite, sillimanite, kyanite	Al_2SiO_5	~5	~10					
Cordierite	$Mg_2Al_4Si_5O_{18}$	~7	~5	~3				
Pyrope	$Mg_3Al_2(SiO_4)_3$	6	4	6				
Chlorite	$Mg_5Al(AlSi_3O_{10})(OH)_8$	4	~3	~7				
Enstatite	$MgSiO_3$	8		8				
Anthophyllite	$Mg_7(Si_4O_{11})_2(OH)_2$	8		7				
Talc	$Mg_3Si_4O_{10}(OH)_2$	8		6				
Serpentine	$Mg_3Si_2O_5(OH)_4$	5		8				
Forsterite	Mg_2SiO_4	6		12				
Staurolite	$Fe_2Al_9Si_4O_{23}(OH)$	4	9		2			
Chloritoid	$Fe_2Al_2(Al_2Si_2O_{10})(OH)_4$	~4	~8		~4			
Almandite	$Fe_3Al_2(SiO_4)_3$	6	4		6			
Cummingtonite	$Fe_7(Si_4O_{11})_2(OH)_2$	8			7			
Wollastonite	$CaSiO_3$	8				8		
Grossularite	$Ca_3Al_2(SiO_4)_3$	6	4			6		
Zoisite	$Ca_2Al_3(SiO_4)_3(OH)$	~6	~6			~4		
Anorthite	$CaAl_2Si_2O_8$	6	6			3		
Diopside	$CaMgSi_2O_6$	8		4		4		
Tremolite	$Ca_2Mg_5(Si_4O_{11})_2(OH)_2$	8		5		2		
Jadeite	$NaAlSi_2O_6$	8	4				4	
Glaucophane	$Na_2Mg_3Al_2(Si_4O_{11})_2(OH)_2$	8	2	3			2	
Albite	$NaAlSi_3O_8$	9	3				3	
Potash feldspar	$KAlSi_3O_8$	9	3					3
Muscovite	$KAl_2(AlSi_3O_{10})(OH)_2$	6	6					2
Phlogopite	$KMg_3(AlSi_3O_{10})(OH)_2$	6	2	6				2

all reduced to a common basis of 24 (O,OH) ions. The figures do not take into account possible ionic substitution, such as magnesium for iron or ferric iron for aluminum. The table provides a useful means of rapid comparison for determining what additions or subtractions are necessary in the transformation of one mineral to another of related composition and for indicating the minerals that may appear in rocks of specific bulk composition.

The significant features of the important silicates with respect to metamorphism can best be discussed with regard to their structural type. Of the common tektosilicates, quartz is present in silica-rich rocks over practically the whole range of metamorphic conditions (at high temperatures and low pressures tridymite and cristobalite may form). The feldspars are common and abundant, but individual species show marked differences in occurrence. Albite is found over a wide range of conditions. Except for the occurrence of paragonite in a few schists, jadeite in eclogite, and glaucophane in glaucophane schists, albite is the principal carrier of sodium in metamorphic rocks. Potash feldspar occurs more usually as microcline than as orthoclase, conditions of metamorphism being generally favorable to the crystallization of the ordered form. Anorthite is rare in metamorphic rocks, the common calcium aluminum silicate being zoisite or epidote. Plagioclase composition is often a sensitive indicator of metamorphic grade; in the lowest-grade rocks it is pure albite, and the calcium content increases as the grade increases.

Minerals of the inosilicate group, that is, the pyroxenes and amphiboles, are common and abundant constituents in metamorphic rocks. Some amphiboles and pyroxenes are virtually confined to these rocks, for example, jadeite among the pyroxenes, and anthophyllite, cummingtonite, tremolite, actinolite, and glaucophane among the amphiboles. In general, the amphiboles are typical of metamorphic rocks formed at low and moderate temperatures, the pyroxenes of those formed at higher temperatures. However, diopsidic pyroxene is often associated with hornblende or with calcite in calcareous rocks at fairly low grade.

Minerals with layer lattices, that is, the phyllosilicates, are especially characteristic of metamorphic rocks. Talc, serpentine, the chlorites and chloritoids, and muscovite and biotite are common and widespread, some of them being practically confined to these rocks. The clay minerals of sedimentary rocks are particularly susceptible to recrystallization, and argillaceous rocks are thus especially suitable for tracing the initial stages of metamorphism. In this connection Yoder and Eugster (1955) have recognized a series of subtle changes in the crystal structure of the clay mica, from a one-layer, randomly stacked polymorph to a one-layer ordered polymorph to a two-layer ordered polymorph during progressive metamorphism. Similar changes can be expected in other phyllosilicates, such as the chlorites.

The nesosilicates are common and abundant in metamorphic rocks, and some of them, such as the garnets, epidote, and the aluminum silicates, are especially typical of such rocks. The nesosilicates in general are closely packed structures and would be expected to show marked stability under high pressure. The aluminum silicates—kyanite, sillimanite, and andalusite—are found in metamorphic rocks with high aluminum content. Stability relations of these three polymorphs are illustrated in Figure 10.1. Kyanite has a considerably higher density than the other polymorphs and indicates conditions of high pressure, although it occasionally occurs as a vein mineral. Andalusite is characteristic of thermally metamorphosed rocks and appears to be unstable under stress. Sillimanite is widespread in high-grade metamorphic rocks. Under extremely high temperatures typified by the contact action of basic lavas the mineral mullite, which resembles sillimanite closely but is of somewhat different composition, may appear. The garnets occur frequently in metamorphic rocks. Their composition is, of course, conditioned

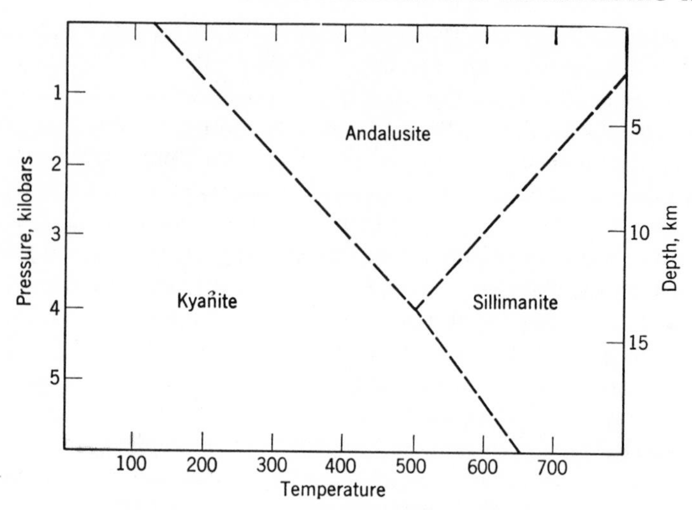

Figure 10.1 Stability fields of the aluminum silicates.

by the bulk composition of the rock, but they are also sensitive indicators of metamorphic grade. In the lowest grades of metamorphism the only garnet found is the manganese variety spessartite. At somewhat higher grades of regional metamorphism almandite may be formed, whereas garnets rich in the pyrope component are formed only at high metamorphic grades. The calcium-iron garnet andradite is characteristic of limestones that have been metasomatically altered by iron-rich emanations. The zoisite-epidote group is a common constituent of metamorphic rocks of low and medium grade. Minerals of this group are important carriers of calcium and aluminum in such rocks.

In general it can be said that the distinctive silicates of metamorphic rocks are the inosilicates and the phyllosilicates. This is not due to chance; they are minerals with fairly high density, the formation of which is favored by increased pressure; their structures generally tolerate considerable atomic substitution and may therefore form under variable bulk composition of the rocks. Some nesosilicates, especially minerals of the garnet and epidote groups, are also characteristic of many metamorphic rocks. Many of the tektosilicates or framework structures, on the other hand, show a marked instability under conditions of metamorphism, probably because of their distinctly open lattices.

The role of aluminum in the silicate minerals of metamorphic rocks deserves mention. We have seen that aluminum is capable of either sixfold or fourfold coordination with oxygen; in sixfold coordination it is structurally similar to magnesium or ferrous and ferric iron, whereas in fourfold coordination it acts like silicon. A distinct correlation exists between the type of coordination of aluminum and the grade of metamorphism. In general, four-coordination of aluminum is promoted by increasing temperature; thus the amount of aluminum replacing silicon in such minerals as the amphiboles, pyroxenes, and micas increases at higher metamorphic grades. The effect of pressure is less obvious, but sixfold coordination is more economical of space and should be favored by high pressures. Minerals containing four-coordinated aluminum are characteristic of igneous rocks and the products of thermal and high-grade regional metamorphism, whereas in sedimentary rocks and low- to medium-grade metamorphic rocks aluminum typically occurs in six-coordination.

Many other minerals besides the silicates discussed here occur in metamorphic rocks, but generally in quite minor amounts. An exception should be made for calcite and dolomite, which make up the greater part of some metamorphic rocks. Their stability is a direct function of the partial pressure of carbon dioxide during metamorphism, and they may thus be stable over a wide range of temperatures and pressures, provided sufficient carbon dioxide is present to prevent decomposition. Magnesium carbonate is much more readily decomposed than calcium carbonate, and therefore calcite is stable to higher metamorphic grades than dolomite.

The calculation of the quantitative mineralogical composition of a metamorphic rock from its chemical analysis is a much more complex procedure than for an igneous or sedimentary rock. Metamorphic rocks may crystallize over a wide variety of physicochemical conditions, and the same chemical composition may correspond to several different mineralogical compositions, depending upon the specific conditions (cf. Table 10.2). Of course, the problem is greatly simplified if the qualitative mineral composition is known from microscopic examination. Barth has developed procedures for calculating normative mineralogical compositions of metamorphic rocks, similar in principle to the norms of igneous rocks; he distinguishes catanorms, mesonorms, and epinorms for rocks of high-, medium-, and low-grade metamorphism respectively.

THE STABILITY OF MINERALS

Stability is a relative property. For example, calcite is said to be stable at ordinary temperatures and pressures. So it is, but only in the sense that under specific conditions it remains unchanged indefinitely. Drop it into dilute hydrochloric acid, however, and it is certainly far from stable. Stability must therefore be defined not only in respect to pressure and temperature, but also as regards chemical environment. Strictly speaking one cannot speak of the stability of a single phase (mineral) in a rock, independent of the associated minerals and the pore fluids. Most discussions of particular minerals or mineral associations assume an indifferent chemical environment, in which stability is a function of physical conditions only.

Three states may be distinguished: stability, metastability, and instability. A stable mineral association is that which has the lowest possible free energy under the particular circumstances and which therefore has no tendency to change. A mineral association is unstable when it is not the association with the lowest free energy under the specific conditions, and when the rate of change to an association with lower free energy is appreciable. A metastable association is one with more than the minimum free energy for the system, but in which the rate of change to an association with lower free energy is so slow as to be undetectable—an input of energy is required to make the transformation take place at a finite rate. Stability therefore involves two independent factors, thermodynamics and kinetics.

THERMODYNAMICS OF METAMORPHISM

The thermodynamics of metamorphism can be expressed in terms of the transformation of a mineral association $A + B + C \cdots$ into a different association $L + M + N \cdots$ according to the equation

$$A + B + C \cdots = L + M + N \cdots$$

For this change to take place the net free energy of $L + M + N \cdots$ must be less than that of $A + B + C \cdots$; that is, the free energy change must be negative. This condition can be expressed in the following way

$$\Delta G = (G_L + G_M + G_N \cdots) - (G_A + G_B + G_C \cdots) < 0$$

However, in general the free energy change is obtained indirectly, usually by the application of the second law of thermodynamics, as expressed in the following form

$$\Delta G = \Delta H - T\,\Delta S$$

in which ΔH is the enthalpy or heat of reaction and ΔS the entropy change.

The application of the simple thermodynamic equation to the consideration of actual reactions is complicated by the variation of ΔS and ΔH with temperature and pressure, and in order to study a reaction over a range of temperatures and pressures one must have data on compressibility, specific heat, and thermal expansion for the different phases. In addition, ΔS and ΔH have been measured for very few reactions of geological significance. Such measurements would provide basic data for significant advances towards the elucidation of fundamental geochemical problems. However, qualitative statements based on Le Châtelier's principle are often possible. Increased pressure will cause a reaction to run in a direction such that the total volume of the system decreases. Increasing temperature favors endothermic reactions. Since the volume change accompanying metamorphic reactions (except those involving gas phases) is generally small in comparison to the heat of reaction, temperature changes are more significant than pressure changes in displacing equilibrium.

One of the few reactions of metamorphism for which adequate thermodynamic data are available is the formation of wollastonite from calcium carbonate and silica, according to the equation

$$CaCO_3 + SiO_2 = CaSiO_3 + CO_2$$

The thermodynamics of this reaction and its significance for the metamorphism of siliceous limestone was first discussed by Goldschmidt, and he derived therefrom information regarding pressure-temperature conditions during metamorphism. With the aid of modern high temperature-high pressure equipment Harker and Tuttle have succeeded in determining the equilibrium curve for the reaction experimentally (Figure 10.2). The general features deduced by Goldschmidt are confirmed, although the reaction takes place at a lower temperature than he predicted. At zero pressure of carbon dioxide the formation of wollastonite may take place at temperatures below 400°. As may be predicted from the law of mass action, this temperature increases with increasing pressure, reaching 750° at 2000 atm. Figure 10.2 shows that for conditions above curve *AB* the formation of wollastonite from calcium carbonate and silica can take place spontaneously ($\Delta G < 0$);

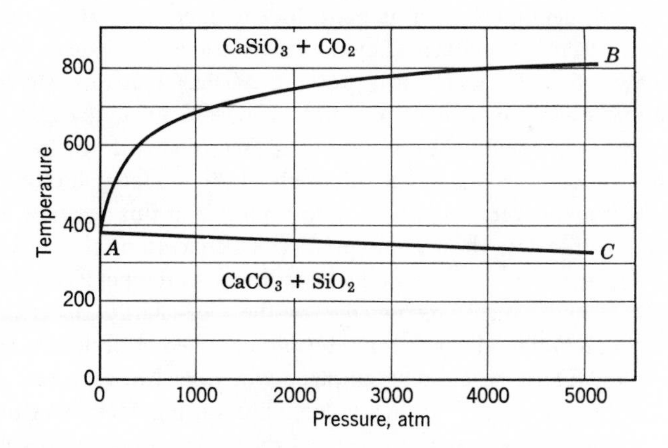

Figure 10.2 Pressure-temperature curves for the reaction $CaCO_3 + SiO_2 = CaSiO_3 + CO_2$.

below curve *AB* the reaction is reversed, and calcium carbonate and silica can be formed from wollastonite and carbon dioxide; curve *AB* itself represents equilibrium conditions ($\Delta G = 0$).

Goldschmidt's computations and Harker and Tuttle's experimental work were performed on a closed system in which the pressure was that of the CO_2. Barth has pointed out that under geological conditions this situation is improbable, since during the reaction of calcium carbonate and silica the carbon dioxide produced will tend to diffuse away. Wollastonite is denser than either calcite or quartz, and so in terms of the solids the reaction proceeds with a decrease in volume, and hence if the CO_2 escapes the temperature of formation of wollastonite will decrease with increasing pressure. Curve *AC* is the quantitative expression of this decrease, as calculated by Barth by means of the Clapeyron equation. This example illustrates the importance of distinguishing between open and closed systems for reactions involving volatile components. Such reactions are common in metamorphism.

Thermodynamic studies have been applied to the investigation of the formation of jadeite in metamorphic rocks. Jadeite is a rare mineral and in spite of its simple chemical composition efforts to make it in the laboratory have only recently been successful. It is, of course, a phase in the system Na_2O—Al_2O_3—SiO_2 and is intermediate in composition between albite and nepheline. The thermodynamic data show that jadeite is a stable phase in this system at 25° and 1 atm pressure. That it is a rare mineral is evidently not because of any lack of stability but probably because of the extreme slowness of the reactions that will form it. The correctness of these deductions from the thermodynamic data has been confirmed by the discovery of jadeite as an important constituent of Californian graywackes which have evidently been metamorphosed at quite low temperatures. In these rocks jadeite has apparently been formed by the reaction albite = jadeite + quartz. For this reaction at 25° and 1 atm pressure ΔG has a small positive value which diminishes with increasing pressure and becomes negative above about 1600 atm; that is, jadeite and quartz would then tend to form from albite. Such a pressure corresponds to a depth of about 20,000 feet, and the jadeite-bearing graywackes were probably metamorphosed at this depth or greater.

Thermodynamic data thus enable predictions concerning the relative stability of possible phases under specified conditions of temperature, pressure, and chemical environment. Knowledge of the free-energy change accompanying any reaction is adequate for determining whether or not the reaction can take place. Those reactions take place which lead to a decrease in the free energy of the system as a whole. Thus during metamorphism the tendency is towards the transformation of the rock material into a mineral association which for the specific conditions has the lowest possible free energy. However, the rate at which such transformation will proceed cannot be determined from the thermodynamic data but is dependent on other factors.

KINETICS OF METAMORPHISM

The significance of kinetics in metamorphism can be illustrated by the simple case of recrystallization of a limestone. A fine-grained aggregate of calcite can alter spontaneously to a coarse-grained aggregate, since the change will result in a reduction of the net surface energy. Yet some fine-grained limestones have remained essentially unchanged for millions of years, often in spite of deep burial and the accompanying temperature increase. Something more than a potential decrease in free-energy content is evidently required to cause a metamorphic change to take place.

The kinetics of heterogeneous reactions are not well understood, but qualitative statements can be made with some degree of assurance. The ionic groupings found in minerals are all relatively stable entities, and before they can be rearranged to give different minerals they must undergo some kind of dislocation. Energy is required for this, and only when energy is available in appreciable amounts can the rate of reaction become significant. This factor is an exponential function of E/RT, in which E is known as the *activation energy*. This function makes the activation energy the most important single factor determining reaction velocities and shows that they are increased by a certain amount per unit temperature rise. The usual statement that a reaction rate is doubled by a rise in temperature of 10° is a rough generalization and implies a uniform energy of activation of about 13,000 cal/mole; measured energies of activation show a considerable range but are of this order of magnitude.

Temperature is thus the most significant factor in determining the rate of a reaction. However, any agent that weakens the bonding within the phases present will give the reacting units greater freedom of movement and thereby decrease activation energies and accelerate reaction. The influence of water, a powerful solvent and ionizing agent, is particularly significant in mineral transformations, which are essentially ionic in character.

Although rise in temperature is perhaps the most important source of activation energy, it is by no means the only one. Laboratory experiments have shown that light, X rays, and gamma rays can accelerate reactions by the input of energy of which they are capable. Although these have little or no significance in metamorphism, the importance of shear in this connection deserves to be emphasized. Shear is an important source of energy in metamorphism and is probably responsible for much reconstitution that would not take place in its absence. This is borne out by field observations which suggest that the formation of low-grade schists has taken place at temperatures that may not exceed those existing in deep geosynclines, where rocks are indurated but not recrystallized. The contrast is evidently a reflection of the differing effects of uniform load and directed pressure.

The activation energy of the reactions taking place during metamorphism may be expected to vary according to the nature of the reaction and to be considerably influenced by the type of structural changes involved. Studies of polymorphic changes have shown that the rate of such transformations depends greatly on the degree of rearrangement of the structural units. A typical example is the rapid inversion of low-quartz to high-quartz compared with the sluggish transformation of high-quartz to tridymite. By analogy it may be expected that the change from chlorite to biotite, for example, probably has a much lower activation energy than the change from chlorite to garnet; in the first change large segments of the chlorite lattice can be directly incorporated in the biotite lattice, whereas in the latter change the sheets of silicon-oxygen tetrahedra must be completely broken down before the garnet structure can be built up from them.

In addition to the above factors, the rate of reconstitution of a rock will also be influenced by the physical condition of the reacting material, particularly the size of the individual grains and the intimacy of their mixture. The reacting units come together more easily the smaller the grain size and the more complete the mixing; that is, fine-grained rocks will respond to metamorphism more rapidly than those of coarser grain. Here stress has an important mechanical effect, by grinding the constituent minerals together and thereby crushing them and bringing them into closer contact.

The importance of reaction rates can be illustrated by a consideration of the kinetics of the wollastonite reaction discussed in the preceding section. The thermodynamic data show that wollastonite can form from calcium carbonate and silica at temperatures as low as 380°. However, in the laboratory the reaction between calcium carbonate and silica is generally not detectable below about 500°. On the basis of the thermodynamic data alone we might conclude that the absence of wollastonite in a limestone of suitable composition indicates that its temperature had never exceeded 380°. This conclusion would not be justified, however, since the rate of formation of wollastonite may not become finite, even under favorable circumstances, until higher temperatures are reached.

The sluggishness of many reactions at low to moderate temperatures undoubtedly results in the persistence of some minerals under conditions in which they should be changed to others with a lower free energy. This is particularly true in rocks that have been metamorphosed at high temperatures. Their mineralogy is more or less "frozen" at these high temperatures and often shows little if any effect of retrogressive metamorphism during cooling. In general, response in rocks to rising temperatures will be more rapid than response to falling temperatures because of the accelerating effect of temperature increase on reaction rates.

METAMORPHISM AND THE PHASE RULE

The most fruitful line of approach to the geochemistry of metamorphism has been through the principles of equilibria in heterogeneous systems as codified by the phase rule. The study of metamorphic rocks indicates that equilibrium is generally attained during metamorphism, although the evidence for it is usually indirect. In igneous rocks it is often possible to decide such a question by direct reference to the results of laboratory experimentation. Experimental observation of equilibria in silicate systems at the temperatures and pressures of metamorphism has not yet progressed to the same extent. In general, the evidence for equilibrium in the phase assemblages of metamorphic rocks rests on such criteria as their simple mineralogical composition and the tendency for certain typical mineral associations to recur in rocks of the same general composition, despite widely different age and locality. By the phase rule the number of minerals capable of existing together as phases of a system at equilibrium is limited by the number of components in the rock. We find that usually even fewer phases than the maximum number allowed by the phase rule are present. Many chemically complex metamorphic rocks, such as amphibolite (hornblende-plagioclase) and mica schist (muscovite-chlorite-albite-quartz), are extremely simple mineralogically. This is evidently because the individual elements that appear in the composition of the rock are not distinct components in the thermodynamic sense; that is, ferrous iron and magnesium often act as a single component and, to a lesser degree, aluminum and ferric iron, and sodium and calcium.

As mentioned above, the approach that has proved so fruitful in the elucidation of the evolution of the igneous rocks, viz., the laboratory investigation of phase relations in silicate systems, has only recently been extended to conditions corresponding to those obtaining during metamorphism. The first extensive research of this kind to appear is that for the system MgO—SiO_2—H_2O (Figure 10.3). This lag is mainly due to the technical difficulties in carrying out equilibrium studies involving volatiles, and to the sluggishness of many reactions at temperatures and pressures corresponding to those of metamorphism. The approach has thus been from observations on metamorphic rocks, largely without the assistance of data obtained from the laboratory. Geologists have studied the *products* of completed reactions, viz., the mineral associations of metamorphic rocks, and much progress has been made, especially in determining compatible and incompatible minerals and mineral associations. Here again, as in the investigation of mineral formation in igneous rocks, real advances date from the introduction of

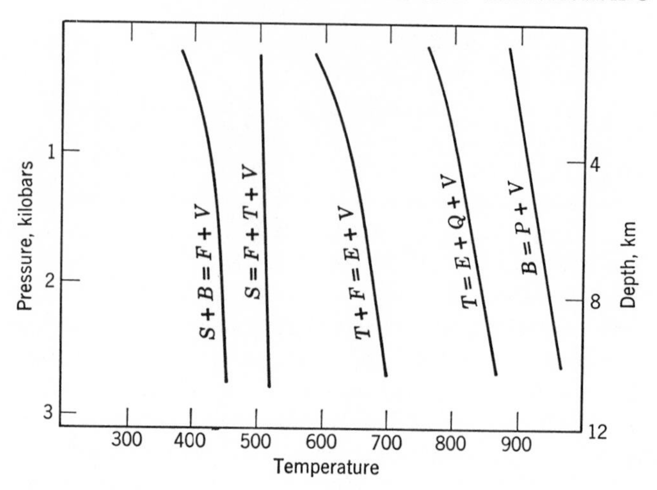

Figure 10.3 Pressure-temperature curves of univariant equilibrium in the system MgO-SiO_2-H_2O. The equation on each curve indicates the reaction to which the curve refers; B = brucite, E = enstatite, F = forsterite, P = periclase, Q = quartz, S = serpentine, T = talc, V = water vapor. (Bowen and Tuttle, *Bull. Geol. Soc. Am.* **60,** 447, 1949)

physicochemical principles, especially the phase rule, as a guide in the study of reactions in rocks. The credit for appying these principles to metamorphic rocks goes in the first place to Goldschmidt, for his classic work on contact metamorphism in the Oslo region published in 1911, and secondly to Eskola, who in 1915 applied the same principles to regional metamorphism in his study of the relationship between mineralogical and chemical composition in the metamorphic rocks of the Orijärvi region. These and other studies led to the formulation of the facies principle for the classification of metamorphic rocks.

THE FACIES PRINCIPLE

Ideally, a genetic classification of metamorphic rocks would be based on the temperature and pressure at which the mineral assemblage originated. It is not yet possible, however, to make precise estimates of the conditions of temperature and pressure at which different metamorphic rocks are formed. Neyertheless, Eskola recognized that the mineral associations are indicative of conditions of formation as well as of the chemical composition of the rock itself. He exemplified this principle by selecting different rocks with similar chemical analyses and showing that the mineralogical composition of each was distinctive (Table 10.2).

*Table 10.2. Mineralogy of Rocks of Gabbroic Composition from Different Facies**

Diabase facies	Pyroxene hornfels facies	Amphibolite facies	Epidote-amphibolite facies	Greenschist facies	Granulite facies	Eclogite facies	Glaucophane schist facies
Diabase (East Karelia)	Hornfels (Oslo region)	Amphibolite (Kisko, Finland)	Epidote-amphibolite (Sulitelma, Norway)	Chlorite schist (Sulitelma, Norway)	Norite granulite (North Finland)	Eclogite (Burgstein, Tyrol)	Glaucophane schist (Scalea, Italy)
Plagioclase 48.4 Pigeonite 37.4	Plagioclase (Mean An_{40}) 48	Plagioclase (Mean An_{43}) 26.5	Plagioclase (An_9) 42.8	Albite 39.9 Chlorite 29.4	Plagioclase 49.5 Hypersthene 25.3	Omphacite 48.5 Garnet 50.5 Others 1.0	Glaucophane 54.3 Lawsonite 26.8 Sericite 15.6 Others 3.1
Hornblende, mica, etc. 14.2	Hypersthene 17 Diopside 18 Biotite, iron ore, etc. 17	Hornblende 71.5 Quartz 2.0	Hornblende 42.2 Clinozoisite 12.3 Others 3.4	Epidote 23.0 Others 7.2	Diopside 9.6 Orthoclase 7.1 Others 7.4		

* After Eskola.

In the rocks of Table 10.2 the bulk composition is much the same throughout, and (assuming equilibrium has been reached in each case) the different mineral associations can only represent different conditions of crystallization. This phenomenon is the basic idea behind Eskola's facies classification. Each of the rocks in Table 10.2 represents a distinct facies. The mineral assemblage of a rock reflects the physical conditions under which it developed and is thus the criterion by which the facies may be recognized. The term is analogous to the stratigraphic facies, which comprises sediments with characters pointing to a genesis under similar circumstances. Eskola formulated the concept of mineral facies in the following words:

> A mineral facies comprises all the rocks that have originated under temperature and pressure conditions so similar that a definite chemical composition has resulted in the same set of minerals, quite regardless of their mode of crystallization, whether from magma or aqueous solution or gas and whether by direct crystallization from solution (primary crystallization) or by gradual change of earlier minerals (metamorphic crystallization).

Ramberg (1952, p. 136), who has done much work on the mineral facies concept, expresses it succinctly: "Rocks formed or recrystallized within a certain (*P*, *T*) field, limited by the stability of certain critical minerals of defined composition, belong to the same mineral facies."

These definitions show that the concept can be applied to any rock at all, sedimentary, igneous, or metamorphic, but in practice it has found its greatest utility in the study of metamorphic rocks. A mineral facies comprises all rocks that have reached chemical equilbrium under a particular set of physical conditions. In effect, the facies concept is an expression of the phase rule stated in the following way: In any system at equilibrium the number and composition of the different phases will depend only on the bulk composition of the system and the temperature and pressure at which equilibrium was reached. Facies are recognized and defined by the occurrence of critical minerals or mineral associations that are characteristic of the facies in question and that do not appear in any other facies. Relatively few minerals are sufficiently sensitive to be critical in this sense; many—quartz, calcite, albite, and others—are stable over such a range of conditions that they may appear in several quite distinct facies.

Eskola (1939) has given a diagram (Table 10.3) enumerating the facies that have been generally recognized and correlating them with the conditions of temperature and pressure under which they originated.

The actual ranges of temperature and pressure characteristic of each facies are not known with any degree of precision. Estimates by different investigators vary considerably, and indeed are frequently somewhat contradictory. Judging from the geological evidence, such as the transition from anthracite

*Table 10.3. Relationship of Mineral Facies to Temperature and Pressure**

	Temperature increasing →			
Pressure increasing ↓	Development of zeolites in igneous rocks and recrystallized sediments			Pyroxene-hornfels facies
	Greenschist facies	Epidote—amphibolite facies	Amphibolite facies *Hornblende-gabbro facies*	Granulite facies *Gabbro facies*
	Glaucophane-schist facies		Eclogite facies *Eclogite facies*	

* After Eskola.
Note: Metamorphic facies are in roman type, igneous facies in italics.

to graphite accompanying the transition from indurated sedimentary rocks to schists of the greenschist facies, the lower temperature boundary of this facies is at least 200°. If we put the lower boundary of the epidote-amphibolite facies at the incoming of biotite, a temperature of 400° may be a reasonable estimate. The boundary between the epidote-amphibolite facies and the amphibolite facies can perhaps be placed near 500°, and that between the amphibolite facies and the granulite facies at about 650°. Granulite facies rocks are notably "dry" in the sense that hydroxyl-bearing minerals are essentially absent; an appreciable amount of available water would probably induce melting under conditions of this facies. The progressive metamorphism of amphibolite facies rocks may result either in partial melting and the formation of a magma (if H_2O is available), or recrystallization in the granulite facies (if H_2O is unavailable or can escape from the system).

Because different facies represent different temperature-pressure fields, their mutual relationships can be represented by a diagram using temperature and pressure as axes (Figure 10.4). If the pressure is due to rock load, it can be correlated with depth of burial, as indicated on the right of the diagram. This figure also serves to illustrate the diffuse boundary between metamorphism and diagenesis, on the one hand, and between metamorphism and magmatism on the other. The minimum melting curves of "granite" and "basalt" define the broad zone within which magmas of different composition crystallize. To the left of the melting curve of "granite" no remelting of normal silicate rocks can be expected. In the region between this curve and the melting curve of "basalt" the amount of remelting (*anatexis*) a rock will suffer will depend largely on its composition and the presence or absence

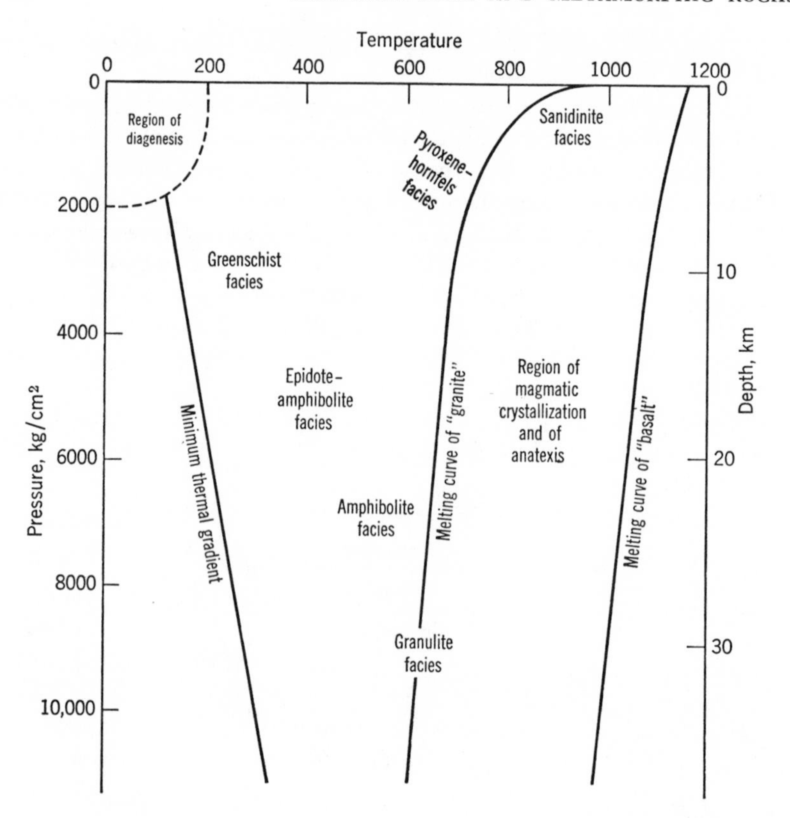

Figure 10.4 The principal metamorphic facies in relation to temperature and pressure.

of volatiles. Pure quartzite and alkali-poor slates may remain essentially solid at even higher temperatures, but other rocks will probably produce some melt in this region. It is clear that no sharp division between metamorphism and magmatism is possible.

In Figure 10.4 the field of metamorphism is limited on the low-temperature side by a line representing the minimum thermal gradient in the crust. This is based on a temperature of 300° at a depth of 30 km. The normal facies of regionally metamorphosed rocks are indicated as lying along a considerably steeper thermal gradient. The reactions accompanying progressive regional metamorphism are generally endothermic, and the process thus requires a considerable input of heat, probably more than that available under the conditions of the minimum thermal gradient. Orogeny, regional metamorphism, and igneous activity are broadly associated in space and time, and they can all be regarded as geological manifestations of an input of energy, mainly thermal.

The greenschist, epidote-amphibolite, amphibolite, and granulite facies may be called the normal facies of regionally metamorphosed rocks, arranged in order of increasing grade of metamorphism. The eclogite facies and glaucophane-schist facies are believed to originate under unusually high pressure, being characterized by minerals with high density (garnet, jadeitic pyroxenes, and lawsonite). A facies which is not included in Table 10.3 is the sanidinite facies, which would be placed to the right of the pyroxene-hornfels facies; it is exemplified by xenoliths in basic lavas, in which the temperature has been very high, leading to incipient vitrification and the development of such minerals as mullite and sanidine.

The characteristic minerals and mineral assemblages of the normal facies are illustrated in Table 10.4. In studying this table it should be borne in mind that mineral transformations, particularly from one solid solution series to another, are conditioned not only by temperature and pressure, but also by chemical environment; for example, the transition from chlorite to garnet evidently takes place at lower temperatures in iron-rich phases than in magnesium-rich ones. Similarly, the stability of calcium and magnesium silicates in the greenschist facies depends in part on a low partial pressure of CO_2, for a high concentration of CO_2 would favor the formation of carbonates. Such considerations are discussed more fully in the following section.

The characteristic mineralogical features of the different facies can most clearly be illustrated by means of triangular composition diagrams, similar in principle to the diagrams used to illustrate three-component systems. In such diagrams each of the apices represents a particular component; those components chosen are mainly responsible for the observed mineralogical variation within the facies. Such diagrams were introduced by Eskola and are usually *ACF* diagrams in which *A* is $Al_2O_3 + Fe_2O_3$, *C* is CaO, and *F* is (Fe,Mg)O. The effect of silica on the possible mineral assemblages can be illustrated by comparing two *ACF* diagrams, one (the most usual case) for rocks with excess SiO_2 (present as free quartz) and the other for rocks deficient in SiO_2. In the same way diagrams have been constructed for rocks of special compositions, viz., *AKF* diagrams in which *K* represents K_2O. *ACF* diagrams illustrating the different facies are given in Figure 10.5.

The critical minerals for the sanidinite facies are sanidine, clinoenstatite-clinohypersthene and their mix-crystals with diopside-hedenbergite (pigeonite) and (at very high temperatures) the aluminum silicate mullite. In limestones and dolomites metamorphosed under these conditions a considerable number of unusual minerals have been described, such as larnite (Ca_2SiO_4), rankinite ($Ca_3Si_2O_7$), and merwinite ($Ca_3MgSi_2O_8$).

A critical mineral combination for the pyroxene-hornfels facies (in common with the granulite facies) is the pair hypersthene-diopside, which do not form mix-crystals under the conditions of this facies. This distinguishes the pyroxene-hornfels facies from the sanidinite facies. Another difference is that $(Mg,Fe)SiO_3$ in this facies is always orthorhombic. Sanidine does not appear, potash and soda feldspar forming distinct phases. The potash feldspar is orthoclase, not microcline, in harmony with the idea that orthoclase is a disordered form and microcline an ordered form.

The granulite facies comprises rocks which have been subjected to high-grade regional metamorphism. It is characterized by the absence of micas; instead of muscovite we find sillimanite (or kyanite) and orthoclase, instead of biotite, garnet and orthoclase. Calcite is probably the stable phase at high lime compositions instead of wollastonite, provided sufficient CO_2 is present.

In the amphibolite facies amphibole always appears, provided the bulk composition allows it. The combination hornblende-plagioclase is critical.

Table 10.4. The Mineralogy of the Different Facies

Facies	Si	Al, Si	K, Al, Si	Na, Al, Si	Ca, Al, Si	Ca, Si, (CO_2)
Greenschist	Quartz		Muscovite Microcline	Albite	Zoisite	Quartz + calcite
Epidote-amphibolite	Quartz	Kyanite	Muscovite Microcline	Albite	Zoisite	Quartz + calcite
Amphibolite	Quartz	Kyanite Sillimanite	Muscovite Microcline	Plagioclase	Zoisite Grossularite	Wollastonite Quartz + calcite
Granulite	Quartz	Kyanite Sillimanite	Orthoclase	Plagioclase		Wollastonite Quartz + calcite
Pyroxene hornfels	Quartz	Andalusite Sillimanite	Orthoclase	Plagioclase	Grossularite	Wollastonite
Sanidinite	Tridymite	Mullite	Sanidine	Plagioclase		Wollastonite (Pseudowollastonite) Larnite Rankinite

Its high-temperature limit is marked by the appearance of diopside and hypersthene in place of hornblende, and its low-temperature limit by plagioclase composition, which changes to albite with decreasing metamorphic grade.

For the epidote-amphibolite facies the combination hornblende-albite-epidote is critical. Under the conditions of this facies the anorthite component of plagioclase is converted to zoisite or epidote; the plagioclase thus becomes high in the albite component. Hornblende is still stable.

At metamorphic grades lower than those for the epidote-amphibolite facies hornblende is unstable, and in its place one finds more chlorite and epidote, or, if the partial pressure of CO_2 is high, dolomite or magnesite plus quartz. Carbon dioxide can decompose silicates under these conditions, forming associations of quartz and carbonates.

The eclogite facies is characterized by pyroxenes of the omphacite type (intermediate between diopside and jadeite) and garnet with a high pyrope

Si, Fe, (Mg), (CO_2)	Si, Mg, (Fe), (CO_2)	Si, Mg, Ca, (CO_2)	Ca, Al, Mg, Si	Mg, (Fe), Al, Si	Fe, (Mg), Al, Si	K, Mg, Fe, Al, Si
Siderite + quartz	Magnesite + quartz Talc	Dolomite + quartz Talc + calcite Tremolite		Chlorite	Chloritoid	Muscovite + chlorite
Cummingtonite	Talc Serpentine Anthophyllite	Tremolite	Blue-green hornblende	Chlorite	Chloritoid Almandite	Biotite
Cummingtonite	Anthophyllite Forsterite	Tremolite Diopside	Green hornblende	Cordierite	Almandite Staurolite	Biotite
Hypersthene	Enstatite Forsterite	Diopside	Augite	Pyrope-almandite		Orthoclase + pyroxene
Hypersthene	Enstatite Forsterite	Diopside	Brown hornblende Augite	Cordierite Pyrope-almandite		Biotite
Clinohypersthene	Forsterite Pigeonite Clinoenstatite	Diopside Melilite Merwinite	Augite	Cordierite		Orthoclase + pyroxene

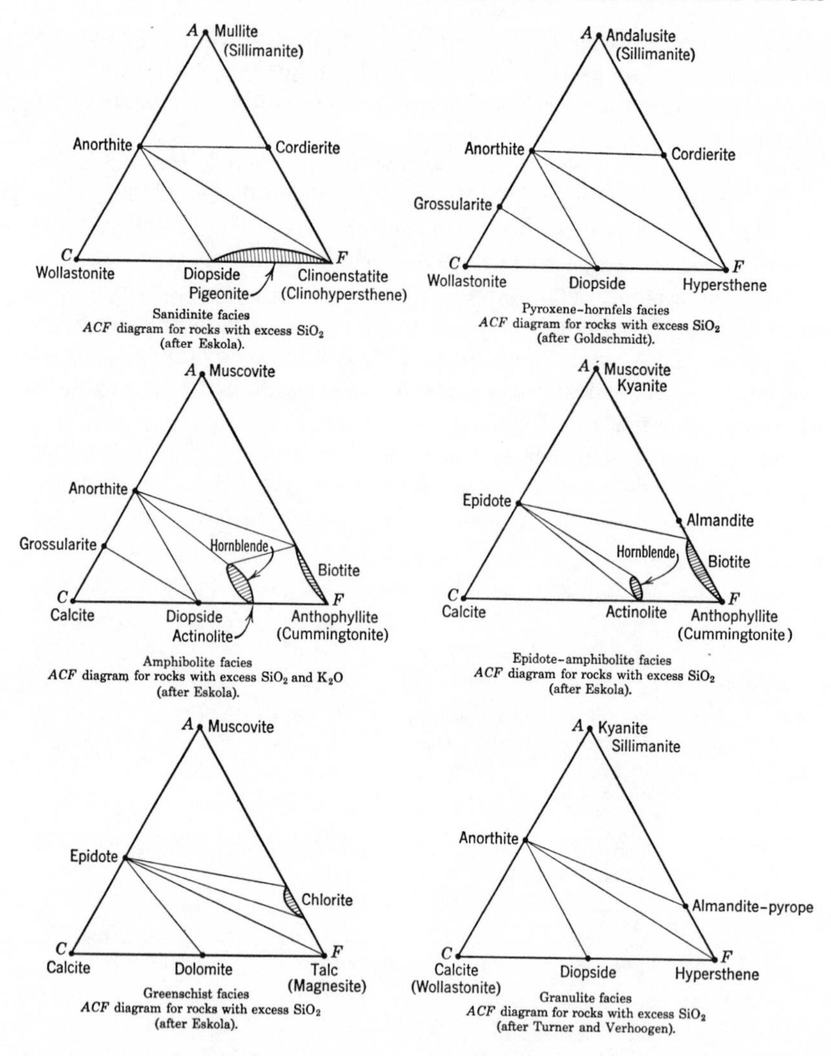

Figure 10.5 *ACF* diagrams for different facies.

content. Other minerals that may be present are kyanite and hypersthene. It is difficult to draw a satisfactory *ACF* diagram for the eclogite facies, since known rocks belonging here have a comparatively small range in composition.

In the glaucophane-schist facies the critical minerals are glaucophane and lawsonite. Muscovite, almandite, epidote, and pumpellyite have also been recorded; these minerals are found in rocks of the greenschist and epidote-amphibolite facies, and glaucophane schists occur in close association with such rocks. This suggests that the glaucophane-schist facies is related to these facies, but probably represents higher pressures, since glaucophane plus lawsonite can be considered as the high-density equivalents of plagioclase plus chlorite.

MINERAL TRANSFORMATIONS AND THE FACIES PRINCIPLE

Much remains to be done to make the facies classification the precise tool that it potentially is. This demands careful study of mineral assemblages and especially the determination of the variability in chemical composition of many of the characteristic minerals. The variability is linked not only with the bulk composition of the rock, but also with the probable temperature and pressure conditions under which it was formed. A mineral facies represents a definite (P, T) interval within which the rock attained its present mineralogical composition. The different facies are characterized by the stability fields of their critical minerals or mineral associations. The boundaries of a facies are delimited by the equilibrium curves for the reactions which produce the critical minerals and mineral associations.

In 1940 Bowen showed that the progressive metamorphism of limestones and dolomites could be considered in terms of a series of reactions of increasing decarbonation taking place at successively higher temperatures at any given pressure. He pointed out that the equilibrium curves for these reactions, when plotted on a pressure-temperature diagram, delimited fields of stability of specific minerals and mineral associations, thus forming what he termed a *petrogenetic grid*. At that time the equilibrium curves for the reactions he considered had not been determined, but he could arrive at approximations for them by evaluating the geological information provided by metamorphosed limestones and dolomites.

The thermodynamic basis for this petrogenetic grid and its relation to the facies principle have been thoroughly discussed by Thompson (1955), who points out that the boundaries between facies are curves of univariant equilibrium. If the equilibria involve only crystalline solids, then these curves have slopes determined by the Clapeyron equation and are virtually straight lines. Many significant equilibria in metamorphism, however, involve gain or loss of a volatile constituent such as carbon dioxide or water, acted on by a pressure equal to or less than that acting on the crystalline solids. If only one volatile constituent is involved in the equilibrium and the pressure on it is equal to the total pressure, the Clapeyron equation still applies, but the curves are concave toward the pressure axis and have a positive slope (since increasing pressure raises the decomposition temperature of a substance containing a volatile component).

These principles are illustrated in Figure 10.6, which gives equilibrium curves for a number of reactions pertinent to conditions of metamorphism.

These curves have been determined either by experimental investigation of the equilibrium or by calculation of the equilibrium curve from thermochemical data. It can be seen that for reactions involving volatiles the curvature of the lines is greatest at low pressures, and they become virtually straight at comparatively moderate pressures.

As mentioned above, these curves represent the condition obtained when the pressure of the volatile component is equal to the total pressure. This condition may not apply in a specific case of metamorphism. For example, if the rock undergoing metamorphism is connected by pore spaces to the surface, the volatile constituent can leak away; its partial pressure may be very low and in the extreme case may approach that at the surface. Under these circumstances the equilibrium curve has a negative slope. This situation is illustrated by the calcite-wollastonite reaction (Figure 10.2). In effect, we have two limiting conditions for any reaction involving volatiles, and the temperature at which the reaction takes place depends upon the extent to which the volatile component is able to escape.

The commonest volatile component in metamorphic reactions is water, and its significance in this connection has been emphasized by Yoder (1955). He points out that many discussions of metamorphism have tacitly assumed that water as a component is always present in sufficient amount to give rise to the most highly hydrated phases stable under the specific temperature and pressure conditions. Anhydrous minerals having hydrous equivalents are then considered to indicate temperatures of formation above the equilibrium curve for reaction to give the hydrous equivalent. Forsterite, for example, reacts with water to give serpentine at about 500°. However, if water is not available, forsterite is stable below 500°, and in the absence of water it can form below 400° from magnesite and quartz. The logical consequence is that the mineralogy of a metamorphic rock depends not only on the composition of the original material and the temperature and pressure of metamorphism, but also on the amount and partial pressure of water and other volatiles. In terms of the facies principle we should be able to distinguish water-deficient and water-sufficient variants of the same facies. For example, in rocks of the greenschist facies the usual calcium aluminosilicate is zoisite or epidote; however, in some tuffaceous graywackes that have suffered low-grade metamorphism prehnite and calcium zeolites, minerals of similar composition except for a higher water content, are found instead. Similarly, the amount and partial pressure of carbon dioxide have a marked effect on the mineralogy of rocks containing appreciable amounts of calcium and magnesium. In the amphibolite and granulite facies both wollastonite and calcite plus quartz are probably stable, their occurrence being conditioned by the partial pressure of carbon dioxide.

A noteworthy feature of Figure 10.6 is that the reaction curves are all steeply inclined to the temperature axis; in other words, they are much more

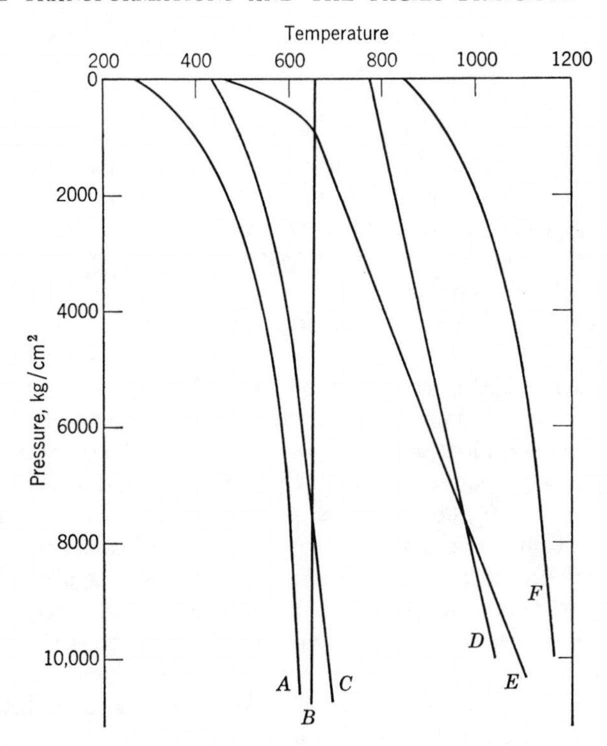

Figure 10.6 Equilibrium curves for some reactions of significance in metamorphism.

	Low Temperature		High Temperature
A.	Dolomite + quartz	=	Diopside + carbon dioxide
B.	Orthoclase + albite	=	Sanidine
C.	Pyrophyllite + corundum	=	Sillimanite + quartz + water
D.	Almandite	=	Hercynite + iron cordierite + fayalite
E.	Muscovite	=	Sanidine + corundum + water
F.	Phlogopite	=	Forsterite + leucite + $KAlSiO_4$ + water

strongly temperature-dependent than pressure-dependent. Such reactions are suitable for assessing temperatures of metamorphism but not for assessing pressure. There is a notable lack of reactions which are strongly pressure-dependent and can thus serve as *geological barometers* and indicators of depth of crystallization. It is known the magnesium-rich garnets characteristically occur in rocks that are believed to have crystallized at great depth, such as eclogites, but the equilibrium relations of such garnets are little known. Diamond is probably a good geological barometer, but its restricted occurrence limits its usefulness. Aluminum substitution in orthopyroxenes may serve as an indication of pressure conditions, but the situation is complicated if other aluminous minerals are present in the assemblage. One relationship that is strongly pressure-dependent is that between kyanite and the other polymorphs of Al_2SiO_5, sillimanite and andalusite, kyanite having a density about 3.6 and the other polymorphs about 3.2. Laboratory measurements confirm that the equilibrium is strongly pressure-dependent and that kyanite is the high-pressure phase at low to medium temperatures (Figure 10.1). Aragonite, the high-pressure phase of $CaCO_3$, has been found in jadeite-bearing metagraywackes of the Franciscan formation in California. Assuming a temperature of metamorphism of 200°, the minimum pressure of formation indicated by the presence of jadeite is 5 kb; aragonite would require a pressure of 7 kb. Under hydrostatic conditions these pressures imply a minimum depth of metamorphism for these rocks of 20 to 25 km.

We are handicapped in our elucidation of the geochemistry of metamorphism by our lack of knowledge of the equilibrium conditions of many important metamorphic reactions, especially those taking place at lower grades. Figure 10.6 points up the absence of equilibrium curves for temperatures below 400–500°. This is a consequence of the difficulty of achieving reaction between solids at lower temperatures in the laboratory and a lack of thermochemical data on minerals from which equilibrium curves can be calculated. In addition, the significant reactions are often very complex. For example, one of the commonest reactions of low-grade metamorphism is muscovite + chlorite = biotite, and the incoming of biotite is an important marker in many regions of progressive metamorphism. This reaction has not been reproduced in the laboratory, and the evidence from the minerals themselves shows that it is not a simple one. Rather than being a direct combination of muscovite and chlorite it seems to be a reaction between muscovite of one composition and chlorite to give muscovite of a different composition and biotite. Most minerals of metamorphic rocks are capable of considerable variation in composition through atomic substitution, and the stability of a mineral and its reactions with other minerals will be affected by the nature and extent of these substitutions. For example, garnet in metamorphic rocks varies from iron-rich almandite in low-grade schists to magnesium-rich pyrope in high-grade eclogites and granulites. Similarly, the anorthite content of plagioclase increases with increasing metamorphic grade, as does the aluminum content of hornblende.

It has long been recognized that the anorthite content of plagioclase in metamorphic rocks often increases with increasing degree of metamorphism. This is a reflection of the independent role of sodium and calcium in these rocks. Sodium is generally present almost entirely in the form of the albite component of plagioclase, whereas calcium may be present in a number of calcium aluminum silicates. In most dynamically metamorphosed rocks of low-grade, albite is associated with zoisite or a mineral of the clinozoisite-epidote series. As the degree of metamorphism increases the plagioclase becomes more calcic and the amount of zoisite decreases, in effect, the zoisite is converted to anorthite, which enters the plagioclase. The reaction can be represented by the equation

$$Ca_2Al_3Si_3O_{12}(OH) = 3CaAl_2Si_2O_8 + CaO + H_2O$$

This is certainly an oversimplification, but it shows that the reaction involves water, and the equilibrium will be affected by the partial water-vapor pressure. If we consider the reaction to take place at some fixed water-vapor pressure, the relationship between degree of metamorphism and plagioclase composition can be represented diagrammatically (Figure 10.7). The relationship is additionally complicated by the peristerite unmixing in sodium-rich plagioclase (Chapter 5). Electron microprobe analyses have shown that plagioclase in rocks of the greenschist and epidote-amphibolite facies is almost pure albite ($Ab_{>99}$), whereas in rocks of the amphibolite facies it is Ab_{80} or more calcic. This is illustrated in Figure 10.7, which shows that albite and zoisite (epidote) are in stable equilibrium up to temperature t_2. At this temperature the albite is entirely converted to plagioclase of composition Ab_{80}, provided sufficient calcium is available; insufficient calcium results in a mixture of two plagioclases, about Ab_{99} and Ab_{80} respectively. As the temperature increases beyond t_2, if calcium is still available, the plagioclase becomes progressively richer in the anorthite component.

Transformations of the type discussed above can thus be used as recorders of the conditions under which metamorphism took place. Careful study of such transformations, both in laboratory preparations and as revealed in

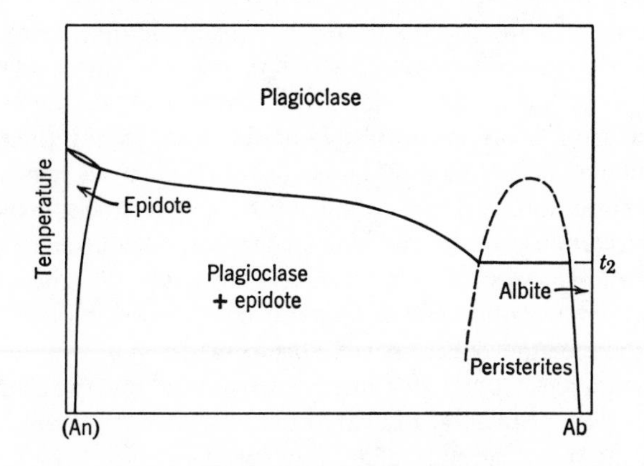

Figure 10.7 Stability relations between zoisite (epidote) and plagioclase. (After Christie, *Norsk Geol. Tidssk.* **39,** 270, 1959)

metamorphic rocks, can contribute greatly to a more precise knowledge of the different facies and their mutual relationship.

METASOMATISM IN METAMORPHISM

In discussing the facies principle we have considered metamorphic rocks as the end products of chemical reaction; the equilibrium assemblage of minerals has been determined by the conditions of metamorphism and the ultimate composition attained by the rock in question. We have not been concerned with the possible changes in composition that the rock may have undergone during metamorphism, that is, whether metasomatism has played a part. It remains to consider some of the problems posed by metasomatism, apart from the mode of transportation of the metasomatizing substances, which was discussed in a previous section.

Demonstration of metasomatic activity depends on chemical evidence that the original rock differed in composition from the final product. This evidence is readily available when we find rocks such as serpentines, talc-carbonate rocks, and the like, having bulk composition corresponding neither with igneous nor sedimentary rocks. However, the situation is seldom so clear-cut. The common types of sedimentary and igneous rocks vary in composition, and analyses of different samples of the same formation must be interpreted with this in mind. In a study of the composition of metamorphic rocks Lapadu-Hargues (1945) presented evidence from which he concluded that there is a progressive change in composition during the metamorphism of pelitic sediments, leading ultimately to rocks of granitic composition. This claim was based on statistical comparison of 302 analyses of shales, slates, mica schists, gneisses, and granites, which shows an increase in potassium and sodium and a decrease in magnesium, calcium, and iron in going from the unmetamorphosed rocks to granite. The validity of such a procedure is doubtful, since individual analyses of all these rocks vary widely, and the conclusion is probably not justified. Even if the trend were a paragenetic one, it could hardly be held to prove that all sequences of progressive metamorphism from sedimentary rocks to gneiss have been subjected to metasomatism. In problems of this sort each occurrence has to be considered individually; in some instances the evidence indicates the introduction and removal of certain elements, in others the essentially isochemical nature of the metamorphism is well documented.

The reactions accompanying metasomatism are governed, of course, by physicochemical principles, of which the phase rule and the law of mass action are the most significant in this connection. However, an additional rule has been found to apply to metasomatism: experience has shown that metasomatism usually takes place without change of total volume; that is, the metasomatized rock generally occupies the same volume as it did before metasomatism. This is highly significant in that metasomatic reactions are not quantitatively represented by the conventional simple equation, which merely balances equal weight of material on the right and left sides, respectively, without taking volume changes into account. Thus Turner and Verhoogen show that the serpentinization of olivine can be represented by the following equations:

$$3Mg_2SiO_4 + SiO_2 + 4H_2O = 2H_4Mg_3Si_2O_9$$

or

$$5Mg_2SiO_4 + 4H_2O = 2H_4Mg_3Si_2O_9 + 4MgO + SiO_2$$

Of these two equations the second better approximates the conversion of olivine into an equal volume of serpentine (assuming excess MgO and SiO_2 are removed in solution) and may be expected to represent more closely the actual process. It is thus unsafe to interpret metasomatism quantitatively in terms of simple equations which actually correspond to reactions involving marked charges of volume, although such equations are useful for indicating qualitatively the general direction in which chemical reaction may have proceeded.

Some instances of metasomatism have already been mentioned, of which the formation of dolomite by the action of magnesium-bearing solutions on calcite, discussed in Chapter 6, is one. Many ore deposits seem to represent a metasomatism of pre-existing rocks, the ore material having replaced them volume for volume. Metasomatism very often accompanies the late stages of magmatism. Many china clay deposits, such as those of Cornwall, are evidently the result of the metasomatic alteration of granite, whereby the feldspar is altered to kaolin. Schematically

$$2KAlSi_3O_8 + 2H_2O = Al_2Si_2O_5(OH)_4 + K_2O + 4SiO_2$$

That the metasomatism takes place volume for volume is shown by the way in which the kaolin occurs as pseudomorphs after the feldspar. Ultrabasic rocks are particularly subject to metasomatism. Conversion to serpentine is the commonest form of alteration, but if calcium is present or introduced, assemblages such as actinolite-chlorite may form. At temperatures corresponding to the greenschist facies CO_2 may replace SiO_2, and talc-dolomite, dolomite-quartz, or magnesite-quartz rock may be the ultimate products. Some of the best-described examples of metasomatism are those accompanying ore deposits, since they are often well exposed by mining operations and their interpretation is important for an understanding of the economic geology. They were first described in connection with ore deposits in the strongly metamorphosed Precambrian of Sweden, and the Swedish term *skarn* has become a general term for such material. Extensive skarn masses generally surround iron ores in limestones; the major skarn minerals are usually garnet (andradite), pyroxene (hedenbergite), and amphibole (actinolite and hornblende). Often the source of the metasomatizing solutions can be traced to neighboring intrusions of granite; sometimes the skarns are due to interchange of material between the ore body and the country rock during regional metamorphism.

An interesting development in the concept of metasomatism is the doctrine of "fronts" (see, for example, Reynolds, 1947). It has been expounded especially in connection with theories of the origin of granite batholiths through the metasomatism of preexisting rocks by processes of ionic migration through solids. This involves not only the addition of elements characteristic of granites, such as sodium and potassium, but the removal of superfluous ones, such as calcium, magnesium, and iron. It is therefore argued that the country rock around a granite should be "basified" by the addition of these superfluous elements, and the wave of migrating material is known as a "basic front," in advance of the encroaching "granitic front." A front occurs wherever there is a diffusion limit marked by a change in the mineral assemblage. On this basis zones of hornfelses around intrusions have been interpreted as evidence of successive fronts. Turner and Verhoogen observe that it is difficult to reconcile the discrepancy in size between some very large granite batholiths and the narrow aureoles bordering them with the doctrine of fronts. The larger the granite batholith, the larger should be the wave of migrating ions pushed ahead of it and the broader the metamorphic aureole. Many hornfels aureoles show that the difference in mineralogy between successive zones is essentially an expression of the temperature gradient in passing out from the intrusion rather than successive changes in chemical composition.

ACCESSORY ELEMENTS IN METAMORPHIC ROCKS

Whereas considerable data on the abundance and distribution of minor and trace elements are available for igneous rocks and, to a lesser extent, for sedimentary rocks, corresponding information for metamorphic rocks is conspicuously scanty. This has led to much speculation on patterns of behavior and their possible geological significance in metamorphic terranes. For example, the occurrence of tourmaline in schists derived from sedimentary rocks has been cited as evidence for boron metasomatism from supposedly deeper-lying granite intrusions; however, since many marine clays show a notable enrichment in boron, this may be sufficient to account for the tourmaline in many schists. Theoretical considerations enable some predictions as to possible patterns of behavior for certain elements. Strongly lithophile trace elements may be expected to follow Goldschmidt's rules for camouflage, admission, and capture by the crystal lattices of the common minerals of metamorphic rocks. The greater variety of minerals and of conditions of crystallization in metamorphic rocks, compared to igneous rocks, will result in more complex relations. The greater possibilities of disorder in crystal lattices at higher temperatures suggest greater tolerance for foreign elements in minerals formed at higher grades of metamorphism. For trace elements of chalcophile or weakly lithophile character electronegativity and the concentration of sulfide ions are probably more significant than ionic size in determining their ultimate deposition.

One of the few investigations of minor and trace elements in metamorphic rocks for which both geological and geochemical data are closely controlled is that of Shaw (1954). He selected the Littleton formation of New Hampshire for examination, because it is a series of pelitic rocks which has undergone progressive regional metamorphism, and its present condition shows a transition from shales through increasingly metamorphosed rocks to sillimanite schists and gneisses. Analyses were made of 63 samples of the formation, representing all grades of metamorphism. The results of this extensive study showed that the concentration of most elements remained constant during the metamorphism. The only detectable changes in trace elements were a slight decrease in nickel and copper and a well-defined increase in lithium and lead, the latter being correlated with some degree of potassium metasomatism.

This work of Shaw has general significance because it shows that a formation of fairly uniform composition retains the primary pattern of minor and trace elements even after extensive and intensive metamorphism. Thus the pattern of minor and trace elements may be a useful guide to the character of the original rock. This principle has been applied to the problem of the original nature of certain amphibolites. Nearly identical amphibolites, consisting essentially of hornblende and plagioclase, may evolve from skarn-like metasomatism of carbonate sediments or from essentially isochemical metamorphism of gabbroic and dioritic rocks. In such amphibolites the most satisfactory clue to their origin may lie in their distinctive inheritance of accessory elements. Dynamothermal metamorphism of gabbroic rocks does not generally remove the initial concentrations of such elements as chromium, cobalt, nickel, and copper; conversely, amphibolites derived by the replacement of marbles tend to be deficient in these elements and show higher concentrations of barium and lead.

Under some circumstances, however, it appears that appreciable changes in pattern of accessory elements are possible. DeVore (1955) has made a large number of analyses of the individual minerals extracted from rocks of different metamorphic facies, and claims that the replacement of one mineral assemblage by another will be accompanied by a redistribution and fractionation of both major and minor constituents. For example, he suggests that the transformation of an epidote-amphibolite facies hornblendite to a granulite facies hornblendite could release large amounts of Cr, Ni, Cu, and Mg, and the reverse transformation could release Pb, Zn, Ti, Mn, and Fe. His data indicate that a cubic kilometer of epidote-amphibolite hornblendite changing to granulite facies hornblendite could release eight million tons Cr_2O_3, four million tons of NiO, and 800,000 tons of CuO, and the reverse transformation could release 11 million tons MnO, 94 million tons TiO_2, 800,000 tons ZnO, and 27,000 tons PbO. This would provide an excellent source of ore-forming elements, but no convincing mechanism for the extraction of this material and its concentration into ore bodies has been proposed, nor have instances of ore deposits formed in this way been described. Under these circumstances this concept must be considered as an interesting speculation rather than an established fact.

ULTRAMETAMORPHISM

If the temperature continues to rise during metamorphism, any rock must eventually melt. In this way a magma is generated, and the further geochemical evolution is no longer part of metamorphism. The regeneration of magma will not, however, take place at a definite temperature and pressure but over a range of temperatures and pressures; the process may not proceed to completion and may be halted at any stage. Mixed rocks are thereby formed with characters partaking of those of both igneous and metamorphic rocks.

The process can be looked upon as the reverse of magmatic crystallization. In the melting of rocks we may expect the first liquid formed to resemble the last liquid fraction of a magma and be rich in silica, soda, potash, alumina, and water. The process of differential fusion or *anatexis* can be conceived as beginning with the "sweating" of the low-melting fraction from the main mass of the rock and its segregation into lenses. In the light of laboratory work on the crystallization of hydrous feldspar melts differential fusion of rocks may be expected to begin at temperatures of 600–700°. If the process ceases at this point, the fused material will crystallize as an aggregate of quartz and feldspar as lenses within the more refractory material, giving rise to the rock type that has been called venite, a particular variety of migmatite. A similar product could be formed by the injection of the last fraction from magmatic crystallization between the layers of a solid rock (lit-par-lit injection); the resulting rock has been called arterite to distinguish it from one formed by differential fusion. In the field it is often difficult to decide whether a migmatite is an arterite or a venite, and indeed both processes outlined above may have been active at the time of formation.

Under conditions of ultrametamorphism the regenerated magma will most likely be granitic. Some granites may well have been formed by the actual remelting of material of suitable composition. However, granites may also originate without remelting. The mineral association of granite—quartz, potash feldspar, biotite and/or hornblende—is typical of the amphibolite facies, and any body of rock with a bulk composition corresponding to that of granite (or which is converted thereto by metasomatism) recrystallizes to give this typical mineral association under conditions of that facies. Generally, as a result of directed pressure during crystallization, the product is a gneiss, but, if directed pressure is weak or lacking, a normal granite may result. Thus granites may be produced in several ways: by the fractional crystallization of a magma; by the crystallization of a melt produced by the differential fusion of a preexisting rock; and by the recrystallization without fusion of a preexisting rock. Any body of granite may include within it representatives of all these types of origin. Deciding the mode of formation of a particular granite may require all the resources, field and laboratory, of a geologist, and even so, the answer may be equivocal. The volume of discussion on the origin of granite is an eloquent expression of the difficulty in determining the boundary between magmatic and metamorphic processes in actual rocks.

SELECTED REFERENCES

Barth, T. F. W. (1962). *Theoretical petrology* (second edition) 416 pp. John Wiley and Sons, New York. Gives a comprehensive and logical presentation of principles and underlying theory of the petrology of metamorphic rocks.

Bowen, N. L. (1940). Progressive metamorphism of siliceous limestone and dolomite. *J. Geol.* **48,** 225–274. A classic paper in which the concept of a petrogenetic grid was first formulated.

Clarke, F. W. (1924). *The data of geochemistry.* Chapter 14.

DeVore, G. W. (1955). The role of adsorption in the fractionation and distribution of elements. *J. Geol.* **63,** 159–190. Discusses the changes in trace-element content between metamorphic facies and the possibility that metamorphic transformations liberate ore-forming materials.

Eskola, P., in Barth, T. F. W., C. W. Correns, and P. Eskola (1939). *Die Enstehung der Gesteine.* 422 pp. Springer-Verlag, Berlin. A comprehensive account of metamorphic petrogenesis by the leading European worker in this field.

Lapadu-Hargues, P. (1945). Sur l'existence et la nature de l'apport chimique dans certain séries cristallophylliennes. *Bull. Soc. Géol. France*, 5th series, **15,** 255–310. Concludes from a study of average compositions of the different rock types that the series shale-schist-gneiss-granite is a paragenetic series brought about by progressive metasomatism.

Packham, G. H., and K. A. W. Crook (1960). The principle of diagenetic facies and some of its implications. *J. Geol.* **68,** 392–407. An illuminating discussion of the relationship of diagenesis to metamorphism.

Perrin, R., and M. Roubault (1949). On the granite problem. *J. Geol.* **57,** 357–379. A statement of the view that granites have originated by ionic diffiusion and reaction in the solid state rather than by crystallization of melts or metasomatism by liquids or gases.

Ramberg, H. (1952). *The origin of metamorphic and metasomatic rocks.* 317 pp. University of Chicago Press, Chicago. A comprehensive account of metamorphism and metamorphic rocks, with emphasis on their interpretation in terms of thermodynamics and crystal chemistry.

Rankama, K., and Th. G. Sahama (1950). *Geochemistry.* Pp. 243–263.

Reynolds, D. L. (1947). The association of basic "fronts" with granitisation. *Science Progr.* **35,** 205–219. A review article discussing the evidence for basic fronts surrounding granite masses formed by metasomatism.

Shaw, D. M. (1954). Trace elements in pelitic rocks. *Bull. Geol. Soc. Amer.* **65,** 1151–1182. Gives a statistical analysis of the content of minor and trace elements in sixty-three rocks from the same formation, representing all grades of metamorphism from shales to sillimanite gneisses.

Thompson, J. B. (1955). The thermodynamic basis for the mineral facies concept. *Am. J. Sci.* **253,** 65–103. A thorough treatment of the thermodynamics of metamorphism and the geological applications.

Turner, F. J., and J. Verhoogen (1960). *Igneous and metamorphic petrology* (second edition), 694 pp. McGraw-Hill Book Co., New York. Contains a comprehensive account of metamorphism and metamorphic rocks, with emphasis on the physical chemistry involved.

Winkler, H. G. F. (1965). *Petrogenesis of metamorphic rocks.* 220 pp. Springer-Verlag, New York. A thorough discussion of the chemical and mineralogical transformations accompanying rock metamorphism.

Yoder, H. S. (1955). Role of water in metamorphism. *Geol. Soc. Amer. Spec. Paper* **62,** 505–524. Criticizes the concept of the ubiquitous occurrence and adequate supply of water in metamorphism and discusses the consequences of a water-deficient environment.

Yoder, H. S., and H. P. Eugster (1955). Synthetic and natural muscovites. *Geochim. Cosmochim. Acta* **8,** 225–280. An exhaustive study of one of the most abundant and widespread metamorphic minerals.

CHAPTER ELEVEN

THE GEOCHEMICAL CYCLE

THE EARTH AS A PHYSICOCHEMICAL SYSTEM

Geochemically, the earth may be considered a closed system, as this term is used in physical chemistry. This concept can be criticized as an oversimplification; we have seen that some material—meteorites and meteoritic dust—is continually being received from outer space, and that some hydrogen and helium is being lost by escape from the upper atmosphere. Nevertheless, these gains and losses are insignificant as compared with the system as a whole. If our interpretation of the geological record is correct, the chemical processes taking place on the surface and within the earth have probably operated with a remarkable degree of uniformity for the last 3000 million years.

By the nature of things the geochemist is mainly concerned with the surface of the earth, since it is the only part accessible to direct examination. In the discussion of the origin and structure of the earth an attempt was made to present a logical account of the probable nature of the interior. The data on which this account was based are indirect, their interpretation speculative. However, the general picture of a nickel-iron core, a mantle largely of magnesium-iron silicate, and a crust in which oxygen, silicon, aluminum, iron, calcium, magnesium, sodium, and potassium are the major constituents gives a consistent interpretation of information gleaned from many independent sources—the study of meteorites, the physics and chemistry of the earth, seismological data, and so on. We may therefore accept such a picture as a working hypothesis, realizing always that it is a hypothesis, but a well-buttressed one.

One of the few subjects upon which universal agreement seems to prevail is that the earth was not created in its present state. The geochemical evidence supports the idea that its internal structure is probably the result of forces originating within the earth itself. The earth is a system with considerable mass and thus exerts a gravitational force on its own components. The resulting gravitational field has affected the distribution of material by concentrating the heavier phases towards the center and the lighter towards the surface. The rate of such a gravitational differentiation clearly depends upon the viscosity of the system; it is more rapid in a gas than in a liquid, more rapid in a liquid than in a solid. The idea of an earth at one time liquid is attractive, for it explains how such gravitational differentiation could take place within a comparatively short time. Geophysical data sustain the theory that the earth has a layered structure due to separation of its material into shells of different density. Goldschmidt termed this the primary geochemical differentiation. It was a differentiation due to gravity acting on a system in which iron, oxygen, magnesium, and silicon were the major components. Iron was the principal component, and the distribution of the elements between a metallic core and a silicate mantle was controlled by their oxidation potential with respect to that of iron. Elements more readily oxidized than iron concentrated in the mantle; the others alloyed with iron to form the core. Hence the fate of an element in this primary geochemical differentiation is in effect a reflection of the number and arrangement of its orbital electrons. Those elements forming ions with a noble gas structure went into the silicate phase; the transition elements, on the other hand, concentrated in the metallic core or in a sulfide phase.

THE CRUST AS A SEPARATE SYSTEM

The separation of crust, mantle, and core enables us to consider the outer part of the earth as a distinct physicochemical system. The migration of material within the crust can then be discussed as an independent phenomenon; it is partly mechanical, brought about by orogenic movements or gravitational forces, and partly chemical. Mechanical movements belong in the field of geology. Geochemistry is concerned with the migration of the elements under the influence of physicochemical forces. This migration has been discussed in terms of the processes of magmatism, sedimentation, and metamorphism. We have seen that the fate of an element during magmatic crystallization is primarily a function of its ionic size. A particular element appears in those minerals in the lattices of which it fits most readily and with the greatest decrease of free energy. The distribution of the elements by ionic size in this way was described by Goldschmidt as the secondary geochemical differentiation. Magmatic crystallization also adds important amounts of a few elements to the atmosphere and hydrosphere.

The processes of sedimentation can be looked upon as leading to a further degree of geochemical differentiation. At the comparatively low temperature of sedimentary processes ionic substitution in minerals is much less prominent, although still significant. Coprecipitation under a particular set of physicochemical conditions is, however, an important way in which certain elements become associated in specific types of sediment. The controlling factors are those pertaining to the properties of ions in aqueous solutions, and the ionic potential is of primary importance. Geochemical differentiation during sedimentation is therefore governed in large part by interrelationship between ionic radius and ionic charge.

Sedimentation also involves an interaction of the hydrosphere and the atmosphere with the lithosphere. Water and carbon dioxide are incorporated in sedimentary minerals; soluble ions, especially sodium, are contributed to the hydrosphere. Processes involving living organisms are intimately associated with sedimentation but can be considered separately, and these processes are even more closely linked with the hydrosphere and the atmosphere. We have seen how photosynthesis has probably been largely responsible for the present composition of the atmosphere and how the balance of dissolved material in the ocean is largely a function of the organic life therein. Thus in the biosphere a further geochemical differentiation takes place through the metabolic action of organisms.

The series of changes so far discussed has led on the whole to an increasing degree of geochemical differentiation. This tendency is reversed by metamorphism. In general, metamorphism tends towards uniformity of distribution of the elements. One can visualize unlimited metamorphism as resulting in an ideal condition in which the whole lithosphere reaches a uniform composition. This may seem an exaggerated view, but that such a tendency exists is evidenced by the comparatively monotonous chemical and mineralogical composition of ancient geological formations. Compared with the chemically diverse rock types of younger ages, the Archaean is dominantly made up of gneisses of relative uniformity, probably due in considerable degree to long-continued metamorphic and metasomatic reactions.

Thus in the beginning the relative abundances of the elements were determined by the stability of their isotopes, that is, by the number of neutrons and protons in the nucleus and the binding energy. The processes by which the earth was formed led to a first separation of elements according to their volatility or ability to form volatile compounds; the earth is evidently impoverished in the more volatile elements and compounds in comparison with the universe as a whole. The separation of the earth into an iron core and a silicate mantle and crust resulted in a strong fractionation of the elements according to their affinity for metallic iron or for silicate; this fractionation was controlled by the number and arrangement of the outer electrons. The next step in the evolution of the earth was the solidification of mantle and crust, which led to a further fractionation, this time determined largely by the role of the different elements in liquid ⇌ crystal equilibria. The major controlling factor was ionic size. During geological time considerable fractionation of the elements has taken place at the earth's surface as a result of sedimentary processes; the fate of an element under these conditions is largely a matter of its ionic potential, the ratio of ionic size to ionic charge. The absolute abundance of an element is conditioned by its nuclear structure; its abundance in a particular part of the universe or of the earth is conditioned by more superficial atomic characters, such as the number and arrangement of the orbital electrons, and the size of the atom or ion. The geochemical behavior of each element depends on its individual properties under the physicochemical conditions at each stage in the geochemical cycle.

THE GEOCHEMICAL CYCLE

This overall picture of the migration of the elements in the outer part of the earth provides us with the concept of the geochemical cycle (Figure 11.1). In the lithosphere the geochemical cycle begins with the initial crystal-

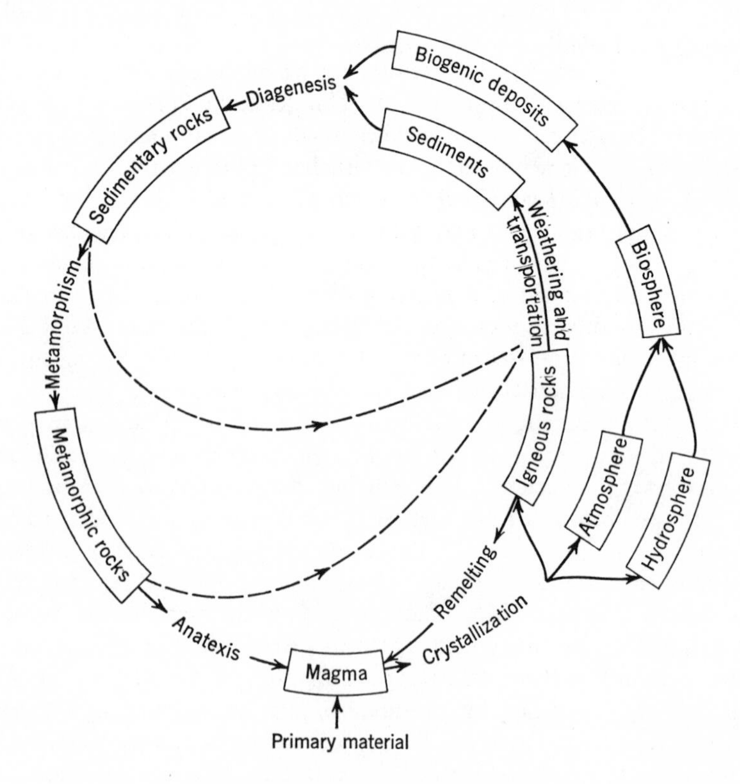

Figure 11.1 The geochemical cycle.

lization of a magma, proceeds through the alteration and weathering of the igneous rock and the transportation and deposition of the material thus produced, and continues through diagenesis and lithification to metamorphism of successively higher grade until eventually, by anatexis and palingenesis, magma is regenerated. Like any ideal cycle, the geochemical cycle may not be realized in practice; at some stage it may be indefinitely halted or short-circuited or its direction reversed. The geochemical cycle is not closed, either materially or energetically. It receives "primary" magma from below bringing energy with it in the form of heat. The surface receives an insignificant contribution of meteoritic matter, which nevertheless is detectable in deep-sea deposits where the rate of sedimentation is very low. The surface receives energy from outside the earth in the form of solar radiation, nearly all of which is, however, reradiated into space.

The geochemical cycle provides a useful concept as a basis for the discussion of many aspects of geochemistry, particularly the course followed by a specific element in proceeding through the different stages. A complete understanding of the cycle in terms of the individual elements is one of the major objectives of geochemistry. An element may tend to concentrate in a certain type of deposit at some stage, or it may remain dispersed throughout the entire cycle. The geochemical cycle of a few elements has been worked out in considerable detail, but for many our knowledge is fragmentary.

ISOTOPIC FRACTIONATION IN THE GEOCHEMICAL CYCLE

The geochemical cycle often results in the concentration of a specific element at some stage in the cycle, as, for example, carbon in biogenic deposits. For elements with more than one isotope a more subtle fractionation, that of the individual isotopes, is also conceivable. For some years it was thought that such fractionation did not occur in nature, except for an element, such as lead, in which one or more isotopes is the end member of a radioactive series; other elements were considered to be of constant isotopic composition regardless of the geological or biological processes they had undergone. With the development of sensitive mass-spectrometric techniques, however, it has been found that the isotopic composition of some elements is changed by such processes. As a result there has been a remarkable development of *isotope geology*, which Rankama defines as the investigation of geological phenomena by means of stable and unstable isotopes and of changes in their abundance.

Isotopic fractionations can arise in several ways, depending on the fact that the thermodynamic and kinetic properties of a system of molecules are mass-dependent, hence show differences on isotopic substitution. The resulting fractionation is most marked when the mass difference on substitution is greatest, that is, for low atomic weight elements substituted in low molecular weight compounds. In nature, fractionation is most efficient in systems involving a gaseous phase, when there is in effect a continuous fractional distillation. Hence isotopic differentiation has been most widely observed in elements with gaseous compounds—H, C, O, S.

The chemical fractionation of isotopes is a reflection of slight differences in the free energies of formation of isotopic molecules from their constituent atoms. In a reaction between molecules containing different isotopes of some element, such as

$$CO_2^{16} + 2H_2O^{18} = CO_2^{18} + 2H_2O^{16}$$

the equilibrium constant may be appreciably different from unity, and the oxygen present as carbon dioxide and as water may differ measurably in isotopic composition. The equilibrium constants for such reactions can be computed by statistical mechanics; they are temperature dependent, and, in general, isotopic fractionation diminishes with increasing temperature. Calculated values of the fractionation factor for the above reaction are 1.055, 1.047, 1.027, and 1.017 at 0°, 25°, 127°, and 227° respectively; the observed values are in good agreement. Temperature is thus the principal control of isotope fractionation, but it may be affected by other factors influencing the chemical behavior of the system, such as pressure, chemical composition, and activity coefficients of the solutions. Kinetic effects may also modify the fractionation factor.

Variations in O^{18}/O^{16} ratios have been applied to a number of geological problems. Actual results are usually reported as δ values, which compare the measured O^{18}/O^{16} ratio in a sample to this ratio in mean sea water, which is taken as a standard. Values of δ are expressed in parts per thousand, positive figures indicating relative enrichment in O^{18}. These measurements can be applied to the study of natural waters and the circulation of water and water vapor. Some typical figures are the following:

Vapor in equilibrium with sea water	−8
River water, Alaska	−20
Snow, Chicago	−17
Rain, Chicago	−7
River water, Mississippi	−4
Snow, South Pole	−50

These figures indicate that snow at the South Pole is remarkably depleted in O^{18} with respect to water vapor from the ocean. Evidently, a marine air mass preferentially loses H_2O^{18} as it loses water through rain or snow and thus becomes progressively richer in H_2O^{16}.

Oxygen isotope measurements have been applied successfully to the measurement of paleotemperatures. For the reaction

$$\tfrac{1}{3}CaCO_3^{16} + H_2O^{18} = \tfrac{1}{3}CaCO_3^{18} + H_2O^{16}$$

the equilibrium constant K reduces to the following expression

$$K = \frac{(O^{18}/O^{16})_{CaCO_3}}{(O^{18}/O^{16})_{H_2O}}$$

The value of K for calcite-water exchange is 1.0286 at 25°, diminishing by 0.00023 per degree increase in temperature. With an analytical accuracy of ± 0.0001 this permits the determination of the temperature of deposition of calcite from water of known isotopic composition to within 0.5°.

Because the O^{18}/O^{16} ratio of ocean water is essentially invariant, the O^{18}/O^{16} ratio of calcite deposited from it will be temperature-dependent. Thus if marine carbonate skeletons grow in isotopic equilibrium with sea water, the O^{18}/O^{16} ratio of the carbonates will be a measure of the temperature at which the carbonate was laid down by the organism. Of course, if the shell material recrystallizes or undergoes isotopic exchange in some later geological event, the O^{18}/O^{16} ratio will be modified, thereby erasing the record of deposition temperature. Apparently reliable paleotemperatures have been measured on fossil carbonates from as far back as the Mesozoic. For example, Upper Cretaceous specimens from the United States, including Alaska, from Siberia, Sweden, Denmark, France, and Germany gave temperatures ranging from 12° to 28°. The lowest temperatures were observed for samples from Alaska and Siberia. The results showed that the climate in the Upper Cretaceous was relatively mild and much more uniform than at present. Moreover the temperature of the oceans in the northern region, and thus probably at depth (deep ocean water coming eventually from the polar regions) was about 14°, considerably higher than the temperature of the present ocean depths.

One of the promising applications of oxygen-isotope measurements is in the field of petrology, including diagenesis and ore formation. Common rock systems are usually composed of two or more coexisting minerals. If these minerals were in isotopic equilibrium with H_2O and with each other at the time of crystallization, the temperature of equilibration can be determined by using the temperature-dependence of several fractionation factors simultaneously. However, the amount of variation of the O^{18}/O^{16} ratio in silicates, carbonates, and oxides in igneous and metamorphic rocks is not well known. The quantitative extraction of oxygen from silicate and oxide minerals is difficult. Early attempts showed that isotopic variations existed, but measurements were relatively crude. Current research activity in this field is concentrated along two lines: the laboratory determination of exchange equilibrium constants as functions of temperature, and measurements of natural mineral assemblages selected so that the isotopically deduced temperatures may be compared with temperatures of formation of the same minerals as estimated from other, nonisotopic data. The interpretation of isotopic measurements must be related to the thermal history of the sample, particularly in rocks which underwent a long, slow cooling following initial crystallization. Studies in areas of regional metamorphism have shown that isotopic equilibrium is probably preserved in low-grade rocks, but that the isotopic distributions in some of the higher-temperature equilibria are probably lost by retrograde effects.

The fractionation of carbon isotopes was discussed in Chapter 9 in connection with the carbon cycle. Another element for which natural isotopic fractionation has been extensively studied is sulfur (Figure 11.2). The reference standard is meteoritic troilite, which shows a remarkably uniform S^{34}/S^{32} ratio of 22.21, and may perhaps be considered to represent primordial

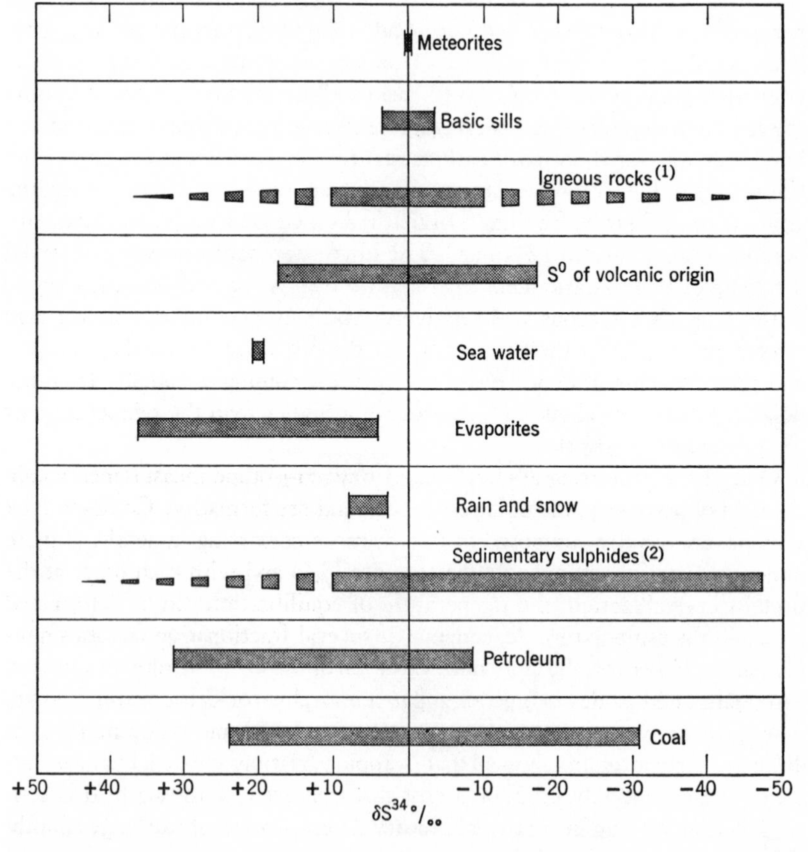

Figure 11.2 Variations of sulfur isotopes in natural materials: (1) including granitized or reworked sediments; (2) excluding rain and snow from industrial areas. (Thode, 1963)

sulfur. The S^{34} content of sulfur in the earth's crust varies by about 5% on each side of this standard, δS^{34} in Figure 11.2 being defined by the equation

$$\delta S^{34} = \left(\frac{(S^{32}/S^{34})_{\text{sample}}}{(S^{34}/S^{34})_{\text{standard}}} - 1 \right) \times 1000$$

Sulfur isotopes, in geochemical and biochemical processes involving oxidation and reduction, are fractionated with the lighter isotope being enriched in the sulfide and the heavier isotope being enriched in the sulfate. The maximum range of S^{32}/S^{34} in nature is from 20.8 for sulfate in the cap rock of salt domes to 23.3 for sulfide from a shale.

ENERGY CHANGES IN THE GEOCHEMICAL CYCLE

We have been considering the geochemical cycle in terms of the material changes which take place during the various processes. Equally significant, if less studied and less well understood, are the energy changes during the cycle. Geochemical processes operate only because of a flow of energy from a higher to a lower potential or intensity; hence energy is no less important than matter in the geochemical cycle.

The earth cannot be considered a closed system in terms of energy, since it receives a large amount of heat from the sun in the form of solar radiation. In an overall view we should consider the whole universe as a single system which is presumably undergoing a spontaneous decrease in free energy and an increase in entropy, as prescribed by the second law of thermodynamics. However, we can deal with the principal aspects of the geochemical cycle in terms of the solar system, since the forces acting on the earth and the energy it receives are primarily the result of its relationship to the sun. Within this limited system the overall development is also toward a decrease in free energy and an increase in entropy. The few apparent exceptions, such as the accumulation of free energy in deposits of coal and oil, represent a decrease in entropy at the expense of the sun and are merely minor fluctuations in a general trend toward a state of higher entropy.

Major sources of energy in the geochemical cycle may be considered under the following heads: (*a*) solar radiation; (*b*) mechanical energy, potential and kinetic; (*c*) reaction energy; (*d*) nuclear energy; and (*e*) heat content of the earth. Transformation of these forms of energy is the common denominator of all geochemical processes and reactions.

Solar radiation is, of course, the product of nuclear reactions whereby hydrogen is converted to helium in the sun. The sun radiates energy in all directions and only a very small proportion, about 5×10^{-8}, impinges on the earth; about half of this is reflected back into space by the atmosphere. The average solar radiation received at the earth's surface is 4.2×10^{-3} cal/cm^2/sec, or about 6750×10^{20} cal annually for the whole earth. It has been estimated that 0.1% is utilized by photosynthesis of plants, terrestrial and marine, about 15% is absorbed by the earth, and the remainder is used to evaporate water from the hydrosphere. Solar energy is thus essential to the cycle of changes involving the interaction of the hydrosphere, the biosphere, the atmosphere, and the crust.

Through the evaporation of water from the hydrosphere solar radiation in converted into mechanical energy. Water vapor in the atmosphere has potential energy, which in turn is transformed into the kinetic energy of rain and running water, the source of most of the energy driving erosion and sedimentation. This kinetic energy is utilized in part in moving solid material downhill. A pound of quartz moved from the top of Pikes Peak to the delta of the Mississippi River loses about 14,000 foot-pounds of energy; its original potential energy is converted into kinetic energy and eventually into heat. The reduction of a mountain range to a peneplain is largely the work of running water and transforms the energy stored in its uplift into heat. Within a comparatively short period of time, geologically speaking, erosion would reduce the whole land surface to a monotonous low-lying plain, were it not for diastrophism, which creates mountains about as fast as they are destroyed. The uplift requires a considerable expenditure of energy. The average elevation of the continents is about 800 meters; to maintain this average elevation against erosion requires an output of 10^{17}–10^{18} cal/year.

Another source of mechanical energy is the kinetic energy of the earth's rotation. A portion of this energy appears as winds and tides. To be sure, tides are brought about by the gravitational effect of the moon and, to a lesser extent, the sun, but the energy dissipated comes from the kinetic energy of the earth and not from these extraterrestrial bodies. The effect of these tides is the slowing-down of the earth's rotation by about one second in 1000 years. The kinetic energy of the earth's rotation has important geological implications, for it acts to maintain the earth's figure against all forces that tend to deform it.

Reaction energy covers all changes of state, physical and chemical, and is the form most immediately associated with the geochemical cycle, because every geochemical process involves such changes. From Le Châtelier's principle we can predict that reactions taking place under conditions of falling temperature, such as the crystallization of magma, will be exothermic; reactions taking place under conditions of increasing temperature, such as progressive metamorphism, will be endothermic. Reactions involving change of state, such as the melting of solid material and polymorphic transformations, are accompanied by important energy changes, and are probably significant in the overall energy balance within the crust and the mantle.

We have already seen that solar radiation is the product of nuclear

Table 11.1. Radioactive Heat Production in Typical Rocks

	Concentration		Heat production (cal/g/10^6 years)			
	U (ppm)	K (per cent)	U	U + Th	K	Total
Granitic	4	3.5	3	6	1	7
Intermediate	2	2	1.5	3	0.5	3.5
Basaltic	0.6	1	0.4	0.8	0.3	1.1
Dunite	0.015	0.001	0.01	0.02	<0.001	~ 0.02

reactions in the sun. Nuclear reactions in the earth—the spontaneous disintegration of radioactive atoms—have been of prime importance throughout the earth's history. Much study has been devoted to this in recent years; a useful summary is provided by Birch (1954). Radioactive decay converts mass to energy, and nearly all the energy appears as heat in the immediate neighborhood of the source. The isotopes of importance for heat generation in the earth are U^{238}, U^{235}, Th^{232}, and K^{40}. Birch gives the figures in Table 11.1 for the radioactive heat production in typical rocks. These figures are based on averages of numerous analyses; however, the individual analyses show a considerable variation, and the sampling problem when dealing with uranium and thorium, present in a few parts per million, associated with minor mineral constituents and irregularly distributed, is a formidable one. Nevertheless, the results show quite clearly the marked concentration of the radioactive elements in the silica-rich rocks of the continental crust.

The figures in Table 11.1 are based on the amounts of uranium, thorium, and potassium in the rocks at the present time. The present amounts of the radioactive isotopes are smaller than those which existed in the past, and for calculations dealing with the whole lifetime of the earth it is important to take the decay into account. The rate of heat production at a time t years ago was $e^{\lambda t}$ times the present rate, for a given present quantity of an isotope with disintegration constant λ. The ratio of the total heat produced over a time t, allowing for the decay, to the amount that would have been produced in the same time if the present quantity had been constant, is $(e^{\lambda t} - 1)\lambda t$. Birch has evaluated this function, with the following results:

t (in 10^9 years)	1	2	3	4	5
$U^{238} + U^{235}$	1.11	1.25	1.47	1.85	2.57
Th^{232}	1.03	1.05	1.08	1.11	1.14
K^{40}	1.33	1.82	2.55	3.65	5.33

These figures point up the significantly greater production of radiogenic heat in the early period of the earth's history, and the increasing fraction contributed by K^{40}.

The heat content of the earth is due to the thermal vibration of its constituent atoms. The internal heat of the earth is enormous, but much of it is essentially static and plays no part in the geochemical cycle at and near the surface. Thermal energy is transferred to the surface by conduction and by igneous activity; much of this energy is generated within the crust by radioactivity. The continental crust averages about 35 km thick. A 35 km column of granite with 1 cm^2 cross section has a volume of 3.5×10^6 cm^3 and a mass of 9.4×10^6 g. The heat generation, from Table 11.1, would thus amount to 65.8 cal/cm^2/year, or approximately 2×10^{-6} cal/cm^2/sec. The average output of heat by conduction is about 1.4×10^{-6} cal/cm^2/sec, or roughly 3×10^{20} cal/year for the whole earth. A granitic crust 35 km thick would thus supply more heat than is actually escaping at the surface, even without allowance either for original heat or for radiogenic heat liberated in the underlying mantle and core.

Precise estimates of the heat brought to the surface by igneous activity are difficult to make. Verhoogen (1960) estimated that about 3×10^{15} g of lava are erupted annually, giving off some 400 cal/g during cooling and solidification, that is, a total of about 10^{18} cal. Not all magma reaches the surface; erosion reveals igneous rocks of all ages which have been intruded in the liquid state and have cooled within the crust. In addition, volcanic heat is also carried to the surface by thermal springs. The total amount of heat contributed by igneous activity is thus of the order of 10^{18}–10^{19} cal/year. Although the contribution of a single volcano may be spectacular, the total heat transferred by igneous activity is only a small fraction of that due to conduction. The average energy released annually by earthquakes has been estimated to be about 10^{25} ergs, or 0.24×10^{18} cal, also a minute fraction of the heat escaping by conduction.

In terms of the geochemical cycle the upward movement of magma brings with it a certain amount of energy. The problem of how magmas originate is still unsolved. Our knowledge of the thermal gradient within the earth indicates that temperatures within the crust and mantle are on the average well below the temperature of melting of the material at any specified depth. The thermal energy of the earth (including that produced by radioactivity) provides an ample source of heat; the problem is to find some mechanism which will cause melting at local "hot spots" which are the centers of igneous activity. Igneous activity is a small-scale, local phenomenon in relation to the earth as a whole, and it is concentrated along zones of general tectonic activity, especially the margin of the Pacific Ocean. In many places igneous activity is related to deep-seated fractures; these fractures may act as channels for gases carrying heat from greater depths, and movements on such fractures will disturb the balance between temperature and pressure on either side. Such mechanisms may produce magma, which by virtue of its lower density than the surrounding solid material tends to rise into the higher and cooler parts of the crust. The thermal energy of the magma is dissipated to the surroundings during cooling and solidification. The kinetic energy of the atoms or ions decreases as the temperature falls; their potential energy also diminishes when they are arranged in an orderly fashion in a crystal lattice. At the next stage of the cycle erosion and sedimentation reduce the material on which they work to a lower energy state. Erosion moves material with the aid of gravitational forces, and a loss of potential energy results. The chemical reactions accompanying weathering are spontaneous and usually irreversible and lead to a decrease in free energy. The only spontaneous process at the earth's surface resulting in an increase of free energy is the conversion of carbon dioxide and water to complex organic compounds by plants utilizing solar radiation. The general trend is reversed during metamorphism, when thermal and gravitational energy is converted into chemical energy by endothermic reactions and the formation of compounds of higher density. A sufficient increase in temperature may increase the kinetic energy of the atoms or ions sufficiently to overcome the forces holding them in crystal lattices, so that the minerals decompose or melt. A magma is thereby regenerated, and the cycle is complete. A quantitative balance sheet for the energy change during the geochemical cycle cannot yet be drawn up, no more than a quantitative balance sheet for the material changes. However, it does appear that much of the energy of the geochemical cycle is contributed by the disintegration of radioactive elements within the crust, possibly supplemented by some of the earth's internal heat.

SELECTED REFERENCES

Birch, F. (1954). Heat from radioactivity. Chapter 5 (pp. 148–174) in *Nuclear Geology*, H. Faul, ed., John Wiley and Sons, New York. A quantitative account of the significance of radiogenic heat in geology.

Craig, H., S. L. Miller, and G. J. Wasserburg (eds.) (1964). *Isotopic and cosmic chemistry*, 553 pp. North-Holland Publishing Co., Amsterdam. This volume, dedicated appropriately to H. C. Urey, contains numerous contributions to isotope geochemistry.

Goldschmidt, V. M. (1922). Der Stoffwechsel der Erde. *Norsk. Videnskapsselskapets Skrifter*. I. *Mat.-Naturv. Klasse*, No. 11, 25 pp. A remarkably far-sighted essay in interpretative geochemistry, written when little quantitative data existed for the ideas expressed therein.

Goldschmidt, V. M. (1954). *Geochemistry*. Part II of this book deals with the geochemical cycle of individual elements and groups of elements.

Rankama, K. (1954). *Isotope geology*. 535 pp. McGraw-Hill Book Co., New York. An exhaustive compilation and discussion of variations in amounts of the isotopes of the elements in geological materials.

———, K. (1963). *Progress in isotope geology*. 705 pp. Interscience Publishers, New York. A supplementary volume to the preceding work.

Rankama, K., and Th. G. Sahama (1950). *Geochemistry.* In Part II of this book the geochemistry of the individual elements is described, including what is known of the geochemical cycle for each of them.

Shaw, D. M. (ed.) (1963). Studies in analytical geochemistry. *Roy. Soc. Canada Spec. Publ.* 6. 139 pp. A number of review articles, including one on sulfur isotope geochemistry by H. G. Thode and one on oxygen isotope geochemistry by R. N. Clayton.

Verhoogen, J. (1960). Temperatures within the earth. *Am. Scientist* **48,** 134–159. A discussion of the intensity and distribution of energy sources within the earth.

APPENDIX I

ATOMIC WEIGHTS AND IONIC RADII

	Symbol	Atomic Number	Atomic Weight (1955)	Ionic Radius* (sixfold coordination)
Aluminum	Al	13	26.98	0.51
Antimony	Sb	51	121.76	Sb^{3} 0.76; Sb^{5} 0.62
Argon	Ar	18	39.944	—
Arsenic	As	33	74.91	As^{3} 0.58; As^{5} 0.46
Barium	Ba	56	137.36	1.34
Beryllium	Be	4	9.013	0.35
Bismuth	Bi	83	209.00	Bi^{3} 0.96; Bi^{5} 0.74
Boron	B	5	10.82	0.23
Bromine	Br	35	79.916	1.95
Cadmium	Cd	48	112.41	0.97
Calcium	Ca	20	40.08	0.99
Carbon	C	6	12.011	0.16
Cerium	Ce	58	140.13	Ce^{3} 1.07; Ce^{4} 0.94
Cesium	Cs	55	132.91	1.67
Chlorine	Cl	17	35.457	1.81
Chromium	Cr	24	52.01	Cr^{3} 0.63; Cr^{6} 0.52
Cobalt	Co	27	58.94	Co^{2} 0.72; Co^{3} 0.63
Copper	Cu	29	63.54	Cu^{1} 0.96; Cu^{2} 0.72
Dysprosium	Dy	66	162.51	0.92
Erbium	Er	68	167.27	0.89
Europium	Eu	63	152.0	0.98
Fluorine	F	9	19.00	1.36
Gadolinium	Gd	64	157.26	0.97
Gallium	Ga	31	69.72	0.62
Germanium	Ge	32	72.60	0.53
Gold	Au	79	197.0	1.37

	Symbol	Atomic Number	Atomic Weight (1955)	Ionic Radius* (sixfold coordination)
Hafnium	Hf	72	178.5	0.78
Helium	He	2	4.003	—
Holmium	Ho	67	164.94	0.91
Hydrogen	H	1	1.0080	—
Indium	In	49	114.82	0.81
Iodine	I	53	126.91	I^{1-} 2.16; I^{5} 0.62
Iridium	Ir	77	192.2	0.68
Iron	Fe	26	55.85	Fe^{2} 0.74; Fe^{3} 0.64
Krypton	Kr	36	83.80	—
Lanthanum	La	57	138.92	1.14
Lead	Pb	82	207.21	Pb^{2} 1.20; Pb^{4} 0.84
Lithium	Li	3	6.940	0.68
Lutecium	Lu	71	174.99	0.85
Magnesium	Mg	12	24.32	0.66
Manganese	Mn	25	54.94	Mn^{2} 0.80; Mn^{4} 0.60
Mercury	Hg	80	200.61	1.10
Molybdenum	Mo	42	95.95	Mo^{4} 0.70; Mo^{6} 0.62
Neodymium	Nd	60	144.27	1.04
Neon	Ne	10	20.183	—
Nickel	Ni	28	58.71	0.69
Niobium	Nb	41	92.91	0.69
Nitrogen	N	7	14.008	0.13
Osmium	Os	76	190.2	0.69
Oxygen	O	8	16.0000	1.40
Palladium	Pd	46	106.4	0.80
Phosphorus	P	15	30.975	0.35
Platinum	Pt	78	195.09	0.80
Potassium	K	19	39.096	1.33
Praseodymium	Pr	59	140.92	1.06
Radium	Ra	88	226.05	1.43
Radon	Rn	86	222	—
Rhenium	Re	75	186.22	0.72
Rhodium	Rh	45	102.91	0.68
Rubidium	Rb	37	85.48	1.47
Ruthenium	Ru	44	101.1	0.67
Samarium	Sm	62	150.35	1.00
Scandium	Sc	21	44.96	0.81
Selenium	Se	34	78.96	Se^{4} 0.50; Se^{6} 0.42
Silicon	Si	14	28.09	0.42
Silver	Ag	47	107.880	1.26
Sodium	Na	11	22.991	0.97
Strontium	Sr	38	87.63	1.12
Sulfur	S	16	32.066	S^{2-} 1.84; S^{6} 0.30

	Symbol	Atomic Number	Atomic Weight (1955)	Ionic Radius* (sixfold coordination)
Tantalum	Ta	73	180.95	0.68
Tellurium	Te	52	127.61	Te^4 0.70; Te^6 0.56
Terbium	Tb	65	158.93	0.93
Thallium	Tl	81	204.39	1.47
Thorium	Th	90	232.05	1.02
Thulium	Tm	69	168.94	0.87
Tin	Sn	50	118.70	Sn^2 0.93; Sn^4 0.71
Titanium	Ti	22	47.90	Ti^3 0.76; Ti^4 0.68
Tungsten	W	74	183.86	W^4 0.70; W^6 0.62
Uranium	U	92	238.07	0.97
Vanadium	V	23	50.95	V^3 0.74; V^5 0.59
Xenon	Xe	54	131.30	—
Ytterbium	Yb	70	173.04	0.86
Yttrium	Y	39	88.92	0.92
Zinc	Zn	30	65.38	0.74
Zirconium	Zr	40	91.22	0.79

* L. H. Ahrens, *Geochim. Cosmochim. Acta* **2,** 168, 1952.

APPENDIX II

THE GEOLOGICAL TIME SCALE*

Eras	Approximate age (in years)	Subdivisions	Approximate duration (in years)
QUATERNARY		RECENT	10,000
	10,000	PLEISTOCENE	3 M
TERTIARY	3 M	PLIOCENE	9 M
	12 M	MIOCENE	13 M
	25 M	OLIGOCENE	15 M
	40 M	EOCENE	20 M
	60 M	PALEOCENE	10 M
MESOZOIC	70 M	CRETACEOUS	65 M
	135 M	JURASSIC	45 M
	180 M	TRIASSIC	45 M
PALEOZOIC	225 M	PERMIAN	45 M
	270 M	CARBONIFEROUS	80 M
	350 M	DEVONIAN	50 M
	400 M	SILURIAN	40 M
	440 M	ORDOVICIAN	60 M
	500 M	CAMBRIAN	100 M
PRECAMBRIAN	600 M	Worldwide subdivisions not well established	2800 M+

Oldest rock dated 3500 M

Age of the earth 4500 M

* These figures, like those in railway timetables, are subject to change without notice.

APPENDIX III

ESTIMATED ANNUAL WORLD CONSUMPTION OF THE ELEMENTS

(in tons unless otherwise stated)

Element	Principal Sources	Amount of Element	Price of Element (U.S. $ per ton)
H	water, methane	2×10^6	150
He	natural gas	3500	7000
Li	petalite, lepidolite, spodumene, lake brines	1500	20/kg (Li) 1000 ($LiCO_3$)
Be	beryl	300	120/kg
B	Na, Ca borates, brines	400,000 (B_2O_3)	330 (B_2O_3)
C	diamond	6 (industrial)	
	graphite	300,000	
	coal	1.8×10^9	
	petroleum	0.9×10^9	
	natural gas	0.4×10^9	
N	air	1.7×10^7	44 ($NaNO_3$)
	soda niter	1×10^6 ($NaNO_3$)	
O	air	2×10^7	30
F	fluorite	1.1×10^6	
Ne	air		
Na	halite	1.0×10^8 (NaCl)	400
Mg	sea water, magnesite	150,000 (Mg) 9×10^6 ($MgCO_3$)	750
Al	bauxite	6.1×10^6	560
Si	quartz	700,000	350
P	apatite (phosphorite)	7×10^6	1000

Element	Principal Sources	Amount of Element	Price of Element (U.S. $ per ton)
S	sulfur, pyrite, natural gas	2.0×10^7	27
Cl	halite	5×10^6	60
Ar	air	20,000	400
K	sylvite, carnallite	1.0×10^7	20/kg (K) 25 (KCl)
Ca	calcite	6.0×10^7 (CaO)	14 (CaO)
Sc	thortveitite	50kg(Sc_2O_3)	2000/kg (Sc_2O_3)
Ti	ilmenite, rutile	10,000 (Ti) 1×10^6 (TiO_2)	2500 (Ti) 500 (TiO_2)
V	U, Pb vanadates	7000	7000
Cr	chromite	1.4×10^6	2500
Mn	pyrolusite, psilomelane	6×10^6	700
Fe	hematite, magnetite	3.1×10^8	50
Co	Co sulfides, arsenides	13,000	3000
Ni	pentlandite, garnierite	400,000	1800
Cu	chalcopyrite, chalcocite	5.4×10^6	800
Zn	sphalerite	3.8×10^6	300
Ga	bauxite	10	1000/kg
Ge	germanite	100	250/kg
As	arsenopyrite, enargite	40,000	1000 (As) 80 (As_2O_3)
Se	byproduct of Cu smelting	1000	9000
Br	sea water, brines	110,000	500
Kr	air		
Rb	pollucite, salt deposits		1000/kg
Sr	celestite	8000	250 ($Sr(NO_3)_2$)
Y	monazite, euxenite	5 (Y_2O_3)	60/kg (Y_2O_3)
Zr	zircon	1000 (Zr) 200,000 (zircon)	10/kg (Zr)
Nb	columbite, pyrochlore	1300	90/kg
Mo	molybdenite	45,000	7000
Ru	platinum ores	120 kg	1500/kg
Rh	platinum ores	3000 kg	4000/kg
Pd	platinum ores	24	900/kg
Ag	silver sulfides	8000	40/kg
Cd	sphalerite	13,000	6000
In	byproduct of Zn and Pb smelting	10	50/kg
Sn	cassiterite	190,000	5000
Sb	stibnite	60,000	1100
Te	byproduct of Zn and Pb smelting	200	12/kg
I	brines, byproduct of Chilean nitrate	4000	2000

Element	Principal Sources	Amount of Element	Price of Element (U.S. $ per ton)
Xe	air		
Cs	pollucite	1000 kg	200/kg
Ba	barite	3.2×10^6 (barite)	100 ($BaCO_3$)
La-Lu	monazite, bastnasite	2000 (oxides)	8000
Hf	zircon	50	150/kg
Ta	tantalite	300	60/kg
W	scheelite, wolframite	30,000	6000
Re	molybdenite	1000 kg	1300/kg
Os	platinum ores	60 kg	1600/kg
Ir	platinum ores	3000 kg	2500/kg
Pt	platinum ores	30	4000/kg
Au	gold, gold tellurides	1600	1150/kg
Hg	cinnabar	9000	15/kg
Tl	byproduct of Zn and Pb smelting	30	17/kg
Pb	galena	2.8×10^6	440
Bi	byproduct of Pb smelting	3000	4500
Th	monazite, byproduct of U extraction	50	50/kg
U	uraninite	30,000 (U_3O_8)	16/kg (U_3O_8)

(Data mainly from 1964 Minerals Yearbook)

APPENDIX IV

SOME QUESTIONS AND PROBLEMS

GENERAL

1. Explain where you would look for concentrations of the following elements: B, Mn, Be, Cr, Au, Hf, F.

2. Write a paragraph about each of the following men, indicating their contributions to geochemistry: Urey, Fersman, Bullen, Pauling, Bowen, Eskola, Day, Laue, Berzelius.

3. Define the following, giving one example of each: incongruent melting, feldspathoid, dispersed element, clarke, exsolution, ionic potential, geochemical fence, coordination number, polymorphism.

4. You are asked to investigate the geochemistry of a specific element. How would you proceed?

5. Write an essay on the geochemical significance of the periodic classification of the elements.

6. Discuss (with dates where possible) some significant events in the development of the science of geochemistry.

7. Write an essay on geochemical prospecting.

8. Discuss three examples of the application of non-radiogenic isotopes in geochemical research.

9. Describe the potentialities and limitations of spectrographic analysis in geochemical research.

10. For each of the following materials name three minerals characteristic of each, giving also their compositions (formulas if possible): metamorphosed siliceous dolomite, gabbro, granite, river mud, marine evaporate.

11. What significance do the following places have in geochemistry: Katmai, Canyon Diablo, Ytterby, Stassfurt, Kilauea, Skaergaard, Kimberley.

12. Write an essay on the geochemistry of the radioactive elements.

13. Write an essay on the geochemical cycle of phosphorus.

14. In the system $CaO—MgO—Al_2O_3—SiO_2$ each of the following associations: (a) wollastonite + cordierite, (b) diopside + andalusite + quartz, (c) anorthite +

enstatite + quartz, could be equivalent to a single bulk chemical composition. Write equations expressing this equivalence, and calculate the composition.

15. Give the names and formulas of the minerals that might form in the following systems, and plot their positions in triangular composition diagrams: SiO_2—Al_2O_3—H_2O; Fe—Mn—O; MgO—Al_2O_3—SiO_2; $CaAl_2O_4$—SiO_2—H_2O; MgO—SiO_2—H_2O.

16. What are the minimum components in the system Fe_3O_4—Mn_3O_4—$ZnFe_2O_4$—$ZnMn_2O_4$? What minerals belong in this system?

CHAPTER 2

1. Discuss the abundances of the elements with respect to nuclear structure.
2. What are the principal classes of meteorites and what is the mineralogical composition of each class?
3. Do elements have an age? How do we know?
4. Write an essay on the geochemical significance of meteorites.
5. Write an account of the chemistry of the sun.
6. How are estimates of the cosmic abundances of the elements obtained?
7. How do you account for the fact that iron is much more abundant than its neighboring elements in the periodic table?
8. Write an essay on tektites and their geochemistry

CHAPTER 3

1. Discuss the relationship between the geochemical properties of the elements and their position in the periodic table.
2. The Earth is believed to have a zonal structure. Summarize this structure, giving the significant chemical and physical properties of each zone.
3. Take the figures for the average composition of the Earth's crust, and convert the data into (*a*) atoms per 10^6 atoms Si; (*b*) atoms per 10^6 atoms O.
4. Assume that there is a discontinuity in the mantle at a depth of 400 km and it marks a transition from olivine (D = 3.3) to a high-pressure polymorph (D = 3.7). If the circumference of the Earth has decreased by 500 km during geological time, and this is due to an increase in the amount of the high-pressure polymorph, calculate the initial depth of the discontinuity.
5. How has the mass of the Earth been determined?
6. What is the geochemical classification of each of the following elements: Li, As, U, C, Ne, Ru.

CHAPTER 4

1. Define isomorphism and solid solution, distinguishing clearly between them. There are different types of solid solution; name them, giving one example of each.
2. Enumerate the different structural types of silicate minerals, giving one example (with formula) of each.
3. Place the following elements in order of decreasing ionic radius: Mg, Si, Ba, B, Ca, Cs.

4. Name the four most abundant elements (*a*) in the Earth's crust; (*b*) in meteorites; (*c*) in the universe.
5. Discuss the application of thermodynamics to geochemical problems.
6. Write an essay on the stability of minerals.
7. What is the phase rule? The mineralogical phase rule? Who was the author of each? When?
8. Give Le Chatelier's Principle, and discuss some examples of geochemical interest.
9. Does the geochemistry of beryllium resemble that of magnesium? If not, why not?

CHAPTER 5

1. In what type of rocks of magmatic origin, and in what minerals of these rocks, would you expect to find the following elements: Ba, Sr, Y, Hf, Rb, Sn, Ge, Ni.
2. Draw the PT equilibrium diagram for the one-component system SiO_2, and discuss its geological significance.
3. Write an essay on the volatile components of a magma.
4. Discuss the significance and interpretation of chemical analyses of igneous rocks.
5. Define magma. Where does it come from? How does it originate?
6. Draw a phase diagram with a reaction curve. Discuss its significance.
7. Would you expect to find aluminum in olivine? If not, why not?
8. Discuss with examples the principles governing the distribution of minor and trace elements during the crystallization of a magma.

CHAPTER 6

1. Explain the concept of the "geochemical fence," illustrating your answer with a diagram showing some important geochemical fences.
2. Outline differences in chemical composition between igneous and sedimentary rocks.
3. Write an essay on minor and trace elements in sedimentary rocks.
4. Write an account of the geochemistry of salt deposits of nonmarine origin.
5. Write an essay in the geochemistry of soils.
6. Calculate the mineralogical composition of some sedimentary rocks from their chemical analyses by using the procedure of Imbrie and Poldervaart.

CHAPTER 7

1. Write an essay on the history and evolution of the ocean.
2. It has been estimated that 600 g of igneous rock has been weathered for each kilogram of sea water. What is the potential supply per kilogram of sea water of each of the following elements: Li, B, S, Na, K, Cl, Sr, Ba, assuming complete solution? What percentage is actually present in sea water? Discuss the significance of these results.

CHAPTER 8

1. Enumerate (*a*) the additions to the atmosphere during geological time, and (*b*) the losses from the atmosphere during geological time.
2. How do we know the mass of the atmosphere?

CHAPTER 9

1. Write an account of the geochemical cycle of carbon. Illustrate your answer with a diagram.
2. Discuss the role of organisms in geochemistry.
3. Write an essay on geochemistry and animal health.
4. Write an essay on the isotopic composition of sulfur, with special reference to its biogeochemistry.

CHAPTER 10

1. Discuss the role of solid-state reactions in metamorphism.
2. Outline differences in chemical composition between igneous and sedimentary rocks, and show how these differences can be applied to determining the original nature of metamorphic rocks.
3. Discuss the application of thermodynamics and kinetics to problems of metamorphism.
4. Enumerate the minerals you would expect to find in the following materials, showing by equations the possible reactions producing them:
   (*a*) a metamorphosed dolomite which contained some quartz,
   (*b*) a metamorphosed limestone containing some kaolinite,
   (*c*) a metamorphosed sandstone cemented with siderite.
5. A metamorphic rock is composed of grossular, wollastonite, diopside, and quartz; its chemical analysis is SiO_2 52.75; Al_2O_3 5.65; Fe_2O_3 1.23; FeO 3.71; MgO 6.33; CaO 30.20. Calculate the weight percentage of each mineral in the rock (assume all FeO is in diopside, all Fe_2O_3 in grossular).
6. Write equations for the following reactions:
   (*a*) muscovite + quartz = sillimanite + orthoclase + water,
   (*b*) pyrope = spinel + cordierite + forsterite,
   (*c*) grossular = anorthite + wollastonite + gehlenite,
   (*d*) anorthite + enstatite + water = zoisite + chlorite + quartz,
   (*e*) tremolite + water + carbon dioxide = serpentine + calcite + quartz.

CHAPTER 11

1. Write an essay on the transformations of energy during geological processes.
2. Calculate (in $cal/cm^2/sec$) the amount of heat generated in a layer of basalt 16 km thick, given that the radioactive elements in basalt generate heat at the rate of 1.1 $cal/g/10^6$ years. Assume the density of basalt is 2.9.

INDEX